OIL CULTURE

OIL CULTURE

Ross Barrett

and

Daniel Worden

Editors

University of Minnesota Press

MINNEAPOLIS • LONDON

The University of Minnesota Press gratefully acknowledges the
generous assistance provided for the publication of this book by the
Hamilton P. Traub University Press Fund.

Chapters 2, 3, 4, 5, 6, 7, 12, 14, 16, and 18 originally appeared in the
Journal of American Studies 46, special issue 2 (May 2012). Copyright 2012
Cambridge University Press; reprinted with permission.

Passages from Ogaga Ifowodo's "The Oil Lamp" are reproduced in chapter 11.
Originally published as *The Oil Lamp* (Trenton, N.J.: Africa World Press, 2005), 38.

Passages from Sheryl St. Germain's "Midnight Oil" are reproduced in
chapter 20. Originally published in *Navigating Disaster: Sixteen Essays of Love
and a Poem of Despair* (Hammond: Louisiana Literature Press, 2012);
reproduced with the author's permission.

Published by the University of Minnesota Press
111 Third Avenue South, Suite 290
Minneapolis, MN 55401–2520
http://www.upress.umn.edu

Library of Congress Cataloging-in-Publication Data
Oil culture / Ross Barrett and Daniel Worden, editors.
Includes bibliographical references and index.
ISBN 978-0-8166-8968-2 (hc : alk. paper) — ISBN 978-0-8166-8974-3 (pb : alk. paper)
1. Petroleum industry and trade—United States—History. 2. United States—
Social and customs—History. I. Barrett, Ross, 1977– II. Worden, Daniel, 1978–
⌐ HD9565.O583 2014
338.2´72820973—dc23
2013050891

Printed in the United States of America on acid-free paper

The University of Minnesota is an equal-opportunity educator and employer.

20 19 18 17 16 15 14   10 9 8 7 6 5 4 3 2 1

We dedicate this book to
Clementine *and* William.
May your futures venture beyond oil.

Contents

Foreword

ALLAN STOEKL

Oil Culture is not just a collection of highly informative essays, but it asks an implicit question: What *is* oil culture? (And are there cultures, more than one?) Along with that, and perhaps conditioning it: How are oil and culture conjoined? What is it for oil to be cultural? And culture to be oily? Why is oil culture and its fragmentation in cultures so fundamentally important?

These are, I think, central questions, for oil in itself is mere matter, but as a number of authors in this collection point out, it contains, when burned, concentrated energy, the power to do work much more effectively—more energy, less mass—than, say, coal, let alone human muscle. Yet there is a tendency on the part of some critics, not represented in this collection, to hold that oil, and the fuels derived from it, is *the* most important element in culture: that economic value, for example, derives solely from energy inputs and "EROEI" ("energy returned on energy invested"), and not, say, from human labor. Perhaps this kind of overstatement is only the flip side of another reduction, one much more commonly seen and endemic to mainstream economics: the source of value is labor, conjoined with invested capital, supply and demand, improvements in efficiency, and so on.

This latter view results in what I would call the invisibility of energy inputs, and, most notably in our era, the invisibility of that ultimate fossil fuel, oil. Oil is invisible, undetectable, when we "take it for granted": we assume needlessly that "peak oil is garbage," whatever that might mean, or that it will be very easy to find substitutes for oil when it becomes scarce and prohibitively expensive. We assume that current civilization, such as it is, is exclusively the result of human ingenuity as applied to technological development, or of human reason as it comes to transcend all material inputs. We assume that the chief problems of the world—overpopulation, mass warfare directed against civilians, climate change—are the result of technology out of control, all the while forgetting that technology is nothing without highly refined energetic inputs. We assume that our current prosperity is the result of our personal decisions, our cleverness at school and our academic degrees, our correct career moves, our sagacity in investing our earnings—all the while forgetting that our luxurious homes, our cars, our clothes are the result of myriad energy slaves whose activities are those of all the joules liberated from the vast quantities each of us depends on, but never see, and never question. Where does oil come from? How is it extracted from the earth? How is it refined? And, most important, as a number of the essays in this collection demonstrates, how does capitalism, in conjunction with our government, foster this energy use, and this invisibility? How is capitalism as we know it today inseparable from the energy regime of oil?

Given this history of invisibility, how then might we think about oil in culture without "reifying" it—that is, without simply turning the current situation on its head and seeing oil as the be-all and end-all of culture in general?

I think the most effective way of refusing such a reification of oil, all the while granting it the visibility it deserves, is to write its history—as many of the essays in this collection do. It's when we think about what "oil history" could mean that we take a *natural* entity and recognize its *cultural* centrality. Oil is natural in the sense that no one put it there in the ground: it is the result of natural processes, the arrested decomposition of plant and organic matter over millions of years. And yet everything that is done with it—the pumping, the refining, the grading, the distribution, the use in transport, manufacture, heating, the generation of electricity—is fully cultural. Oil is a natural element—and one that has a history. What's more, its history, at least over the last two hundred years or so, is inseparable from the history of the advance and development—if that's the word—of modern industrial civilization. It is thus not just a commodity; it is *the* commodity. Much as we try, embracing nuclear, solar, wind, tides, whatever, we still cannot find a real substitute for oil. It is dumb matter, a natural offshoot of natural processes, gunk in the ground, that we use, and that uses us, keeping us prisoner to our energy slaves, to the rich energy inputs that we find so hard—even impossible—to derive from any other energy source.

Oil is, then, the poster child for the argument that the natural, today, is fully cultural. It is futile to imagine a pristine natural space, some deep-green realm whose preservation will save us. Even nature preserves, even spaces seeming (for the moment) beyond our reach, are at this late date natural artifacts: if we manage to "preserve" the Amazon, it too will be simply an artifact, the consequence of decisions we have or have not consciously made. Oil is the ultimate natural-cultural artifact. Once it ceases to be gunk in the ground, it is what we do with it. And yet it is an agent as well: it calls to us, and we respond to its entreaties, its interpellations. It is the ultimate siren's song, the ultimate Eurydice we reach for—as it disappears. We are the slaves of our energy slaves, in a surprising (to us) revision of the Hegelian master-slave dialectic. We work to further oil's consumption, imagining ever more wasteful "uses" for it—from lavish oil-heated homes to gas-guzzling Hummers. We cannot help but realize that in some sense we *are* oil: all those grain-fed cattle we eat, themselves fed with cereals grown with fossil-fuel–derived fertilizers, and protected by oil-based pesticides; all the water purified and pumped with the use of fossil fuels; all the wars powered and won with fossil fuels (not least the wars for control of oil resources). And we are the disappearance of oil. We are human, social, cultural—that is, we are not formed by a god, we form, make, ourselves—but we are also what we have done with oil. And oil is natural in the sense that it is present, then absent: it is finite, and that ultimate finitude is well beyond our control.

Back in the days of existentialism, we could write of the finitude of Man, all the while ignoring the prime locus of human finitude: the very fuel that serves to make "his" modernity. At a certain point, if we recognize our finitude, we recognize, ironically, not what is cultural but what is "natural." If the existentialists posited a human finitude based on the independence of man—no god in the sky, only the self-creation of our own meaning, our own destiny—now we can start to see a finitude deriving precisely from what is natural in man. If we can say that today the natural is now fully cultural—just think of oil and the ways in which its myriad uses have transformed the world—we can also say, just as convincingly, that the cultural is relentlessly natural. It is natural less in a positive way than in the subtractability of oil. But if the "cultural" as we know it today is inseparable from rich oil-based energy inputs, the horizons of culture will appear to be inseparable from the horizons of the limited availability of oil. The most advanced human culture will recognize its natural status when it finally *recognizes* oil, when it really sees it, and when it recognizes, consequently, the limits not just to growth but to any and all room for maneuver. Of course advanced technology can always pull a little more oil out of the ground, derive a little more energy from what there is—*but there are limits*. Per capita wealth, population density, physical fitness and comfort, levels of education, and all this in a temperate world unaffected by the long-term effects of the combustion of fossil fuels—all are highly fragile artifacts dependent on naturally conditioned but temporary energetic inputs. The "Death of Man"—or at least Man's radical reconfiguration into a species barely recognizable today—may be less a function of semiotic sleight-of-hand, as it was in the 1960s, and more the result of a natural limitation, the wall toward which the species is headed at top speed.

Thus the need, I would argue, for the essays in this collection. We cannot know oil (and hence ourselves) if we know it only in geophysical or chemical terms: where it is to be found, how it is to be refined. And we cannot know it if we see it only as something manipulated by greedy corporations, or by speculators working for private profit. Those elements are part of the story of course, but they, in different ways, contribute to oil's invisibility: it is just another commodity, just another thing we buy and use, necessary to the American Way of Life, sure, but mainly just another way of getting ripped off. We need to do more: we need to understand, given the centrality of oil, its weird natural/cultural status, what oil has been in history, and what it will be: politically, culturally, aesthetically, historically. All the ways that oil has been, and will be, consumed, and all the ways it has, and will, consume us, exercising its material agency—all this constitutes a central task for the cultural historian, the literary critic, the geographer, the philosopher. The stimulating and brilliant essays in this collection are an essential first step in this direction, sure to be followed by many more.

Acknowledgments

Oil Culture began as a panel at the American Studies Association's Annual Meeting and has grown beyond our expectations. We would like to thank Nancy Quam-Wickham, who served as respondent to that initial conference panel, as well as William Lucas, John Horne, the editorial board of the *Journal of American Studies,* and Marc Anderson at Cambridge University Press. We are fortunate to have had the support and guidance of Richard Morrison and Erin Warholm-Wohlenhaus at the University of Minnesota Press. We owe a special note of thanks to all of the contributors to this volume for their incisive, expansive, and thoughtful essays. We also thank Cat and Danielle, for so many things.

Introduction

ROSS BARRETT AND DANIEL WORDEN

Oil may come and go at any time and it may not be visible.

—Alabama Department of Public Health, from a sign posted on
public beaches following the BP Deepwater Horizon Oil Spill

The National Commission on the BP Deepwater Horizon Oil Spill and Offshore Drilling's Report to the President reproduces an Alabama Department of Public Health sign warning residents of the presence of oil on public beaches.[1] Along with more practical advice, such as "do not handle tar balls," the sign makes the above claim, both a practical warning and a canny statement about oil's ontology. Oil is not entirely visible to us as a commodity, a fuel, a resource, or a political and economic agent, yet it is also not invisible. Events like the BP Deepwater Horizon Oil Spill— "the largest peacetime offshore oil spill in history"[2]—are shocking not only because they are ruptures in the seemingly smooth, technologically sophisticated transport of fossil fuel, but also because they reinforce the already visible presence of oil in our everyday lives. Oil is not invisible to us as much as it is contained—in our cars' gas tanks, in pipelines, in shale, in tar sands, in distant extraction sites.[3] We might not interact with oil as a viscous material very often, but we are aware of it as such. This gives oil a curious valence in the cultural imagination—it is foundational and ever present, yet it is also secreted away.

Oil's signature cultural ubiquity and absence is the occasion for this book. *Oil Culture* analyzes, interprets, and explicates oil's presence in culture. This presence manifests in multiple ways: oil is material, mystical, historical, geological, and agential. Oil does things, connotes meaning, and is leveraged by nations, corporations, and individuals. Only in recent years, beginning with conversations about peak oil and transforming into debates about deepwater offshore drilling, fracking, and bitumen, has oil become the subject of cultural analysis for a critical mass of scholars. Wishfully, we think that oil's emergence as a subject of analysis signals a shift in its hold on our world. More skeptically, we also acknowledge that this new scholarly interest in oil is so vital because oil is everywhere, and it shows little sign of being eclipsed by another energy source in the near future.

The BP Deepwater Horizon Oil Spill is but one in a long series of disasters—environmental, political, economic, military—that underwrite petroleum's more than 150-year history as a modern fuel. Understood from the beginning as a finite resource, petroleum is nonetheless routinely thought of as a necessity to modern life—without oil, we are often told, the economy will simply collapse, and our modern way of life with it. As Allan Stoekl has noted, energy discourse is saturated with an "uncritical faith in the capacity of human genius": "Fossil fuels entail a double humanism: they are burned to serve, to magnify, to glorify the human . . . and they are produced solely through the free exercise of the mind and will."[4] It is difficult to think of oil as a material substance that is not entirely the product of and subject to human will, and thus oil's necessity to modern life appears less as a problem than an ontological given. The lack of oversight, regulation, and cleanup technology that led to and exacerbated the Deepwater Horizon spill inspired even more reaffirmations of oil's necessity to human life, often under the guise of arguments for increased regulations and explorations into alternative energy. Even in the face of environmental disaster, and an environmental disaster that has been and will continue to be repeated on sliding scales as risky deepwater drilling continues to be practiced in the Gulf of Mexico and in even more treacherous parts of the world such as the Arctic, we are unable to give up oil. As the National Commission's report states, "Given Americans' consumption of oil, finding and producing additional domestic supplies will be required in coming years, no matter what sensible and effective efforts are made to reduce demand."[5]

In contemporary culture, no better a symbol of oil's persistence can be found than the return of the television drama *Dallas* in 2012.[6] One of the most successful television dramas of all time, the first version of *Dallas* ran from 1978 to 1991, and its revamped version stars cast members from the original *Dallas* alongside younger actors who play the next generation of the show's focal point, the Ewing family. In its original incarnation, *Dallas* staged its major tension around Bobby and J. R. Ewing, with Bobby representing morality and ranching to J.R.'s avarice and drive to strike oil. In its contemporary iteration, *Dallas*'s major tension is between Bobby's son Christopher, a proponent of deepwater drilling for methane hydrate, and J.R.'s son John Ross, who wishes to drill for oil the old-fashioned way on the Ewing's Southfork ranch. While rhetorically positioning this conflict as one between clean and dirty energy, the show represents the massive earthquakes and damage to the ocean floor caused by deepwater methane drilling as a merely technological problem, one that will inevitably be solved as the free market demands more fuel than can be produced through traditional drilling. Like the Deepwater Horizon's pursuit of new fossil-fuel deposits, *Dallas's* vision of "alternative energy" is an even more treacherous version of traditional drilling, and moreover, it represents not an alternative to but a continuation of our dependency on fossil fuels.

However melodramatic, the family tension in *Dallas* resonates with popular audiences and signals just how integral oil is to contemporary life in the United States and to the American cultural imagination. When the actor Larry Hagman, best known for his role as J. R. Ewing, passed away while filming the second season of the relaunched *Dallas* in November 2012, newspapers and other media outlets memorialized him as the character he was so well known for playing. Hagman's *New York Times* obituary relayed that "at the height of the show's popularity, he handed out fake $100 bills with his face on them."[7] The popularity of J. R. Ewing, "one of television's most beloved villains," is evidence of the hateful, yet pleasurable, proximity of oil to our lives and culture more broadly.[8] An underwriter of nearly every major museum, and whose petroleum products are used to make film, ink, paint, and countless other tools used to produce art today, the oil industry is as ubiquitous and necessary to contemporary life as money, the point of Hagman's funny one-hundred-dollar bills. Even more than money, though, oil is finite, and as demonstrated in *Dallas's* own fantasy of methane hydrate as a viable alternative energy, it is unclear how we can imagine a world without fossil fuels that remains a livable environment and not a postapocalyptic wasteland.

This stalemate between oil as a nonrenewable resource and oil's centrality to modern life requires more than just a commitment to alternative energy, more than just individual consumer choices such as buying local or driving a hybrid car. While these acts of personal conservationism are a welcome deviation from traditional oil profligacy, they do not grapple with a deeper problem—how oil has shaped contemporary ways of being and imagining not only in North America and Europe but also the world. Like vegetarian "hamburgers" or vegan "cheese," many "green" visions of the world entail maintaining our lives and practices as they exist with petroleum and simply swapping oil out for a different energy source that magically takes its place and replicates precisely its roles.[9] We believe that this failure of imagination can be partially remedied by understanding how oil works in culture.

This volume seeks to remedy the relative silence that scholars in the humanities have maintained about oil. This silence stems not from some imperative within any given discipline, but rather from a deep-rooted reluctance among humanistic scholars to use oil, a crude material indeed, as a framework for thinking of anything beyond economics, energy, and politics. As we will see, the customary reluctance has begun to wane in recent years. During the twentieth century, however, academics did relatively little work to assess oil's saturation of modern social life and thought. In a well-known 1992 review of Abdelrahman Munif's *Cities of Salt* novels, the Bengali novelist Amitav Ghosh identified a possible reason for this pattern of evasion, noting that "the history of oil is a matter of embarrassment verging on the unspeakable."[10] Whether inspired by an unspoken (and perhaps unspeakable) Western shame about the ruinous legacy of oil capitalism or not, humanistic scholars have until recently abnegated the historical and theoretical analysis of oil to industry insiders, energy policymakers, environmental historians, and journalists.

To study the development of the oil industry, one can turn to a number of industry-centered histories, including Arnold Daum and Harold Francis Williamson's *The American Petroleum Industry* (1959–63), Daniel Yergin's best-selling *The Prize: The Epic Quest for Oil, Money, and Power* (1991), and Roger M. Olien and Diana Davids Olien's *Oil and Ideology: The Cultural Creation of the American Petroleum Industry* (2000). These works offer industry-focused, and in some cases explicitly triumphalist, assessments of the rise and development of oil capitalism. There are also a number of texts that offer crucial case studies of oil's material development in particular nations, such as Robert Vitalis's *America's Kingdom: Mythmaking on the Saudi Oil Frontier* (2006), and more general introductions to oil energy, geology, economics, and politics, such as Gavin Bridge and Philippe Le Billon's *Oil* (2012), Morgan Dowley's *Oil 101* (2009), and Vaclav Smil's *Oil: A Beginner's Guide* (2008). In political theory, Timothy Mitchell's *Carbon Democracy: Political Power in the Age of Oil* (2011) analyzes the ways that oil has both facilitated and frustrated democratic organization, expression, and action in the Middle East. Along with these books, there is also a large body of books about the energy crisis, war, and environmentalism that deal explicitly with oil, which includes Kenneth S. Deffeyes's *Beyond Oil: The View from Hubbert's Peak* (2005), Peter Maass's *Crude World: The Violent Twilight of Oil* (2009), Daniel Yergin's *The Quest: Energy, Security, and the Remaking of the Modern World* (2011), and Steven Coll's *Private Empire: ExxonMobil and American Power* (2012). While these studies are all integral to any understanding of oil, they do not engage specifically with oil as a cultural material, a force not only in economic and political life but also in everyday experience and aesthetics. As such, this work has done little to illuminate the resource's hold over common ways of thinking and acting in the oil-guzzling West.

As Imre Szeman has influentially argued, we are largely incapable of imagining a world without oil: "It is not that we can't name or describe, anticipate or chart the end of oil and the consequences for nature and humanity. It is rather that because these discourses are unable to mobilize or produce any response to a disaster we know is a direct result of the law of capitalism—limitless accumulation—it is easy to see that nature will end before capital."[11] Focused entirely on the very industry that has led to this impasse, existing work on the oil economy has little to say about alternatives, and the hard-nosed pragmatism common in industry-based histories of oil leaves little room for imaginative, let alone utopian, thinking. By analyzing oil as both an industry and a culture, a business and a set of aesthetic practices, a natural resource and a trope, this volume aims to find some room at the edges of oil's hold on us, to see the ways in which oil's dominance might, in and of itself, be a part of oil culture in the first place rather than an economic or physical necessity.

On the Cultural Analysis of Oil

Oil Culture draws inspiration from an emerging body of scholarship concerned with the cultural forms of oil capitalism. In the past twenty years, a growing number of scholars have investigated how creative texts, images, performances, and other productions have worked to stimulate and sustain the oil economy. As suggested above, this body of work has sought to fill a significant lacuna created by the material emphases of early oil studies and to challenge those assumptions and ambivalences that previously undercut humanistic inquiries into oil's cultural expressions. In literary studies, for example, it was long assumed that the ascendance of American oil capitalism left few traces in the fiction and poetry of the United States. In his above-cited 1992 review, Amitav Ghosh offered perhaps the best-known articulation of this idea, arguing that "the Oil Encounter . . . has produced scarcely a single work of note" and, more emphatically, that "there isn't a Great American Oil Novel."[12] In making this argument, Ghosh reinvigorated a claim that had received frequent expression in earlier criticism. In the introduction to a 1967 anthology of southwestern writing, for example, Martin Shockley declared that "with minor exceptions . . . oil culture is not represented in literature"—an assertion that ignored at least a dozen widely read period novels on Texan and Oklahoman oil, let alone the many novels, poems, and short stories devoted to other aspects of the petroleum business that had appeared in the first half of the twentieth century.[13]

In the 1990s and 2000s, scholars across the humanities began to rethink these assumptions and explore the cultural dimensions of oil capitalism. A variety of intellectual and historical developments likely spurred this reorientation. The violent crises that beset the late-century oil economy—crises triggered by the depletion of proven fields and the petroleum industry's increasing reliance on crude produced by risky and violent oil frontiers—inspired new interest in the mechanisms by which petroleum had been established as the world's primary energy commodity. The contemporaneous growth of cultural geography, environmental history, and ecocriticism as distinct fields of inquiry likely worked to further encourage critical analyses of oil capitalism's spatial and symbolic forms. Working within this charged climate, turn-of-the century historians, geographers, and anthropologists turned new attention to the broad array of cultural forms that have propelled the oil economy.

Much of this work concentrated on the specific cultural dynamics of the oil boom. This subject received an important early treatment in Polish journalist Ryszard Kapuściński's *Shah of Shahs* (1982), an imaginative account of the Iranian Revolution that includes a brief but evocative chapter on the petro-mythologies that flourished during the golden age of Iranian oil. A well-known passage in this chapter considers the power of promotional images of oil, focusing on photographs that "preserve the moment when the first oil spurts from the well: people jumping for joy, falling into each other's arms, weeping," and in so doing conjure "the illusion of a completely changed life, life without work, life for free."[14] In its brief but suggestive evocation of the power of visual representation to win popular support for oil, Kapuściński's chapter laid the groundwork for subsequent scholarship on the cultural dynamics of the oil-boom landscape.

Building on this precedent, several scholars have examined promotional representations of the American oil industry's first flush field: the explosively productive oil region of northwestern Pennsylvania. Brian Black's *Petrolia: The Landscape of America's First Oil Boom* (2000), for example, explores how reporters and illustrators used the framework of the sublime to legitimate the disastrous overproduction that unfolded in the Pennsylvania oil fields.[15] Others have concentrated on the cultural dimensions of the mid-twentieth-century Venezuelan oil boom. Fernando Coronil's *The Magical State: Nature, Money, and Modernity in Venezuela* (1997) advanced an influential account of the ways that state-organized spectacles and promotional narratives helped secure popular allegiance to the Venezuelan oil economy and the tenuous government that administered it. More recently, Miguel Tinker Salas's *The Enduring Legacy: Oil, Culture, and Society in Venezuela* (2009) analyzes the oil camp as a cultural laboratory that transnational companies used to promote forms of citizenship and social life attuned to the petroleum economy. The Nigerian oil boom has similarly inspired a variety of cultural analyses. Andrew Apter's *The Pan-African Nation: Oil and the Spectacle of Culture in Nigeria* (2008) addresses the 1977 Second World Black and African Festival of Arts and Culture as a spectacle of national community built on oil-based prosperity and pan-ethnic unity. In a series of important essays, Michael Watts has analyzed the fetishizing discourses, oil mythologies, and booster narratives that arose around Nigeria's booming oil economy of the 1970s.[16]

While some critics and historians have focused on the imaginative representations and ideological constructs of the flush field, others have concentrated on the cultural practices underpinning petroleum-based lifestyles in the United States and other gas-guzzling Western societies. John Urry, Cotten Seiler, and others, for example, have studied the popular discourses of autonomy, speed, and adventure that have sustained the modern American system of automobility.[17] Various scholars have also begun to consider the ways that popular representations have forestalled efforts to understand oil's ubiquity and unavoidable transience. Peter Hitchcock has recently argued that oil's entrenchment in American society has depended in part on cultural strategies of evasion and erasure, and Stephanie LeMenager has charted the aesthetics of and affects clustered around oil in film, literature, and other media.[18] In three important articles published between 2007 and 2012, Imre Szeman identified three dominant contemporary discourses of oil that frustrate efforts in North America and Europe to envision alternative energy economies and analyzed present-day cinematic and photographic attempts to reopen dialogues about a post-oil future.[19]

Recent work in ecocriticism and the environmental humanities, finally, has sought to develop a conceptual vocabulary that can better capture the concerns posed by climate change and global environmental movements. These efforts have yielded a range of ideas and terms that promise to inspire new ways of conceiving oil. Rob Nixon, for example, has recently used the concept of "slow violence" to explain the gradually compounding environmental ruin produced by industry and war and to challenge disaster discourses focused on sudden crises (explosions, oil spills, embargoes, etc.).[20] Timothy Morton's discussion of the "hyperobject" as a material that exists outside of time likewise offers a useful framework for reimagining dominant cultural interpretations of oil's historical presence and temporal nature.[21]

The past twenty years have, in sum, seen the emergence of a new body of multidisciplinary research attentive to oil's cultural presence in the various social, economic, and political realms that constitute the global petroleum system.[22] Drawing energy from this body of work, we have developed the present volume with the hope that it might extend and complicate ongoing dialogues about petroleum's cultural ramifications. *Oil Culture* aims to advance the field of oil studies, more specifically, by critically engaging the insights of existing work on oil's symbolic life, exploring new approaches to this subject, and expanding the historical, theoretical, and thematic horizons of cultural scholarship on oil. The essays collected here accordingly present the work of established scholars and figures new to the field, and investigate oil's imaginative manifestations in diverse cultural media and sociohistorical climates. In so doing, the essays bring a wide array of historical, interpretive, and theoretical methods to bear on representations of fossil fuels and the petroleum industry. The total body of scholarship will, we hope, trace the preliminary outlines of a newly comprehensive account of oil's symbolic existence over the past 150 years.

This account relies on a particular model of "oil culture." We use the phrase to designate a dynamic field of representations and symbolic practices that have infused, affirmed, and sustained the material armatures of the oil economy *and* helped to produce the particular modes of everyday life that have developed around oil use in North America and Europe since the nineteenth century (and that have since become global). We first understand cultural signification to be a fundamental process of the expansive economic system that has arisen around oil. This proposition builds on the insights of several recent scholars. Diverging from accounts that would identify oil as the object of a discrete "industry," historians and theorists have increasingly understood the resource as the central concern of a vast network or "assemblage" of interlinked technological, commercial, financial, and political initiatives.[23] Others (including, most prominently, Imre Szeman) have understood oil as the foundation for a whole phase of capitalism premised on cheap energy, petrochemical goods, and risky modes of accumulation.[24] These accounts have increasingly considered the possibility that cultural signification might constitute a central function within the multivalent structures of the oil economy. In so doing, they have turned to a specific theoretical model for inspiration: Guy Debord's well-known account of the "society of the spectacle." Debord's 1967 treatise identified the spectacle as a new phase of capitalism defined by the hyperproduction of commodities and of the signs of commodified value.[25] Suffusing every last corner of everyday social experience, this web of images reorganized modern life around commodified appearances and the ideal of market exchange. Drawing on this theoretical model, contemporary oil scholars such as Andrew Apter and Michael Watts have interpreted various oil economies (and especially the OPEC petrostates of the 1970s) as spectacular edifices built on images of petroleum-based progress and prosperity.[26]

Following the work of Apter and Watts, we embrace Debord's spectacle as a useful model for thinking through the deep ties between oil capitalism and cultural representation. Indeed, we argue that the global oil economy is at root a spectacular system, built on and sustained by proliferating cultural significations. This symbolic imperative arises in part from the peculiar physical properties of oil, which more than any other commodity demands the unceasing generation of imaginative interpretations of its value (the sort of mystifying signs that constitute the spectacle in Debord's account). As a material whose utility is largely realized through its own destruction, oil requires creative accounts of its worth that depart from its physical form. As a substance that can (at least initially) be extracted without much work, moreover, oil encourages fetishistic representations of its value as a magical property detached from labor. And as a fluid generally recognized to be dirty, sensually offensive, volatile, and transient, oil has long required especially elaborate affirmations of its economic and social benefits. In order to remake the difficult physical material of oil as a viable market commodity, then, petroleum producers and distributors have been continually forced to make intensive efforts to imaginatively recode the resource since its nineteenth-century rediscovery. Articulating oil's value in innumerable creative ways that transcend (and obscure) its actual material constitution, the resulting images, narratives, and discourses have contributed to the formation of an oil spectacle that has sustained industrial and financial commitments to the expanding system of petrocapitalism.

This vast economy of cultural representations has worked in turn to accelerate oil capitalism's reorganization of everyday life in North America, Europe, and an expanding array of non-Western industrial nations (including, most significantly, China) around the maximally intensive consumption of oil and its chemically engineered derivatives.[27] Scholars in a variety of fields have recently grappled with the cultural processes that have helped to make oil ubiquitous within modern life. Moving beyond simple economic or technological accounts, Szeman and Watts have explained oil's reorientation of daily life by theorizing broader structures of experience and understanding that have arisen around petroleum use. Szeman has used the term "oil ontology" to describe the emergence of a new phase of being in which oil functions as the "structuring real of contemporary social-political imaginary"; Watts has suggested that urban development patterns have organized a "regime of living" defined by oil use.[28] Building on these accounts, we emphasize the critical role that cultural representations play in the establishment of comprehensive systems of being, belief, and knowledge keyed to the priorities of oil capitalism. Oil culture, we argue, has helped to establish oil as a deeply entrenched way of life in North America and Europe by tying petroleum use to fundamental sociopolitical assumptions and aspirations, inventing and promoting new forms of social practice premised on cheap energy, refiguring petroconsumption as a self-evidently natural and unassailable category of modern existence, and forestalling critical reconsiderations of oil's social and ecological costs.

Though we have thus far emphasized the hegemonic effects of oil culture, the broad field of oil-inspired cultural activity also includes significant countercurrents of doubt, uncertainty, ambivalence, and critique. Since oil's nineteenth-century rediscovery, there has existed a steady chorus of protest and lamentation aimed at the petroleum industry, a chorus sounded in reformist treatises, muckraking investigative reports, conservationist essays, environmentalist manifestos, political protests, theoretical critiques, and other cultural texts that have attempted to check the onrush

of oil. And promotional imaginings of oil have long carried their own internal ambivalences and contradictions, moments of conflictedness that suggest the fragile
tenuousness of the ideological constructs (techno-utopianism, financial adventurism, and so on) on which the unlikely oil economy and its turbulent markets have
depended.

As we envision it, then, oil culture encompasses the fundamental semiotic processes by which oil is imbued with value within petrocapitalism, the promotional discourses that circulate through the material networks of the oil economy, the symbolic
forms that rearrange daily experience around oil-bound ways of life, and the many
creative expressions of ambivalence about, and resistance to, oil that have greeted the
expansion of oil capitalism. We understand the study of oil culture as a project that
traces the various threads of petroleum's symbolic life in pursuit of two overarching
historical and theoretical objectives, each freighted with significant political implication. On the one hand, the cultural analysis of oil aims to reconstruct those symbolic
forms and practices that enabled the emergence, development, and entrenchment
of oil capitalism. On the other hand, by attending to those moments in which oil's
ceaseless colonizing expansion has engendered uncertainty, ambivalence, or resistance, the cultural analysis of oil can also uncover symbolic materials that may be
useful to the construction of alternative perspectives on fossil fuels and the energy
economy. The former mode of inquiry promises to yield a fuller understanding of the
mechanisms underpinning oil capitalism's dominance, an understanding that might
be used to confront that system's hegemonic self-presentations. The latter approach
may in turn help to open space from which to begin to conceive a post-oil future.

PLAN OF THE BOOK

The contributors to this volume pursue these two overarching projects from a variety
of disciplinary angles and train their attention on a diverse selection of cultural forms.
We have organized the essays into five sections that track the cultural interpretation
of petroleum across several successive phases in the development of oil capitalism,
moving from imaginative engagements with the emergence of that system in the
nineteenth century to present-day efforts to contend with the global oil economy. In
an effort to produce a narrative that is attentive to the contradictory spatial dynamics
of oil capitalism (its local concentrations and global extension) and oil's own tendency to confound established territorial demarcations, we have assembled a group
of essays that examine oil culture both within and across national boundaries. The
bulk of the chapters concentrate on oil's multivalent existence within the American
cultural imagination. As the historical fountainhead for the modern petroleum industry, the contemporary epicenter of global oil capitalism, and the society in which the

most intensely realized fossil-fuel lifestyles have taken shape, the United States has proven to be an especially productive incubator for cultural reckonings with oil, and the essays gathered here explore some of the many impulses (promotional, propagandistic, creative, educational, critical) that have informed these imaginative projects. To complement this account of American oil culture, we have included a handful of essays that examine petroleum's cultural dimensions in Europe and the various extractive frontiers on which oil-thirsty Western societies have increasingly depended (including the Niger Delta, the Gulf of Mexico, and the Alberta oil sands). These chapters shed new light on the role that cultural representations have played in enabling and resisting the colonizing drive of American and European oil corporations.

Part I, "Oil's Origins and Modernization," explores promotional narratives and interpretive frameworks that arose around the early American oil industry. Examining advertising, popular religious tracts, public monuments, and creative fiction, Heidi Scott, Rochelle Zuck, Ross Barrett, and Frederick Buell explore how late nineteenth- and early twentieth-century representations of oil helped to establish the conditions necessary for the rise of a dynamic oil economy. Studying promotional accounts of the lighting commodity that held sway before petroleum, Heidi Scott shows that mid-nineteenth-century imaginings of whale oil established a set of cultural frameworks that deeply shaped how late-century marketers and consumers understood petroleum. Using the widely reported exploits of the medium-turned-oilman Abraham James as a case study, Rochelle Zuck explores how early publicity appropriated spiritualist themes to legitimize the fledgling oil industry. Ross Barrett addresses an early sculptural monument commissioned by the Standard Oil Trust as the centerpiece of a turn-of-the-century publicity campaign designed to rehabilitate the public image of the petroleum industry and an early expression of a promotional discourse (petro-primitivism) that would inform much twentieth-century oil boosterism. Frederick Buell traces the development of two cultural discourses that have informed literary interpretations of oil's meaning since the nineteenth century; while the first of these (exuberance) has given rise to promotional imaginings of oil as a vehicle of personal and societal progress, the second (catastrophe) has inspired critical reconsiderations of petroleum's ruinous potential. Taken together, these essays suggest that early cultural accounts of oil worked both to accommodate Americans to the tumultuous petroleum economy and open space for critical and protestatory responses to oil.

Part II, "Oil's Golden Age," addresses cultural projects that fueled and contested the formation of a modern system of oil capitalism in the mid-twentieth century, a system that linked national spaces of intensive oil consumption (chiefly, the United States) to a new array of productive frontiers. Sarah Frohardt-Lane and Daniel Worden examine cultural texts that affirmed the oil-based lifestyles that emerged in the United States at midcentury. Frohardt-Lane shows that Office of War Information propaganda strengthened Americans' commitment to private automobile travel by celebrating carpooling as a patriotic act, denigrating public transportation, and presenting limitless driving as a reward for the sacrifices required by World War II. Analyzing the 1952 novel and 1956 film *Giant* as test cases, Worden identifies an ideological configuration (fossil-fuel futurity) that informed a wide variety of twentieth-century American representations of oil and that celebrated petroleum use as a means by which a normative familial life and an ideal liberal-pluralist society might be realized. Hanna Musiol and Georgiana Banita address the efforts of writers and filmmakers to assess the relationship of Western oil-consuming nations to the period's new extractive frontiers. Musiol examines John Joseph Matthews's 1934 novel *Sundown* as an account of the Osage oil boom that uses the bildungsroman form to dramatize the disastrous effects of unchecked production on the social life of an oil frontier within the continental United States (the Osage reservation in Oklahoma). Banita analyzes Bernardo Bertolucci's 1967 corporate-sponsored documentary *La via del petrolio* as an experimental attempt to visualize the flow of oil from Iranian fields to Italian refineries, an attempt that ultimately figures petroleum as a wondrous vehicle of transnational unity reconnecting post–World War II Europe.

The essays in Part III, "The Local and Global Territories of Oil," extend the investigations of the previous section by exploring representational practices and cultural projects that enabled, affirmed, and resisted the crystallization of a global system of oil capitalism during the second half of the twentieth century. Chad Parker, Michael Watts, and Jennifer Wenzel accordingly analyze the roles that history writing, creative fiction, mapping, and other spatial representations played in facilitating and contesting Western corporate efforts to establish production zones in the Middle East, Africa, and the Gulf of Mexico. Parker uses a particular instance of creative remapping—a campaign undertaken by Aramco in the 1940s and 1950s to redefine the potentially oil-rich Buraimi oasis (in present-day Oman) as a historical territory of Saudi Arabia—to address the cultural practices that propelled the expansion of corporate oil production in the Middle East. Employing the turbulent fields of the Niger Delta and Gulf of Mexico as case studies, Watts analyzes the legal, political, and cultural practices that have established the oil frontier as a primary productive realm of global petrocapitalism and a highly tenuous space defined by primitive accumulation, social disorder, and ecological ruination. Wenzel addresses Nigerian author Ben Okri's 1987 short story "What the Tapster Saw" as an illuminating example of petro-magic-realism, a literary mode that arose during the Nigerian oil boom and that uses tropes drawn from Yoruba narrative tradition to contend with the state violence and environmental degradation that defined oil exploration in the Niger Delta. While Parker,

Watts, and Wenzel study the cultural dimensions of the violent production sprees that have unfolded in oil borderlands, Matthew Huber and Sheena Wilson address the contemporary ideologies and representational practices that have enabled the ever-more-intensive infiltration of Western life by petroleum. Huber analyzes an enduring promotional construct of oil-based existence ("entrepreneurial life") that propelled the neoliberal refashioning of American social life, institutional politics, and urban settlement during the last several decades of the twentieth century. Interpreting a range of industry advertising and press imagery, finally, Wilson illuminates the ways that contemporary commercial photography uses constructs of femininity to promote the continued expansion of domestic oil industries, delegitimize women's anti-oil activism, and derail post-Deepwater protests against offshore drilling.

The essays in Part IV, "Exhibiting Oil," focus on the multivalent interpretations of oil offered by the exhibitions and public displays of aquariums, art institutions, and science museums. Offering broad audiences clear glimpses of the otherwise inconceivably vast or impossibly dispersed structures of oil capitalism, the public exhibition has exerted considerable pressure on twentieth- and twenty-first-century imaginings of petroleum. Dolly Jørgenson, Catherine Zuromskis, and Stephanie LeMenager demonstrate that the exhibition has served as an important vehicle for the dissemination of official booster arguments about oil *and* a significant venue for the articulation of more complicated perspectives on oil capitalism. Jørgenson examines strategies employed by contemporary Gulf Coast aquarium displays to refigure deepwater petroleum extraction as an ecologically beneficial activity. Analyzing two significant photographic exhibitions—the 1975 *New Topographics* show at the George Eastman House and a 2009 traveling exhibition of Edward Burtynsky's photographs of oil landscapes, entitled *Burtynsky: Oil*—Zuromskis explores the quiet resonances of unease and anxiety that structure contemporary landscape photographers' visions of petromodernity. Seizing on the institution's still-resonant displays of oil's prehistoric roots, LeMenager reconceives the George C. Page Museum in Los Angeles (known colloquially as the La Brea Tar Pits museum) as a space in which existing conceptions of environmental history and human dominion over the natural world might be productively reoriented, and in which contemporary Americans might begin to conceive a post-oil future.

The essays in Part V, the final section, entitled "The Futures of and without Oil," take up this last problem raised by LeMenager's essay and address cultural efforts to understand, critique, and look beyond the transient structures of oil capitalism. Chapters by Gerry Canavan, Imre Szeman, Melanie Doherty, and Ruth Salvaggio explore twentieth- and twenty-first-century attempts to develop cultural languages for the representation of oil that might satisfactorily account for the impossibly dense entanglements of contemporary life and petroleum consumption, overturn existing discourses that seek to delay or defuse such evaluations, and foster alternative perspectives on the past and future of oil-based modernity. Gerry Canavan begins the section by addressing historical efforts to grapple with the inescapable transience of oil. Carefully reading a series of twentieth-century science-fiction novels, he tracks the development of two interpretive perspectives that continue to inform understandings of petroleum's end: a techno-utopian belief in the capacity of modern invention to supplant oil with more sophisticated energy systems, and an apocalyptic terror of oil exhaustion as the end point of human history. Imre Szeman investigates three contemporary oil documentaries to assess the difficulties posed by representing petrocapitalism on film and the capacity of the documentary medium to depart from dominant narratives of the oil industry. Melanie Doherty addresses Reza Negarestani's 2008 novel *Cyclonopedia* as an experimental bricolage of literary and theoretical writing that uses a creative engagement with the insights of Speculative Realist philosophy to open new perspectives on oil capitalism. Drawing inspiration from powerful environmentalist icons of the twentieth century (most prominently, Rachel Carson's famous evocation of a "silent spring"), finally, Ruth Salvaggio uses post-Deepwater photographs of the oiled bird to theorize a new mode of resonant imagery that might inspire imaginative engagements with the hidden social and ecological histories of oil extraction and spur the formation of a new sort of sensory engagement with the natural world that would displace the authoritative gaze of technological modernity.

Oil Studies: Present and Future

This volume is an attempt to consolidate the field of "oil studies" as a major component of not just environmental and energy studies but also cultural studies more generally. It is our hope that the contributions to this volume will lead to further work in anthropology, art history, film studies, geography, history, literary studies, museum studies, visual cultural studies, and related fields. Many dimensions of oil's cultural existence call for further study. These include oil's relationship to the seemingly abstract signs of digital media, petroleum's symbolic life within the traditional fine arts, the interpretations of oil advanced by television and print advertisements, cultural promotions produced by industry organizations such as the American Petroleum Institute, poetic and theatrical engagements with oil, the oil-themed programming of advocacy and public-interest groups such as the Center for Land Use Interpretation, and many other representations and symbolic practices.

For a scholar, the framework of oil studies can be both exciting and daunting. Oil studies is exciting because it challenges us to think about material culture in a way that we might not be used to, and it forces us to consider economic, environmental, and political realities that might otherwise seem only tangentially related to our fields of inquiry. Oil studies is daunting for these same reasons—it asks us to think about the real world in a way that forces us to move outside familiar periodizing terms and national boundaries.[29] Oil is global, and oil is older than our modern world. The effects of oil, too, will outlast us all. We believe that these questions of scale and scope can productively and critically reorient us in relation to culture, and make cultural studies all the more relevant to the pressing questions of climate change.

It is perhaps wishful thinking to believe that the positing of "oil studies" will somehow hasten the end of oil's hold on us. Nonetheless, the contributions to this volume all serve as evidence that thinking of oil culturally can help to envision a future without oil. We hope that the future of oil studies is in the study of the past, and we regret already that this will not be the case for some time.

NOTES

1. National Commission on the BP Deepwater Horizon Oil Spill and Offshore Drilling, *Deep Water: The Gulf Oil Disaster and the Future of Offshore Drilling, Report to the President* (January 2011), 190, http://www.oilspillcommission.gov.

2. William R. Freudenberg and Robert Gramling, *Blowout in the Gulf: The BP Oil Spill Disaster and the Future of Energy in America* (Cambridge: MIT Press, 2011), 13.

3. As Rob Nixon notes, "Niger Delta communities have suffered the equivalent of an Exxon-Valdez sized spill annually for half a century." Occurring gradually over time, these spills do not register as much as the more spectacular, sudden spills in or near the United States. See Rob Nixon, *Slow Violence and the Environmentalism of the Poor* (Cambridge: Harvard University Press, 2011), 274.

4. Allan Stoekl, *Bataille's Peak: Energy, Religion, and Postsustainability* (Minneapolis: University of Minnesota Press, 2007), xv.

5. National Commission, *Deep Water*, 305.

6. The new *Dallas* premiered on June 13, 2012, on the TNT network.

7. Enid Nemy, "Larry Hagman, Who Played J.R. Ewing in 'Dallas,' Dies at 81," *New York Times*, November 24, 2012, http://www.nytimes.com.

8. Ibid.

9. For an account of these narratives of a world without oil that looks exactly like a world with oil, see Imre Szeman, "System Failure: Oil, Futurity, and the Anticipation of Disaster," *South Atlantic Quarterly* 106 (Fall 2007): 805–23.

10. Amitav Ghosh, "Petrofiction: The Oil Encounter and the Novel," *New Republic*, March 2, 1992, 29.

11. Szeman, "System Failure," 820–21.

12. Ghosh, "Petrofiction," 29, 30. These claims were meant to underscore the singularity of Munif's works and to redirect the reader to the problematic character of global oil as an object of literary exploration.

13. Martin Schockley, *Southwest Writers Anthology* (Austin: Steck-Vaughn, 1967), xvi. These novels include: Edna Ferber, *Cimarron* (New York: Doubleday, 1929); Karle Baker's *Family Style*

(New York: Coward-McCann, 1937); Mary King O'Donnell, *Quincie Bolliver* (Boston: Hough-ton Mifflin, 1941); Edward Lanham, *Thunder in the Earth* (New York: Harcourt, Brace, and Company, 1942); Carl Coke Rister, *Oil Titan of the Southwest* (Norman: University of Okla-homa Press, 1949); Jewel Gibson, *Black Gold* (New York: Random House, 1950); William Owens, *Fever in the Earth* (New York: G. P. Putnam's Sons, 1958); and Tom Pendleton, *The Iron Orchard* (New York: McGraw-Hill, 1966).

14. Ryszard Kapuściński, *Shah of Shahs*, trans. William R. Brand and Katarzyna Mroczkowska-Brand (New York: Harcourt Brace, 1985), 35.

15. Historian Paul Sabin and art historian Rina Youngner have similarly analyzed how rapa-cious promotional discourses and oil-field photography sustained local investments in the Penn-sylvania petroleum business. See Paul Sabin, "A Dive Into Nature's Great 'Grab-Bag': Nature, Gender, and Capitalism in the Early Pennsylvania Oil Industry," *Pennsylvania History* 66 (Fall 1999): 472–505; and Rina Youngner, *Industry in Art: Pittsburgh, 1812–1920* (Pittsburgh: Uni-versity of Pittsburgh Press, 2006), 41–55. Other work in history has similarly focused on late nineteenth- and early twentieth-century oil fields, both in Pennsylvania and in California. See Brian Frehner, *Finding Oil: The Nature of Petroleum Geology, 1859–1920* (Lincoln: University of Nebraska Press, 2011); and Paul Sabin, *Crude Politics: The California Oil Market, 1900–1940* (Berkeley: University of California Press, 2004).

16. See Michael Watts, "Oil as Money: The Devil's Excrement and the Spectacle of Black Gold," in *Money, Power, and Space,* ed. Stuart Corbridge and Ron Martin (Oxford: Wiley-Blackwell, 1994), 406–45; "Petro-Violence: Community, Extraction, and Political Ecology of a Mythic Commodity," in *Violent Environments,* ed. Nancy Peluso and Michael Watts (Ithaca: Cornell University Press, 2001), 189–212. See also Michael Watts, ed., *Curse of the Black Gold: 50 Years on the Niger Oil Delta* (New York: Powerhouse, 2010).

17. See John Urry, "The System of Automobility," *Theory, Culture & Society* 21 (Octo-ber 2004): 25–39; Cotten Seiler, *A Republic of Drivers: A Cultural History of Automobility in America* (Chicago: University of Chicago Press, 2009). See also David Gartman, "The Ages of the Automobile: The Cultural Logics of the Car," *Theory, Culture & Society* 21 (October 2004): 169–95; John Ott, "Landscapes of Consumption: Auto Tourism and Visual Culture in California, 1920–1940," in *Reading California: Art, Image, and Identity, 1900–2000,* ed. Stephanie Barron, Sheri Bernstein, and Ilene Susan Fort (Berkeley: University of California Press, 2000), 51–68.

18. Peter Hitchcock, "Oil in an American Imaginary," *New Formations* 69 (Summer 2010): 81–97; Stephanie LeMenager, "The Aesthetics of Petroleum, After *Oil!,*" *American Literary His-tory* 24, no. 1 (Spring 2012): 59–86; and LeMenager, "Petro-Melancholia: The BP Blowout and the Arts of Grief," *Qui Parle* 19, no. 2 (Spring–Summer 2011): 25–56.

19. See Szeman, "System Failure"; Imre Szeman, "The Cultural Politics of Oil: On *Lessons of Darkness* and *Black Sea Files,*" *Polygraph* 22 (2010): 3–15; Imre Szeman and Maria Whiteman, "Oil Imaginaries: Critical Realism and the Oil Sands," *Imaginations* 3 (2012): 46–67.

20. Nixon, *Slow Violence.*

21. Timothy Morton, *Hyperobjects: Philosophy and Ecology after the End of the World* (Min-neapolis: University of Minnesota Press, 2013); Morton, *The Ecological Thought* (Cambridge: Harvard University Press, 2010).

22. This burst of scholarly energy has also given rise to a variety of other research initiatives. These include the Petrocultures research cluster at the University of Alberta, founded by Imre

Szeman and Sheena Wilson, and the inaugural Petrocultures Conference (held in September 2012), as well as the *Oil in American History* special issue of *Journal of American History* 99, no. 1 (June 2012).

23. See Michael Watts, "Crude Politics: Life and Death on the Nigerian Oil Fields," *Niger Delta Economies of Violence Working Papers* (2009): 9–12. See also Gavin Bridge, "Global Production Networks and the Extractive Sector," *Journal of Economic Geography* 8 (2008): 389–419. For an influential account of the assemblage that uses the electrical grid as its chief example, see Jane Bennett, "The Agency of Assemblages and the North American Blackout," *Public Culture* 17, no. 3 (Fall 2005): 445–66.

24. See Szeman, "System Failure," 805–9.

25. See Guy Debord, *The Society of the Spectacle*, trans. Donald Nicholson-Smith (New York: Zone Books, 1994): 11–34.

26. See Andrew Apter, *The Pan-African Nation: Oil and the Spectacle of Culture in Nigeria* (Chicago: University of Chicago Press, 2005), 14–45; Watts, "Oil as Money," 408–13.

27. For a statistical analysis of the global expansion of intensive oil consumption, see the U.S. Energy Information Administration, *Short-Term Energy Outlook* (June 11, 2013). http://www .eia.gov/forecasts/steo/report/global_oil.cfm.

28. Szeman, "Cultural Politics of Oil," 34–35; Watts, "Crude Politics," 11–12.

29. This challenge to periodization is also explicitly posed in Patricia Yaeger, "Editor's Column: Literature in the Ages of Wood, Tallow, Coal, Whale Oil, Gasoline, Atomic Power, and Other Energy Sources," *PMLA* 126, no. 2 (2011): 305–26.

PART I

Oil's Origins of Modernization

1

Whale Oil Culture, Consumerism, and Modern Conservation

HEIDI SCOTT

In 1859, the first gush of fossil oil from the ground in Titusville, Pennsylvania, opened a new era for energy production in an energetic young America. Within ten years, a clamorous jumble of drilling towers, pipelines, railroads, and refineries assembled to supply the thirsty markets of the industrial Northeast. Quietly senescing in the background, localized mainly to the New England ports of New Bedford and Nantucket, was an aging titan of American enterprise: the whaling industry. In 1861, *Vanity Fair* ran an illustration of a "Grand Ball Given by the Whales in Honor of the Discovery of Oil Wells in Pennsylvania," a dapper black-tie and ball-gown affair decorated with banners: "We wail no more for our blubber," and "Native Land may they never secede."[1] Indeed, the Civil War would deal another blow to the whaling industry as its ships were appropriated by the Union navy and lost in battle, or in the case of the Great Stone Fleet, purposely sunk in southern harbors as (unsuccessful) blockades.

As a niche market, whaling continued into the twentieth century. The products were updated for twentieth-century automobile and domestic markets, including transmission and brake fluid, plant food and pesticides, and laundry detergent (Figure 1.1). The romance of the perilous sail-driven hunt was lost in the efficient industrialized fleet of iron, steam-powered ships equipped with pneumatic harpoons, which by the mid-twentieth century pushed some whale populations beyond replenishment rates. In addition to the direct threat of species extinction, ecologists have established the ecosystem benefits of healthy whale populations, which seed the deep oceans with carcasses (whale falls), and through defecation add nutrients to shallow waters.[2] Recognizing animal rights arguments and the ecosystem services provided by healthy

FIGURE 1.1. "Pursuit to Preservation" exhibit in the New Bedford Whaling Museum, 2012. Photograph by the author.

whale populations, in 1986 the International Whaling Commission banned commercial whaling.

There are a number of reasons it is worth going retrograde, all the way back to the whalers of colonial America and the early Federal period, in order to gain perspective on consumerism and conservation in today's oil cultures. Whaling was the first American industry to make global economic impacts. Whaling cultivated the entrepreneurial spirit of early Americans, and the whale ship captured the American spirit in microcosm: it required risk, violence, individualism, bravery, and carried a sense of destiny borne on divine favor. The crews of American whale ships were often melting pots of global cultures, as depicted on Melville's *Pequod*. The men were compensated not according to their class in society but their office on the ship. The epic hunt was a difficult, dangerous, and often disgusting venture, but it was also the most exotic and romantic pursuit available to young men seeking to make their fortune and see the world. Whaling also provided an essential point of engagement for evolving ethical perspectives on animal rights and ecological theories of population endangerment due to overhunting. These early environmental qualms place whale oil culture in a historical continuum with the landscape- and ecosystem-level hazards of petroleum extraction. In effect, the eco-ethical protests against whaling that began to emerge in the second half of the nineteenth century prefigure modern conservationist arguments that seek to preserve potential drilling zones for their inherent, cultural, and ecosystem service values.

In this volume, Frederick Buell describes the age of petroleum through a tense dialectic of exuberance and catastrophe echoed in capitalist triumph and oppression, environmental domination and destruction, and human liberation versus physical and psychological shackling. Whale oil culture provides a premonition of this petroleum-era paradox of exuberance and catastrophe. Each of thousands of whaling missions was a boom or a bust according to myriad uncontrollable factors. The marketing of whale oil products exploited the adventure of the hunt, but the market itself was unsteady, based on unreliable whale oil supplies and competing plant- and fossil-based oils. The global oceans, landscapes of the quarry, were sublime and often bountiful, but they were also violent, lonely, and increasingly void of sperm whales in the latter half of the nineteenth century. It is worthwhile to explore how the emotions and ethics of whale oil culture foreshadow the more dominant and ubiquitous culture of petroleum and its counterculture of environmentalism today. On the side of exuberance, the marketing of whale and petroleum products has capitalized on the adventure, risk, and acquisitiveness inherent in the American psyche. On the side of catastrophe, a continuum emerges between the ethical arguments against whaling and those against oil drilling based on the desecration of the source: the whale's body and the drilling landscape. Whaling literature and the marketing of whale oil products reveal a historical antecedent to the decadence, dependence, and degradation characteristic of our petroculture.

In the late eighteenth and early nineteenth centuries, whale oil was nearly as essential to a functional society as petroleum is to our own. Until kerosene (or "coal oil") hit the market in the second half of the nineteenth century, whale oil was the leading source of lamp oil used for illumination, and it provided the best lubrication for the hulking machines of the industrial age. From the ports of New England, American whaling ships of that era would compete on remote oceans with the great global powers of the day—among them England, France, and Japan—and emerge victorious, hulls riding low in the water with a burden of spermaceti. Yankee success in whaling unnerved the British, who in the 1780s placed an embargo on American whale oil to support its own competing industry, knowing full well that its refusal to buy would foster a trade relationship between upstart America and rival France. Ambassador and future president John Adams wrote with sarcasm to British Prime Minister

William Pitt the Younger: "The fat of the spermaceti whale gives the clearest and most beautiful flame of any substance that is known in nature, and we are all surprised that you prefer darkness, and consequent robberies, burglaries, and murders in your streets, to the receiving as a remittance, our spermaceti oil."[3] That same year, the French Marquis de Lafayette signed a contract that filled the streetlamps of the City of Light with American sperm oil.

Whale oil culture is deeply immersed in the adventure of its acquisition. As a commodity, whale oil was sold using the romantic danger of whaling as a marketing strategy that actively promoted its harrowing intrigue. The hunt, the harpoons, the Nantucket sleigh-ride,[4] and the final vanquishing of the beast were all salable images associated with whale products. In our era of animal rights and Save the Whales campaigns, the violence of whale oil culture in this heady era may seem terribly antiquated and grotesque. Will the taboo on whale slaughter promoted by adventure-conservation television shows like *Whale Wars* in the future become a taboo on petroleum extraction, due to its environmental impacts? As we will see, the nascent environmentalism based on whale suffering and diminishing populations is continuous with modern eco-centrism seeking to defend landscapes of petrochemicals against the degradation of intense drilling.

WHALE BODIES AND OIL BARRELS

Commonalities between whale oil culture and petroleum culture are both symbolic and economic. In our time, the increasingly desperate search for new reserves of petroleum drive oil corporations to more remote and lower-quality reserves, such as offshore deepwater drilling and tar sands production. By Melville's time of the mid-nineteenth century, the heyday of whaling, more ships competed for dwindling populations of sperm whales, and the outfits were driven into the far hinterlands of the Pacific and Indian oceans to seek their quarry. The American whaling culture of the seventeenth and early eighteenth centuries merely took day trips off the beaches of Nantucket to reach great pods of migrating whales, make kills, and render the beasts on shore. Barely a century later, professional whaling outfits built floating factories that held all the necessary equipment and expertise to seek, kill, retrieve, and render a whale into tidy barrels below decks using an on-board try-works, or rendering stove. These floating factories remained on the seas for years, and for profit's sake would not return home until the ship sailed heavy in the water. The whaling boat's technical efficiency and self-sufficiency previewed the kind of strategic competition for profit now displayed in offshore oil-drilling rigs. Indeed, the improved technology of whaling, particularly the on-board try-works, enabled the markets to expect a regular supply of whale oil and entrepreneurs to find more uses for its products.

Petroleum and whale oil are both made of the bodies of dead organisms. The "body" of petroleum, mostly ancient zooplankton and algae, is obscured by the deep time and geologic pressure required to convert the body into fuel, as opposed to the explicitly embodied source of whale oil. Like petroleum, whale oil came in many graded varieties, from the highest-quality spermaceti extracted from the head of the sperm whale, to the rendered blubber of the sperm whale, to the blubber of the right whale, and so on downward to impure and rotten oil. Spermaceti was to the whale man what light sweet crude is to the oilman.[5] A contemporary historical survey of the Whale-Fishery by William Scoresby (conducted in 1857) enumerates the many uses of whale products developed by manufacturers:

> The oil produced from the blubber of the whale, in its most common state of preparation, is used for a variety of purposes. It is used in the lighting of the streets of towns, and the interior places of worship, houses, shops, manufactures, etc; it is extensively employed in the manufacture of soft soap, as well as in the preparing of leather and coarse woolen cloths; it is applicable in the manufacture of coarse varnishes and paints, in which, when duly prepared, it affords a strength of body more capable of resisting the weather than paint mixed in the usual way with vegetable oil; it is also extensively used for reducing friction in various kinds of machinery; combined with tar, it is much employed in ship-work, and in the manufacture of cordage, and either simple, or in a state of combination, it is applied to many other useful purposes.[6]

One crucial distinction between whale oil and petroleum is that, because of its value and scarcity, the former was not used as a fuel to drive machinery. Both heavy and light machines were engineered to run on coal steam. Whale oil in industrial settings was strictly a lubricant and illuminant. Since petroleum and whale oil had distinct uses in nineteenth-century industry, the assumption that the rise of petroleum single-handedly slayed the whalers is misleading, though the introduction of kerosene extracted from coal undercut the market for lamp oil from whales. Whale oil in its various grades was infinitely useful to industrial-era machinists, and secondary products less commonly pursued eventually found their own markets. The baleen from the mouth of the right whale became a leading material for ladies' corset stays, whips, and umbrellas because of its dual rigidity and flexibility. By the end of the nineteenth century, baleen fell out of the competition when challenged by steel, and later in the twentieth century, by plastic manufactured from petroleum products.

Greater mythos is associated with the occasional discovery of ambergris in the gut of sick whales, the ultimate conflation of the sublime pleasure and disgust associated with whaling. As Melville writes, "Suddenly from the very heart of this plague, there stole a faint stream of perfume, which flowed through the tide of bad smells without being absorbed by it . . . and this, good friends, is ambergris, worth a gold guinea an ounce to any druggist." The first mate Stubb merely cries, "A purse! A purse!"[7]

Speaking of the whale's body in economic terms was commonplace. As a petroleum reserve is valued for its projected quantity of barrels, so whale-men also described their catches. The discourse allowed whales to be simultaneously described as living creatures and as tanks of profit. It was conventional to "kill 140 barrels of sperm oil" or land a "$4000 sperm."[8] Melville's Ishmael examines the head of a dead sperm whale suspended over the deck of the *Pequod* in simultaneous terms of pathos and profit: "Look at that hanging lower lip! What a huge sulk and pout is there! A sulk and pout, by carpenter's measurement, about twenty feet long and five feet deep; a sulk and pout that will yield you some 500 gallons of oil and more. . . . This particular tongue now before us; at a passing glance I should say it was a six-barreler."[9] Melville claims that large sperm whale's head case would yield about 500 gallons of spermaceti, the most valuable purse.[10] The first mate Stubb makes a cheeky economic conversion as a threat to a black boy still finding his sea legs: "'We can't afford to lose whales by the likes of you; a whale would sell for thirty times what you would, Pip, in Alabama. Bear that in mind, and don't jump any more.' Hereby perhaps Stubb indirectly hinted, that though man loved his fellow, yet man is a money making animal, which propensity too often interferes with his benevolence."[11] Whale-men made easy conversions among individual whales, barrels of their oil, and the value of those barrels, which, as Graham Burnett describes, "meant that a whaleman could look at the surface of the Pacific and see cash breaching, just as he could look at the kegs of oil offloaded at Sag Harbor and see an angry sperm whale on the Brazil Banks."[12] Whales were creatures of commerce, quite simply, and the discourse of money coexisted with romantic rhapsodies on their bestial grandeur.

Melville's Ishmael gets misty over the beatific qualities of the whaler's prize, the spermaceti. His raptures are their own advertisement that translates easily to the commercial market:

> It was our business to squeeze these lumps back into fluid. A sweet and unctuous duty! No wonder that in old times this sperm was such a favorite cosmetic. Such a clearer! Such a sweetener! Such a softener! Such a delicious mollifier! After having my hands in it for only a few minutes, my fingers felt like eels, and began, as it were, to serpentine and spiralize . . . as I bathed my hands among those soft, gentle globules of infiltrated tissues, woven almost within the hour; as they richly broke to my fingers, and discharged all their opulence, like fully ripe grapes their wine; as I snuffed up that uncontaminated aroma,—literally and truly, like the smell of spring violets; I declare to you, that for the

time I lived as in a musky meadow; I forgot all about our horrible oath [to seek Moby Dick]; in that inexpressible sperm, I washed my hands and my heart of it; . . . all the morning long I squeezed that sperm till I myself almost melted into it; I squeezed that sperm till a strange sort of insanity came over me; and I found myself unwittingly squeezing my co-laborers' hands in it, mistaking their hands for the gentle globules. Such an abounding, affectionate, friendly, loving feeling did this avocation beget, that at last I was continually squeezing their hands, and looking into their eyes sentimentally; as much as to say,—Oh! My dear fellow beings, why should we longer cherish any social acerbities, or know the slightest ill-humor or envy! Come, let us squeeze ourselves universally into the very milk and sperm of kindness.[13]

His ecstasy of the spermaceti builds to erotic undertones supported by the rich, multisensual experience of the whale's bounty. Such ardent affection for his unbathed shipmates shows the alchemical nature of the spermaceti, which renders body fluids into the "very milk and sperm" not only of kindness, but of a sense of well-being built upon a thriving profit. The material substance corrects physical ills with its capacities to clear, sweeten, soften, and mollify. But it does not stop there. Spermaceti transcends the material to animate the civic and philosophical ideals of the young America, where such exceptional, abundant natural resources absolve any need for acerbity, ill humor, and envy. Ishmael's ecstasy is the promise of success through diligent hard work: the American dream.

What we see in the commercial whaling imagery of the mid-nineteenth century is congruent with the oil cultures of later decades, where the grandson of the salty harpooner may well have been a "wildcatter" drilling his claim in Texas. The difference is one of degree, not of kind. The risk and reward was extreme for the whaleman, measured by the inches in thrusting harpoons and dodging flukes, tried by the stench of a half-rendered whale on deck and the years of privation on board, and rewarded by the fragrant, pellucid, precious spermaceti that became high-quality lubricants, illuminants, and cosmetics in the marketplace. Those somewhat mundane products were the spoils of an epic and sublime adventure, filled with a dual share of pleasure and horror.

Despite the romance of the hunt, whale ships like the fictional *Pequod* were commissioned, built, outfitted, staffed, and floated as pure business enterprises. Its purpose was to produce spermaceti at a profit, so the whale ship served as a prototype for the American oil corporation that would rise to prominence with Standard Oil by the early twentieth century.[14] Pure business enterprise does not make for great mythological literature, however, and *Moby-Dick* is celebrated as a raw exploration of man versus nature precisely because its crazed captain Ahab disdains any practical application of means toward an end. For Starbuck and the other first mates, the *Pequod* is sailing under an austere dictate to maximize profits for its stakeholders—a corporate mentality. For Ahab, she sails the waters solely to wreak revenge on the white whale—an egocentric rationale. This disagreement within the "board of directors" emerges very early in the text and continues to pluck at the tense lines between characters of unequal power. Starbuck is the only mate who directly challenges Ahab's ambition, and early in the sailing he says, "I came here to hunt whales, not my commander's vengeance. How many barrels will thy vengeance yield thee even if though gettest it, Captain Ahab? It will not fetch thee much in our Nantucket market." To which Ahab tersely replies, "Nantucket market! Hoot!"[15] Much further from their home shore, Starbuck discovers that some of their spermaceti barrels are leaking below decks, so by the dictates of business, they must seek a port to "up Burtons and break out." Ahab responds, "Let it leak! . . . Let the owners stand on Nantucket beach and outyell the Typhoons. What cares Ahab? Owners, owners? Thou art always prating on me, Starbuck, about those miserly owners, as if the owners were my conscience. But look ye, the only real owner of anything is its commander; and hark ye, my conscience is in this ship's keel."[16] At this moment Starbuck fully understands that the mission for profit is moot compared to Ahab's mission of revenge, and that the *Pequod* is no average whaler, and no business operation.

Whale Oil Products: Marketing Romance

In addition to epic literature, there is considerable emotive content in whale oil product advertising that partakes of the same grandeur and glory that Melville extolls. The heroic imagery of many nineteenth-century advertisements can be captured in one of the hundreds of images used to market whale oil (Figure 1.2).

This advertisement for Mitchell and Croasdale of Philadelphia, dating from 1857–60, shows the tranquil whale ship floating upstage, while the action of the scene centers downstage on two harpoon boats and an enormous sperm whale.[17] The whale is impossibly buoyed on top of the water to show its fabulous bulk in comparison to the men in the harpoon boats, and it mightily froths the water from tail to blowhole. One of the harpoon boats is scuttled on the whale's back dangerously close to its smashing flukes, and at least four men are thrashing in the water. A pair of legs flails helplessly in the air, a synecdoche of the lost whale-man symbolic of the uselessness of legs against a sublime sea beast. The other harpoon boat is in better position, with its men at oars following orders and its harpooner poised to sink another iron in the whale's flank. Score one for the whale, one for the men. The whale is still spouting clear water, so the men have yet to deliver the fatal blow when the whale begins to spout blood. The image, *in medias res*, lingers on the violent excitement of irresolution.

FIGURE 1.2. Advertisement for Mitchell and Croasdale of Philadelphia, c. 1857–60.
Courtesy of the New Bedford Whaling Museum.

Below the tumult, in boldface, proprietors Mitchell and Croasdale hock their wares:
different grades of oil for tanning and lubrication, and candles made of spermaceti—
cozy, mundane consumer goods equally available in dozens of other shops. The rhetoric of the image seems, to contemporary sensibilities, to contradict the rhetoric of
the writing. In effect, the advertisement has two messages: (1) These goods were
acquired through a remote, dangerous, and deadly encounter between courageous
predatory man and mighty whale; (2) The whale's blubber is now available in bottles
and candles for your convenient purchase above Arch 37 in the wharves of Philadelphia. The implication of this strange marriage of image and content is that consumers
ought to feel the excitement of the hunt as they light their evening candles, and feel
the sleek power of the whale as its oil keeps their machines lithe. The lives of the men
and cetaceans sacrificed for the market embellish the rhetoric of consumption rather
than being effaced from it. It is an advertisement that belongs to an age when frontier
adventure, risk, and danger were commonly praised in popular discourse.

This flavor of advertisement was favored in exuberant pro-petroleum rhetoric in the first half of the twentieth century. Once the sheer use-value of gasoline had established thirsty markets and spawned the American romance with the automobile, petroleum companies deployed the Texas wildcatter as an embodiment of the risk-hungry, wily, egotistical, free-enterprising, self-made man so readily worshipped in public discourse. This characterization gelled with gasoline ads that commonly depicted juiced-up consumers in thrall to their powerful cars, the new proselytes of the open road. The February 13, 1950, cover of *Time* magazine shows wildcatter Glenn McCarthy with a frontier suntan and tousled hair glancing back at a drilling tower.[18] "Diamond Glenn" is wearing a Houston-brass pinstripe suit, and the anthropomorphic drilling tower sports a Stetson, cowboy boots, and a pair of bulging biceps. McCarthy comes across as a wildcatter for the corporate postwar twentieth century: dressed like a businessman but harboring the adventurous soul of a cowboy. The *Time* story begins: "If it were possible to cap the human ego like a gas well, and to pipe off its more volatile byproducts as fuel, Houston's multi-millionaire wildcatter Glenn McCarthy could heat a city the size of Omaha with no help at all. Whether he would allow his rampant psyche to be dedicated completely to so prosaic a project, however, is doubtful—several million cubic feet would undoubtedly be diverted to a McCarthy Memorial Beacon which would nightly cast its glare as far west as El Paso."[19] So popular was his charm-smacking, egotistical persona that he was replicated in Edna Ferber's 1952 novel, *Giant,* adapted into the 1956 film. The appeal of this iconic wildcatter figure has not diminished in the intervening half-century: a similar, though toned-down, version of the "Diamond Glenn" persona became a two-term American president in the twenty-first century.

Petroleum advertising, however, has shifted over the last half-century from the Texas frontier to the technocratic refinery laboratory. This change from risky adventure to controlled expertise echoes the evolution of whale oil advertising in the second half of the nineteenth century. Whale oil advertising grew to rely on more staid claims for the exceptional quality of their products as compared to competitors, rather than reveling in the danger of whaling. According to one advertisement, Fulcrum Lubricants "positively will not gum . . . will not evaporate . . . will not become rancid or discolor . . . is absolutely free from acids."[20] The only image is a row of bottles that helps the buyer recognize the brand in stores. This clean style is more familiar in today's oil rhetoric, where the geographical source of petroleum extraction is almost never disclosed, but we commonly hear how one brand of gasoline has better detergents and molecular-level engineering that lead to better performance and fuel economy. For example, Shell touts how its "patented cleaning agents shield and protect critical engine parts from performance-robbing gunk" and how "Shell Nitrogen Enriched Gasolines actively protect fuel injectors and intake valves."[21] We never hear

Shell advertising the environmental status of its oil fields in the Niger Delta, in analogy to Mitchell and Croasdale's heroic imagery of the harpooning scene on the high seas. Any information about the people and landscapes involved in petroleum extraction is reserved for defensive press releases or filtered through corporate feel-good initiatives that provide small grants for environmental and humanitarian projects. Oil corporations today carefully promote images of philanthropy and responsibility, often advertising issues unrelated to their business, such as Chevron's AIDS advocacy.

Why the marked shift in these parallel rhetorics of oil consumerism? Perhaps one reason for the changed tune of petroleum rhetoric is that the very volatility of the product is now seen as dangerous on both human and environmental scales. When corporations emphasize the hyperscientific process of petroleum extraction, refinement, and engineering, they underscore the rhetoric of control and mastery that reassures consumers about a dangerous and ubiquitous substance. There is also growing consumer distaste for the suffering caused by the oil products they purchase. This distaste has environmental origins, with concern for whale welfare and, later, the degradation of the landscapes of petroleum extraction. It also links with modern environmental justice, as the human rights of whale-men, oilmen, and natives of the locales that oil extraction has transformed have gained attention. Environmental economics has become an essential risk communication tool that shows in dollar terms the value of the "work" done by ecosystems (and the whales within them), therefore justifying regulations that limit the exploitation of these systems, especially for private profit. The concept of ecosystem services aids in the transition from a cultural mind-set of exploiting oil resources (whales and drilling zones) for the production of profitable consumer goods to a mind-set of preserving oil landscapes because they provide calculable benefits to the society that leaves them intact. Oil resources are also potentially cultural and ecological resources that enable ecotourism, biodiversity, robust food webs, and pollution remediation.

ETHICAL EVOLUTION AND EXTENSION

Whaling literature constantly mediates between economics and emotions elicited by the hunt. Alexander Starbuck (not Melville's character) wrote an essay called "The Story of the Whale" for *Harper's New Monthly Magazine* (1856), which shows how the eco- and therio-centrism of the conscience-stricken whale-man is beginning to emerge in the early waning period of the industry, the middle of the nineteenth century. In this Romantic aesthetic, the whale's mighty body is wrecked by the avaricious hunt:

Of all game pursued by the destructive hand of man [the whale is] the most sublime. . . .
Imagination cannot conceive any thing more awful than the butchery that now takes

place. Terrified, the whale plunges from wave to wave—springs with agony out of the water, and covers the surrounding ocean with blood and foam. He dives downward, leaving a whirlpool in his path—he rushes upward, and the fatal lance enters some still untouched spring of life—whichever way he turns, the cold iron goads him to desperation . . . the once mighty breathing mass, now dead, is tossed contemptuously in the troughs of the sea.[22]

This description could apply to the Mitchell and Croasdale advertisement discussed above, but its pathos is clearly aimed to elicit sympathy for the whale, who suffers as human quarry, and anger toward the heartless hunters and their cold harpoons. Starbuck extends man's destruction of the mighty and peaceful beast to its entire habitat: whale-men are "destroyers of the sea"; their actions toward sea dwellers have an eviscerating effect on the whole landscape.[23] True to the ambivalence of his period, Starbuck goes on to describe the heroic bravery of the harpooner involved in mortal peril, and he relishes stories of whale violence and seemingly premeditated vengeance like the story of the *Essex*, the ship splintered by a monster whale that sparked Melville's epic. However, the final pages of Starbuck's essay are plaintive ones, where he weighs the intelligence and adaptability of the whale against the onslaught of the market-driven hunt:

> Circumstances favor the probability that the time will eventually come when the great leviathan of the deep will be exterminated. In the course of two centuries it has been driven from sea to sea; and now, with the scientific discoveries of Wilkes and Maury added to perseverance of the whaleman, it has no resting place. It is estimated that ten thousand are annually slain, and the increase can not equal the destruction. The number of ships engaged in the pursuit is constantly increasing.[24]

Starbuck finds hope in the "intellectual faculties" and potential self-awareness of the great mammal, and imagines behavioral adaptation toward preservation through a combination of greater stealth, wariness, and aggression toward whalers.[25] Starbuck invokes both the evolution of the animal and the preservation of its habitat, the ocean, which is "indeed richer in treasure than the land—its great characteristic is abundance," as necessary bulwarks against the increasing sophistication of whaling science.[26] As goes the animal, so goes the landscape, and both deserve protection in the face of purely economic drives.

Today's environmentalism has picked up Starbuck's thread. Animal rights activism is gaining strength from Enlightenment-era declarations of the rights of man based not upon shared species (humanity's unalienable rights), nor on the ability to suffer (as in Bentham's and Singer's animal rights philosophy), but most recently upon the demonstrated intelligence of cetaceans. A recent article in *The Economist* discusses an American Association for the Advancement of Science (AAAS) conference panel that used comparative anatomy to defend the rights of whales as "people."[27] Their brains have similar anatomical complexity as humans and contain spindle cells, a kind of nerve cell that we associate with abstract cognition. Initial steps toward establishing the legal personhood of whales include using the word "population" (a human-linked word) instead of "stock" to refer to cetacean groups, and researching specific populations of whales to discover their unique "culture" that distinguishes them from other groups in the same species.

Philosophy and anatomy have the power to make inroads on the legal standing and protective status of certain species. But this line of argument has more reach. Legislation like the Endangered Species Act protects not only particular species threatened by human activity, but also, by ecological necessity, their entire habitat. Thus landscapes, by "ecological extension," can receive special protection status against human invasion based upon the species that live within them. The Spotted Owl, to take one controversial example, has long been silencing bulldozers in Pacific Northwest forests. According to Starbuck, man without conscience is a destroyer of the sea, but with an eco-conscience that foreshadows major tenets of Deep Ecology, even the whaler can perceive that his quarry is a protective mother, or a helpless calf, or a stoic old bull with war scars to prove his mettle. The "abundance" of the ocean is its leading characteristic, and Starbuck uses whales as the charismatic mega-fauna that make the whole seascape sublime and worth appreciation. Deep Ecology holds that all species, and even all land-/seascapes, are equal in intrinsic worth, and should therefore all be treated with the same ethical status. Ecological extension implies the rethinking of physical boundaries from the skin of the animal body to the entire landscape upon which the body depends; this eco-spiritual identification originates in European Romanticism, notably Rousseau and Wordsworth. Across the Atlantic and a century later, Aldo Leopold's Land Ethic became an important part of this tradition as it proposed that the land-/seascape should be considered a member of the community, not a substrate and resource base that human groups possess for their utility.

With this strategy environmentalists have approached the protection of landscapes that hold petroleum. Oil drilling is not overtly slaughter, as whaling is, so the ethical line is necessarily a subtler one. The argument about the "body" of the landscape deserving protection can be made by submerged metaphor, as when the National Resources Defense Council describes how the Arctic Wildlife Refuge "continues to pulse with million-year-old ecological rhythms" and includes "a fragile swath of tundra that teems with staggering numbers of birds and animals."[28] The landscape itself has a "pulse," a "rhythm," and in effect a metabolism or homeostasis regulated by its complement of native species. Against this evolved, haven-like biodiversity, the NRDC characterizes the apocalyptic hell-scape of Prudhoe Bay, already exploited for oil, as "a landscape defaced by mountains of sewage sludge, scrap metal, garbage, and more than 60 contaminated waste sites that contain—and often leak—acids, lead, pesticides, solvents, and diesel fuel." Their rhetorical strategy is to characterize the former landscape as alive and homeostatic in ecological balance, and the latter as sick or dead, trampled under the treads of industrial exploitation. While an Endangered Species Act assessment might argue to protect the entire landscape on the premise "save the Musk Ox!," this NRDC strategy uses the macrocosmic vision of a living, dynamic ecosystem as a superorganism. The landscape is a popular level of analysis and intervention in today's environmental ethics.

Oil landscapes are often described in terms of an original eco-dynamic purity lost to the smeary onslaught of industrial production. Nearly 150 years after Alexander Starbuck's plea for the landscape of the whale, Joshua Hammer's 1996 essay in *Harper's* focuses on Nigeria in familiar terms of opposition, with biblical undertones to invoke the "sin" of ecosystem degradation that follows rampant drilling:

> To fly into the Niger Delta is to fall from grace. From the air, the silvery waters seem peaceful. Dubbed "The Venice of West Africa" in 1867 by British explorer Winwood Reade, the Delta stretches 290 miles along the Atlantic coast from the Benin River in the west to the Cross River in the east. In between, the powerful Niger feeds an intricate network of tributaries and creeks that partition sandbars and mangrove islands into cookie-cutter shapes as they meander toward the Gulf of Guinea.
>
> Yet far beneath the belly of the airplane, oil fields mottle the landscape, their rigs ceaselessly pumping crude and natural gas from deep underground. The gas burns incessantly in giant geysers of flame and smoke, and at night the flares that ring the city of Port Harcourt and fishing villages deep within the mangrove swamps cast a hellish glow. As the smoke from the flares rises above the palm trees, methane and carbon dioxide separate from the greasy soot. The gases rise but the grime descends, coating the trees, the laundry hanging on lines, the mud-daubed huts, and the people within. There is nothing pure left in Nigeria.[29]

This vision of a lost Eden is reminiscent of Dickens's Coketown, where the maniacal machines "ceaselessly" and "incessantly" pump the landscape to the glow of sooty fires. The impact of modern industrialism has moved far afield from urban centers like London or Pittsburgh into the nineteenth-century colonial frontier of wild Africa, eliciting Hammer's retrospective dialectic of purity and contamination, Eden and

Hell, innocence lost. From his God's eye view, Hammer analyzes the aesthetic conversion from the intricate ecological "network of tributaries and creeks" to the descent of universal grime even on remote fishing villages: the ecological body's descent from healthy heterogeneity to sickly homogeneity. This shroud of death suffocates the entire landscape, showing no favorites between human and nonhuman, animal, plant, or mineral. Environmental justice, which focuses on the oppression of people in degraded landscapes, is signaled by his focus on huts and laundry, and the implication that humans who suffer the conditions of oil extraction do not benefit from their sacrifice. Still, the passage is one of eco-centric rather than anthropocentric injustice, with a common sad fate for all the "people" of this Nigerian landscape. Like the great whale grossly skinned at the ship's side, this ecosystem bears the stamp of ethical perversion that cannot be justified by economic gain once it is painted in Hammer's vivid terms. In addition, any economic calculus that favors exploitation must consider the imbedded services lost by transforming the landscape from a productive ecosystem into a productive oil system.

The advertising rhetorics of whale and petroleum oil have followed a similar trajectory across their discrete historical arcs. From advertising literature based on blind risk and wild adventure, both whale oil and petroleum evolved in popular discourse toward a calculated "responsible" rhetoric of technological expertise, quality, and control. This evolution is the result of the parallel shift from an initial culture of exuberance to an emergent culture of catastrophe that came to surround each oil source. For whale oil, the shift came toward the end of the nineteenth century, and petroleum followed about a century later. In both cases, nascent environmental awareness and protest necessitated the emergence of "quality control" assurances, as the impacts of too much extraction became clear in the damage to oil landscapes, both sea and land. Of course, global cultures are nowhere near renouncing petroleum because of its environmental impacts, from landscape to climate; with ample substitutes, weaning cultures from whale oil was a much less formidable task. Industrial consumerism often proceeds from celebration to concealment of the origins and means of its oily products, and environmental ethical writing about whales and wells exposes these elisions to a newly conscious audience.

Notes

1. "Grand Ball Given by the Whales in Honor of the Discovery of Oil Wells in Pennsylvania," *Vanity Fair* (April 1861), 186.

2. Joe Roman and James J. McCarthy, "The Whale Pump: Marine Mammals Enhance Primary Productivity in a Coastal Basin," *PLoS ONE* 5 (October 2010), n.p., www.plosone.org.

3. Eric J. Dolin, *Leviathan: The History of Whaling in America* (New York: W. W. Norton, 2007), 168.

4. The term "Nantucket sleigh-ride" refers to the consequence of a successful harpoon-thrust: when the small harpoon boats skewered a whale with harpoons, the injured animal usually bolted and pulled the boat and its men violently through its wake. During the "sleigh-ride" the men held on until the whale exhausted itself.

5. Peter Tertzakian, *A Thousand Barrels a Second: The Coming Oil Break Point and the Challenges Facing an Energy Dependent World* (New York: McGraw-Hill), 16.

6. William Scoresby, *The Northern Whale-Fishery* (London: The Religious Tract Society, c. 1857), 170–71.

7. Herman Melville, *Moby-Dick* (New York: Longman Critical Editions, 2007), 362.

8. D. Graham Burnett, *Trying Leviathan: The Nineteenth-Century New York Court Case That Put the Whale on Trial and Challenged the Order of Nature* (Princeton: Princeton University Press, 2007), 110.

9. Melville, *Moby-Dick*, 300–301.

10. Ibid., 305.

11. Ibid., 366.

12. Burnett, *Trying Leviathan*, 137.

13. Melville, *Moby-Dick*, 368–69.

14. Robert D. Wagner Jr., *Moby-Dick and the Mythology of Oil* (self-published, 2010), 99.

15. Melville, *Moby-Dick*, 159.

16. Ibid., 416.

17. Advertisement for Mitchell and Croasdale's whale oil products. New Bedford Whaling Museum, c. 1857–60.

18. "Texas: King of the Wildcatters," *Time,* February 13, 1950, 18–21.

19. Ibid.

20. "Fulcrum Lubricants," Fulcrum Oil Company. New Bedford Whaling Museum, n.d.

21. "Shell Nitrogen Enriched Gasolines," Shell Oil Corporation, http://www.shell.us/products-services/on-the-road/fuels.html.

22. Alexander Starbuck, "The Story of the Whale," *Harper's New Monthly Magazine* 12 (March 1856): 466, 473–74.

23. Ibid., 474.

24. Ibid., 480–81.

25. Ibid., 481.

26. Ibid., 466.

27. "Whales Are People, Too," *Economist,* February 25, 2012, 92–94.

28. "Arctic Wildlife Refuge: Why Trash an American Treasure for a Tiny Percentage of Our Oil Needs?" Natural Resources Defense Council, 2012. http://www.nrdc.org/land/wilderness/arctic/asp.

29. Joshua Hammer, "Nigeria Crude: A Hanged Man and an Oil-Fouled Landscape," *Harper's* 292 (June 1996): 58.

2

The Wizard of Oil

Abraham James, the Harmonial Wells, and the Psychometric History of the Oil Industry

ROCHELLE RAINERI ZUCK

On October 31, 1866, four men were traveling by buggy a few miles south of Pleasantville, Pennsylvania, which at the time was little more than a crossroads on the edge of the oil fields. One of the men, Abraham James, was no stranger to the region, although this was the first time he had set foot, physically, in Venango County. According to biographer and fellow spiritualist James Martin Peebles, James had "frequently visit[ed] these Pennsylvania oil regions as a spirit, accompanied by his spirit-guides."[1] On this particular visit, James, along with C. P. Easton, George Porter, and George McBride, was looking at a piece of property owned by Porter. Suddenly, James was forcefully possessed by his spirit-guides and conveyed out of the buggy and over a fence on the east side of the road. He did not know if he was "in the body or out" and was moved around a field and thrown violently upon the ground. James then stuck a penny into the ground and fell into a "psychometrical condition," during which "Indian spirits controlled his body mechanically, while wisdom spirits induced the trance condition."[2] Speaking through James, spirits communicated that the men were standing on an immense oil deposit. Harmonial no. 1 was sunk on the William Porter farm, exactly on the spot James had indicated. Drilling began on August 31, 1867, and the event was written about in newspapers across western Pennsylvania and New York and featured in spiritualist publications. Oil was struck in February 1868 at a depth of approximately 835 feet and Harmonial no. 1 began to produce more than one hundred barrels

FIGURE 2.1. John A. Mather, Harmonial Well. Courtesy of the Drake Well Museum, Pennsylvania Historical and Museum Commission.

a day (three other Harmonial wells would later be drilled in the area). The Pleasant-ville excitement had begun, and Abraham James and his spirit-guides stood at the convergence of two nineteenth-century American phenomena: spiritualism and oil.

American spiritualism and the oil industry developed around the same time and in relatively close geographic proximity. By the mid-nineteenth century, spiritualism had anywhere from half a million to a rather generous estimate of eleven million adherents, and had garnered immense popular attention as the public learned of the spirit rappings of Kate and Margaret Fox at their home in Hydesville, New York, in 1848.[3] Reflecting on the popularity of American spiritualism, theologian Theodore Parker remarked that "in 1856, it seemed more likely that spiritualism would become the religion of America than in 156 that Christianity would become the religion of the Roman Empire, or in 756 that Mohamedanism would be that of the Arabian pop-ulation."[4] In addition to lectures, spiritualist conventions, and circles held in private homes, there were numerous books, pamphlets, and periodicals devoted to spiritual-ist philosophy, and spiritualism circulated throughout many of the popular reform movements of the nineteenth century. The oil boom also generated popular enthusi-asm and curiosity. With the success of Col. Edwin Drake and the Pennsylvania Rock Oil Company (later Seneca Oil) in 1859, people began to flock to the western Penn-sylvania oil fields hoping to find wealth and opportunity. An era of pond freshets, exploding wagons full of nitroglycerin, and wild financial speculation had begun.[5]

As this essay will suggest, American spiritualism and the early oil industry had more in common than just temporal and geographic proximity. These two nineteenth-century phenomena were both invested in a belief in the unseen, whether in the form of deceased loved ones or underground oil reserves. Spiritualists such as James turned to the oil industry because of its lucrative financial opportunities and because of its potential to demonstrate the "practical" applications of spiritualism and Harmonial philosophy (as advanced by Andrew Jackson Davis, the "seer of Poughkeepsie," whose work James read and admired).[6] Spiritualism offered an alternative to evangelical Christian and classical republican conceptions of industry, and a vibrant communication network ready to relay sensational accounts of the oil region to the general public.[7] Reading accounts of James's work as an "oil wizard" reveals the industrial aspirations of spiritualism and the psychometric aspects of the oil industry, both of which have been largely erased from twentieth-century historiography.[8] This essay focuses on biographical accounts of James and accounts of his work in Chicago (1863–64) and Pleasantville, Pennsylvania (1866–68), to show how spiritualism shaped and was shaped by nineteenth-century oil culture. Throughout the nineteenth century, spiritualist publications, newspapers, technical manuals, and popular accounts of the oil industry produced James as a new kind of male medium, capable of meeting the exigencies of the oil fields. He proved infinitely reproducible as an agent of "practical spiritualism" and was discussed alongside the other drillers, operators, laborers, teamsters, and investors at work in the oil region. As petroleum geology began to establish itself as a discipline in the early twentieth century, accounts of the early oil industry began to reframe James, along with other practitioners of divination, as an amusing, if somewhat embarrassing, anomaly in an attempt to distinguish the modern "scientific" oil industry from its chaotic and superstitious beginnings. While later historians have offered a more sympathetic reading of divination's role in the oil fields, James and his Harmonial Well have largely disappeared from the historical record. Yet despite scientific innovation and revisionist history, the oil industry still bears traces of its psychometric past and must contend with the ways in which its future is dependent on successfully channeling the unseen.

ABRAHAM JAMES: THE MEDIUM AND THE MESSAGE

Like the oil that he was credited with locating and the spirits with whom he apparently communed, James provided a kind of rhetorical availability and functioned as a site onto which discourses, desires, and anxieties of various kinds could be mapped. Just as Edwin Drake's personal history was slightly altered to meet the exigencies of the oil fields—he was given the honorary title of "Colonel" by James M. Townsend, then president of the Pennsylvania Rock Oil Company, to establish his credibility in western Pennsylvania—details of James's life were adapted to produce him as someone who could traverse the spiritual and the industrial. Various and sometimes contradictory pictures of James emerge from nineteenth- and twentieth-century accounts of his life and work. Descriptions of his physical appearance, age, background, personality, and the identity of his spirit guides frequently differed; not even his name was consistent.[9] Certainly, James himself may have been partially responsible for these changes; he seems to have answered to both "Abraham" and "Abram" and may have given different accounts of his life and work in the various places that he visited. Merely viewing James as a confidence man, however, obscures the ways in which, in the early 1860s, James and the oil he sought to locate were commodities in search of value, a value that then had to be produced and marketed to the public.[10] This section will focus on the ways in which biographical accounts of James, while they may have differed on some of the details of his life, represented him as a liminal figure who moved between the spiritual and the physical realms. Pamphlets written by Chicago spiritualist George Shufeldt in the 1860s and Peebles's *The Practical of Spiritualism: A Biographical Sketch of Abraham James* (1868) constructed James as a new kind of male medium, a practical spiritualist who brought both "Indian" wisdom and geological knowledge to his exploration of subterranean frontiers.

Who was Abraham James? A survey of the various accounts of his life suggests that he was born in Pennsylvania and may have worked for the railroad and traveled to California and the Pacific Northwest as a young man. He was involved with the Spirits' Well in Chicago in 1863–64 and arrived in the oil region of Pennsylvania around 1866. The rest of his biography is harder to pin down. An obituary, published in the *New York Times* on November 29, 1884, claims that James was "reported dead" in Oregon (the precise date is unknown) at the age of seventy-seven, which suggests that he was born sometime around 1807.[11] However, this account is not substantiated by any of the other narratives of James's life and work. Peebles's account of James's life indicated a birth date of June 16, 1827, a date supported by the 1900 Census, which shows an Abraham James born on that date who resided in Fredonia, New York, at the time of the census. That Abraham James was married to a Martha Ann James, had several children, and practiced vitapathy in Fredonia, New York, while also working in the oil business.[12] The Abraham James listed in the 1900 Census also had a twin brother, William James, a fact that dovetails with the account of James's family given in Peebles's biography.[13] The "facts" of James's life, including verifiable birth and death dates, familial connections, and places of residence have proved somewhat elusive, which has allowed historians and biographers to project certain traits onto him, while still maintaining an air of mystery that proved rhetorically useful for promoting his spiritual powers.

James was positioned as an explicitly Christian spiritualist, in the style of his biographer, J. M. Peebles, himself a well-known nineteenth-century spiritualist and author, and Andrew Jackson Davis. Several accounts of James published in the 1860s note that he was of Quaker origins, often as testament to his veracity and the validity of his communion with the spirit world. Presenting James as a Quaker linked him with the broader philosophical and religious history of spiritualism.[14] Moreover, associating James with a particular religious tradition served as a means of emphasizing spiritualism's connections with, rather than opposition to, Christianity and American religious history. Peebles, who published the only full-length biography of James in 1868 as part of an attempt to prove "that religion and science are a unit," went further in proclaiming James's connections with mysticism and establishing a genealogy for his spiritual powers.[15] He wrote of James's birth: "Under the sign of Gemini, Jupiter the reigning planet, and Venus exercising a controlling influence, Abraham James, a twin brother, was born on the 16th day of June 1827." James's mother, according to Peebles, was also gifted with the "second sight."[16] Peebles repeatedly doubled James, giving him all the familiar markings of a mystic while still maintaining his connection to Christian tradition and material reality; he was a twin born under the sign of the twins to a mother with two kinds of vision and lived a kind of "dual existence," simultaneously inhabiting spiritual and earthly realms.[17] Later discussions of James expanded on connections between the medium and Christian prophecy. In *The Early and Later History of Petroleum* (1873), written by J. T. Henry, the medium's experience is likened to that of Paul on the road to Damascus. Harry Botsford's *The Valley of Oil* explicitly linked James with the Old Testament prophets, and many accounts of his well placement note his religious background so as to distinguish him from predatory "sharpers" and duplicitous oil firms such as those satirized by a popular song of the 1860s ("There's Ketchum & Cheatum, and Lure'um and Bleed'um, and Swindle'um all in a row").[18] Within the context of nineteenth-century industrialism, which was framed as immoral and irreligious by ministers and social reformers, James's particular brand of spiritualism offered a model of moral industry that combined practicality, individuality, and empiricism with the ability to transcend one's personal concerns and focus on the greater good. Such a model resonated not only with mainstream Protestant discourse but also with republican notions of the ideal male citizen.

Yet James's masculinity was itself informed by spiritualist discourse and stood in contrast with that embodied by the rough and uncouth oilmen described in many nineteenth-century accounts of the industry. Shufeldt, Peebles, and other writers de-emphasized or feminized James's physical presence, using what was to spiritualists a familiar discourse associated with male mediums.[19] In a pamphlet entitled *Petroleum Near Chicago* (1864), Shufeldt described James as "mild and soft in his manners,

[and] gentle in his disposition."[20] Despite his weak constitution, James had been forced by material circumstances, Shufeldt recalled, to work for the railroad and perform other forms of physical labor to which he was unsuited. In a later pamphlet, Shufeldt reiterated his assertion that James's physical appearance did not attract attention, characterizing him as "unostentatious."[21] Peebles further highlighted James's physical weakness and asserted that James had spent his early years in California for health reasons. His work as a conductor on the railroad had been very taxing for someone so fragile. According to Peebles, James had a "spare slender figure" and a countenance that phrenologists would interpret as marked by candor and integrity. He was gentle and retiring, and his gait was "easy and unstudied."[22] James's physical body was unremarkable even to the point of weakness, which further emphasized his sensitivity and spiritual, rather than physical, capacities. And, according to Peebles, James inherited his spiritual powers from his mother, who continued to encourage his mediumship after her death.[23] These powers were presented as a maternal legacy that James, because of his gentle and receptive nature, was able to wield. As Molly McGarry and other historians of spiritualism have noted, young women played a crucial role in spiritualist practice, but "women and—sometimes less often—effeminate, 'gentle' men were called to spiritualist service."[24] Yet narratives about James balanced traits that were coded "feminine"—mildness, gentleness, and a slender physical form—with James's business sense, work experience, and the scientific knowledge of his spirit-guides, offering readers a vision of nineteenth-century manhood that traversed the spiritual and the scientific/industrial. Moreover, his work with the oil industry functioned to create a new vision of the male medium as uniquely suited to the spiritual and material exigencies of nineteenth-century capitalism.

Discussions of James's education and work experience reinforced this sense of a man who moved easily between the spiritual/cerebral and the industrial/physical, in accordance with Harmonial philosophy. Shufeldt in particular highlighted his lack of formal education, initially framing James as a man "of good natural abilities, [and] fair ordinary education."[25] His second account of James, published in *History of the Chicago Artesian Well* (1865), went a bit further, calling James "simple-minded, in the sense that he knows nothing of frauds, trickery, or imposture—perfectly truthful and upright in his character, unostentatious, and seeking no publicity or notoriety." In short, according to Shufeldt, James was "a natural, honest man."[26] Peebles presented a slightly different view of James's education, claiming that his family could not afford a great deal of schooling, but noted that James did attend Unionville Seminary and taught school briefly in the South. Several accounts suggest that James worked as a conductor for the railroad and used his powers as a medium to prevent accidents. According to Peebles, it was James's "mediumship" that gave him the reputation of being a safe conductor.[27] It was while visiting his brother, who was "tunneling for gold" in California, that James was introduced to Andrew Jackson Davis's *The Principles of Nature, Her Divine Revelations, and a Voice to Mankind* (1847).[28] Through James's engagement with Davis's Harmonial philosophy, "he clearly grasped the theory that this visible world was but the phenomenal exhibition of that superior spirit-realm which comprehends the energizing forces, the primal forms and the eternal laws of the world."[29] Peebles wrote that James, like Davis and other spiritualists and Transcendentalists who were influenced by Emanuel Swedenborg, embraced a vision of the spirit realm that "existed in a relationship Swedenborg called 'correspondence' with the material realm" and that "the two realms operated according to parallel laws and spiritual realities had physical expression."[30] Like his work on the railroad, James's search for oil was framed in the language of correspondence and Davis's Harmonial philosophy rather than the language of the individual pursuit of wealth. With the exception of his observations of his brother's work in the California gold rush, Peebles reported, James was untrained in geology.[31] His placement of wells was not attributed to his own knowledge but to that of his spirit-circle, made up of American Indians and geologists who were said to be eager to demonstrate through James that human beings do not lose their interests and experience after death. As Peebles wrote, James's experiences proved that "changing worlds does not change identity" and "activity and progress" continues after human beings pass into the next realm.[32] Although Shufeldt emphasized James's intellectual blankness (even calling him "simple-minded"), Peebles framed James as educated by experience and as part of a larger effort in which the specific geological knowledge needed to produce wells was supplied not by James, but by his attending spirits. In the hands of Peebles, this group effort reinforced nineteenth-century discourses of individualism by demonstrating that people maintain their "self" even after they die and both the spiritual and physical realms are characterized by "progress."

The equation of progress with the communication of Native wisdom from Indian spirits to a white medium echoed colonialist rhetorics of Manifest Destiny, in which American Indian nations were imagined to give way before the advance of white settlers, and framed the subterranean oil reserves as a kind of "frontier" to be explored by the practical spiritualist. The account of James's work published in the *New York Times* obituary explicitly framed James as an explorer: "Presently the earth opened and an immense cavern yawned before them. Into this James was led by the spirit. They journeyed down into the earth for a long distance, and finally the spirit brought James to the margin of a lake of petroleum of unknown depth and extent. Speechless with amazement, James gazed on that apparently boundless store of wealth for a few minutes, when the spirit led him back to the surface. The mouth of the cavern closed, and the spirit vanished."[33] In this scene, which is reminiscent of images of western expansion, James functioned as what Mary Louise Pratt called the "seeing man," an agent of colonialism who visually surveys and possesses rhetorically available territory.[34] In terms that resonated with nineteenth-century ideologies of Manifest Destiny, the oil fields were constructed as a new kind of frontier, where once again Native people were imagined to pave the way for white "progress." Yet it was the practical spiritualist, not the woodsman or the calvary officer, who was positioned at the forefront of this exploration.

Spiritualist and nonspiritualist publications rehearsed James's biography in order to establish his credibility as a medium and challenge the notion that religion was diametrically opposed to science and industry. Particularly in Peebles's biography, James embodied Swedenborgian notions of correspondence between the spiritual and material realms; reading the Harmonial philosophy of Davis allowed him to better understand his dual nature and apply it to his work on the railroad and in the oil fields. He appeared in the historical record as the product of his familial and religious background, his own experience, and the knowledge of his spirit-circle. James was packaged and sold to nineteenth-century readers in spiritualist publications, secular newspapers and journals, and book-length accounts of the early oil industry. As the concrete details of his life remained sparse and he himself produced few surviving writings, James functioned as a medium through which notions of practical spiritualism and ideological constructions of the oil industry could be communicated to the public.[35]

CHICAGO: THE OILDERADO THAT WASN'T

Chicago is not often included in histories of the early American oil industry, but it was the site of Abraham James's first attempt to place an oil well using psychometrics and the first time that what we might call the spiritualist network engaged the topic of petroleum. While the presence of oil had been noted for some time by workers in the limestone quarries, which provided building materials for many of Chicago's noted landmarks of the period, there was no "oil industry" as such near the city in the early 1860s.[36] Drake's success in Titusville set off a chain of events that led to the famous oil boom in western Pennsylvania and attempts to duplicate, or even surpass, his efforts in locations further west such as Chicago. During this period, Chicago had a well-established spiritualist community that sought to both communicate the practical side of spiritualism and cement the city's place as a modern cultural and economic center. Their efforts led to the drilling of the aptly named Spirits' Well, which was located on a forty-acre tract of land on what was then the outskirts of Chicago, at the corner of Chicago and Western Avenues, about three and a quarter miles from the courthouse.[37] Rival accounts of James's involvement with the Spirits' Well published in the 1860s reveal not only the medium honing his craft but also the internal divisions among practical spiritualists, which included disagreements about the proper relationship between spiritualism and public enterprises like the oil industry. As the following section will suggest, Chicago was a test of sorts, not only of the emergent oil technology but also of the ways in which practical spiritualism could intervene in the pursuit of petroleum.

In the 1860s, Chicago was the site of significant spiritualist activity. Spiritualists had formed a society there in the 1850s and held meetings at "Harmony Hall," located on Clark Street. Andrew Jackson Davis came to the city in 1856 to give lectures, and historian Alfred Theodore Andreas has noted that there were at least fifteen mediums in the city at this time.[38] The first national spiritualist convention was held in Chicago in 1864 and another convention was held the following year.[39] The city was also a center of spiritualist publishing; two spiritualist journals, *News from the Spirit World* (1868–1870) and the *Religio-Philosophical Journal* (1865–1905), were produced in Chicago.[40] *The Spiritual Magazine* described Chicago in an 1866 article about James and the Spirits' Well as "the newest and most go-ahead city of the world."[41] Reflecting on the relationship between Chicago and spiritual philosophy, Dr. William Cleveland argued, " Chicago is new, Spiritualism is new, they commenced life about the same time, and the line of progress continues about equal."[42] Less religiously orthodox than some eastern cities, Chicago was positioned by Cleveland and others as the center of American religious thought, a city that could eclipse New York in terms of cultural influence. Shufeldt argued that Chicago, with its central location and diverse population, could function as a stage for a spiritual performance that could be communicated to the entire world.[43] James came to the city in the early 1860s and was part of a larger group of mediums and spiritualists who formed the Chicago Rock Oil Company. This company intended to demonstrate the power and veracity of spiritualism by the placement of oil wells and promote the advancement of Chicago on the national stage. Accounts of the Spirits' Well suggest that rather than being opposed to capitalism and industrial development, James's spiritual communications were framed as an effect of capitalism: to speak to the interests of nineteenth-century Americans, the spirits were forced to communicate through languages of industry and commerce. Members of the Chicago Rock Oil Company hoped that oil itself could function as a kind of medium and persuade nonbelievers that spiritualism could make practical contributions to modern life.

James was not the first medium involved with the Spirits' Well, but he would come to be celebrated as the one who placed it. In the summer of 1863, Mr. Thomas J. Whitehead and Mr. A. E. Swift, two spiritualists, visited Chicago on business and met in a spiritualist circle with Mrs. Caroline Jordan, a relative of Mr. Whitehead. Through the mediumship of Mrs. Jordan, the group received a communication from the spiritual realm that a further communication of an important nature would soon be given. According to Shufeldt, Mrs. Jordan declared that "Petroleum, or Rock Oil, existed in large quantities near Chicago, and the particular tract or piece of land was pointed out on which it would be found." The communication was said to have come from "E. P. Hines," a noted geologist from Maine and the friend of Mr. Swift. James then joined the group and produced several spoken revelations as well as a geological map, completed in a trancelike state.[44] But questions remained about where exactly to place the well. According to Shufeldt, in "the most prolific oil regions of Pennsylvania, not more than one well in ten strikes the oil," and the odds in Illinois remained even slimmer.[45] Surface indications were notoriously unreliable, and direct communication from the spirits represented an alternative to the often-misleading sensory evidence that could be gleaned from ordinary human observation. Shufeldt described how James determined the location of the well. He was traveling with Swift, Whitehead, and A. F. Croskey across the tract of land that the spirits had indicated in October 1863. They were just about to return home when James became entranced by spirits and indicated that the party was standing on an immense oil reserve.[46] This performance prefigured in several ways the one that he would enact in Pleasantville. James took three investors out in the month of October to tour the site of the proposed well. Entering into a trancelike state, he was physically moved by the spirits, predicted the presence of oil, and argued for drilling to commence.

The Chicago spiritualists were divided, however, by ideological differences and the material challenges of separating the oil and water produced by their wells. James became a locus of these internal divisions. In her account of the operation, which was written as a rebuttal to Shufeldt's histories of the well, Adaline Buffum, editor of *News from the Spirit World,* noted the many parties who were involved in the operation. The drilling "was protected by various bands of spirits: mechanical, scientific, geological, financial, and political; also, ancient bands of Hebrews and Chaldeans, and philosophers, prophets, poets and heralds from the most remote sphere, with Roger Bacon, of the thirteenth century, as sentinel, to drive away all the myriads, and myriads of spiritual and material substances that lurked about, not necessary to the enterprise of boring for oil, and the whole was surrounded by a wisdom sphere."[47] Spirits helped to place the well and were also invoked to ward off the usual dangers that accompanied drilling. Despite the efforts of Bacon and the other spirits, the Rock Oil Company struck water instead of oil and the mediums (and spirits) began to disagree with one another as to how to proceed. Twenty-four spirits were consulted; twelve voted to keep drilling for oil, while the other twelve voted to pursue the water. Conveniently, the spirits could not be held to any one position on the issue. According to Buffum, Jordan and another female medium, Mrs. Genung, challenged James's placement of the well. Genung channeled the spirit of Tecumseh, who apparently trumped James's unnamed spiritual advisers. Although he had been framed in other accounts as gentle and almost effeminate, in Buffum's account James invoked his masculinity when answering Jordan's challenge, telling the women that drilling for oil was "no boy's play" and that they should listen to his advice.[48] Shufeldt's accounts, which largely wrote women out of the history of the Spirits' Well, reinforced the idea that while women were central to the overall project of spiritualism, the placement of oil wells was the work of male mediums such as James.

When the Spirits' Well failed to produce enough oil to be profitable, the Chicago Rock Oil Company quickly transformed itself into the Chicago Artesian Well Company and James soon left for the oil regions of Pennsylvania. Buffum was critical of the Well Company, and her critique registered the imbrication of spiritualism and capitalism. She recounted the spirits' promise of a profitable oil well "for they were obliged to make demonstrations through the power of gold, owing to the present condition of society, since all mankind seemed to be guided through that channel."[49] Returning to an earlier republican discourse that equated industry with corruption, she concluded that the enterprise failed because those involved with the oil well were too "selfish" and "carnal minded" for the operation to be successful.[50] Her conclusion was that spiritualism needed to be cleansed of "imposters" (James is implicated here but not named) and "great public enterprises like the *Chicago Rock Oil Co., alias Artesian Well Co.*"[51] Peebles, however, claimed that Davis's theory of "correspondence" between the spiritual and the physical accounted for the discovery of water. Since human beings are largely made of water, he argued, "sympathy" exists between human bodies and water, which explains why James's early mediumistic efforts resulted in the location of water rather than oil.[52] The spirits meant oil to be a "second lesson" for James, one that he would learn in western Pennsylvania.[53] The Chicago Artesian Well Company incorporated in 1867 and Buffum's warnings, which were informed by republican associations between industry and greed, were largely overwritten by the Harmonial rhetoric of Shufeldt and Peebles.

Spiritualism had already proven useful to the development of "public enterprises" such as the early oil industry by providing an established print network and rhetorics of correspondence between the physical and the spiritual that undercut equations of industry with corruption and binaries of success and failure. Despite a lack of oil, the Spirits' Well became a tourist site, largely due to the marketing abilities of the more "carnal minded" Chicago spiritualists; James's geological drawings were displayed and Shufeldt's *History of the Chicago Artesian Well* was available for purchase by interested visitors. At least three histories of the well were published during the 1860s, the first of which was distributed at the 1864 spiritualist convention, and the well was also featured in periodicals as diverse as *Scientific American, The New York Evangelist, Friends' Intelligencer,* and *The Round Table.*[54] The intersection of spiritualism and the oil industry also created new opportunities for male mediums, who positioned themselves as the forefront of practical spiritualism. James touted his placement of the "celebrated Artesian well" while on the lecture circuit and likely used his connection to the well to establish his reputation as a medium in Pennsylvania.[55]

PLEASANTVILLE: PSYCHOMETRICS AND PETROLEUM

After he left Chicago, James took his brand of practical spiritualism to the oil fields of western Pennsylvania, reportedly arriving on October 31, 1866, and bringing with him the celebrity that he earned in Chicago and the resources of the spiritualist communication network. While he would emerge as the most famous medium to visit the oil fields, James was part of a broader culture of divination that already had a toehold in the region. Various forms of divination, including those practiced by oil dousers, who used Y-shaped rods to lead them to drilling spots, and oil smellers, who relied on their olfactory abilities, were employed in the oil region in the 1860s.[56] The following section will situate James within the broader context of oil diviners and chart the ways in which accounts of James's life and work dovetail with larger literary and historiographical trends related to oil industry. Early accounts of divination and James's psychometrics framed the use of spiritualism as proof of the pragmatic approach of the early oil operators, who used all available means at their disposal to avoid the costs associated with an unproductive well. Yet, as oil geology became a more established and "professional" discipline in the early twentieth century, Abraham James was recast as an object of derision and a curious relic of a bygone era.[57] The focus shifted from James himself to his spirits, who became increasingly sensationalized as stereotypical Indian figures meant to further emphasize the "premodern" features of James and his work. Ultimately, James and the Harmonial wells that he placed seemed to disappear from studies of the history of the Pennsylvania oil fields and the development of the oil industry.

James came to the oil fields at a volatile period between the failure of Charles Vernon Culver's banks in 1866 and the Pleasantville excitement that was produced by Harmonial no. 1.[58] Financial panic swept the region in 1867 as a result of the bank failures. During this period the estimated cost of sinking a well was $3,300, and half of all wells were unproductive. For the 50 percent that did produce, the average life span was six months.[59] As Peebles suggested, "There were little system and no science exhibited in putting down these wells. Hundreds of wells were sunk to the proper depth and no oil obtained."[60] Concerns about the unpredictability of the enterprise intensified when prices fell in 1867 and productivity was comparatively low.[61] Summing up the feelings of many who came to the region in search of oil, one writer for the *New York Tribune* reflected that "the business is all lottery, say the operators."[62] People throughout the region were in search of a means of rendering the process of locating wells intelligible and predictable. According to Roger M. and Diana Davids Olien, "In the oil fields the laws of man were both bent and broken" and "the new petroleum industry was unscientific, unpredictable."[63] Tracing the development of the oil industry from a speculative venture to a scientific enterprise, Edgar Wesley Owen concluded: "In the nineteenth century, they followed hunches and set up their rigs by seepages or located places to drill according to 'creekology' and topographic 'breaks' and 'belt lines,' or followed a douser ('doodlebug'). Later the wildcatters adapted geology almost as readily as did the major companies."[64] When James arrived in the oil fields of Pennsylvania, he worked alongside a host of what Brian Frehner has called "vernacular prospectors," men who generally did not have formal training in geology and engaged in the business of locating oil long before petroleum geography became an established discipline.[65] These vernacular prospectors relied on sensory data, surface indications, and tools such as the dowsing rod or the doodlebug (a term that could refer to the gadget or its user). According to Black: "The divining rod became the most acceptable method of well location [in the 1860s]. Using a forked twig from a witch hazel shrub or a peach tree, a diviner would tightly hold the wood as he passed around the property. A swift downward dip of the wood, often imperceptible to the naked eye, would show where one should locate a well."[66] Other oil-finding devices were more elaborate, including one doodlebug device that "consisted of bells, whistles, and dials attached to a large black box that a man operated while seated and covered with a shroud while four men carried him and the device around prospective oil fields."[67] Although he did not mention James specifically, Black noted that during the 1860s, the industry went "metaphysical" and "spiritualists and oil smellers also made a fine living in the oil fields."[68] James and his spirit-guides were part of a larger cast of characters who populated the early oil fields and derived their authority not from geological expertise, but from spiritual, sensory, or extrasensory sources. Yet the success of the Harmonial wells, which were credited with starting the Pleasantville excitement, combined with James's connections to a broader spiritualist network, served to distinguish him from numerous other "vernacular prospectors" who lived and worked in the oil regions of Pennsylvania and garnered him widespread fame.

Throughout the nineteenth century, James, his Harmonial wells, and his psychometric approach were regularly featured in accounts of the oil region published in handbooks and technical manuals, newspapers and periodicals (produced by spiritualists

FIGURE 2.2. John A. Mather, General View of the William Porter Farm. Courtesy of the Drake Well Museum, Pennsylvania Historical and Museum Commission.

and nonspiritualists alike), and in book-length accounts of the "discovery" of oil written for popular audiences. Spiritualists used their literary connections to keep James in the public eye, celebrating James's work in Pleasantville in periodicals such as *Buchanan's Journal of Man* and books such as Emma Hardinge's encyclopedic *Modern American Spiritualism* (1870).[69] James himself gave lectures on the topic. Accounts written by nonspiritualists offer psychometrics as one of a number of available approaches to what seemed like an arbitrary endeavor. In an 1865 account of his travels in the oil region, William Wright conceded that "the diviners have a power in Petrolia, among a people as keenly inquisitive and practical as are to be found" because their services cost very little as compared to the overall expense of sinking a well, and any sort of advice was better than drilling blind.[70] Writers such as J. T. Trowbridge, whose "A Carpet-Bagger in Pennsylvania" was published in the *Atlantic Monthly* in 1869, claimed that James's success spoke for itself in the oil regions: "I hardly know what effect this practical argument of the spiritualists will have on the minds of unbelievers. I talk with some of these, who smile at it, saying that, although James's enterprise succeeded, many similar attempts to find oil or treasure through spirit agencies have failed, and that consequently nothing is proved. Still I perceive that they talk of James with respect. 'There is one good thing, success.' Everybody appreciates that."[71] James's success, which included several wells that produced more than one hundred barrels per day, overrode skepticism about his approach.[72] An 1879 *History of Venango County* echoes similar sentiments: "There are many operators in whose minds are yet fresh the implicit confidence placed in 'Oil Wizards,' and their power to successfully locate wells."[73] Nearly all of the accounts published in the nineteenth century also note that whether or not other operators believed in James and his spirits, many placed wells very close to Harmonial no. 1, hoping to capitalize on James's success.[74]

Accounts varied as to the extent of James's financial success and what he did after the collapse of the Pleasantville oil boom in 1869. Some framed James as one of the great success stories of the region, while others assert that he made nothing from the endeavor or lost his fortune through a series of bad investments. Cone and Johns suggested that by the time of the Pleasantville collapse, James was one of the wealthiest oilmen in the area, and an article in the *Spiritual Magazine* described James's philanthropic support of spiritualist halls, libraries for children's lyceums, and individual lecturers and mediums.[75] According to *The Derrick's Hand-Book*, after the Pleasantville oil field was exhausted, James purchased land near President, Pennsylvania (roughly halfway between Oil City and Tionesta on the Allegheny River), and attempted to replicate his success with Harmonial no. 1. Yet another account claimed that James tried to sink a well in the middle of the Clarion River and lost $6,000 in the process. The *New York Times* obituary claimed that James left the oil region worth an estimated $500,000.[76] The pages of nineteenth-century stories of the region are filled with examples of drillers, operators, and investors who, like James, made and lost money in the oil fields. That James and his spirits may not have been infallible initially did not detract from the fame of the Harmonial wells and the Pleasantville boom. It was only later that stories of his failed attempt to place a well in the middle of the Clarion River came to eclipse his achievements.[77]

By the early decades of the twentieth century, James and his Harmonial wells had, like the doodlebugs, become an amusing footnote, a reminder of an earlier time before the oil industry became serious and scientific. Such accounts provided more detail about his spirits, often framing them as American Indian leaders, than they did about James. In a 1915 account of a commemoration of Drake's strike, an anonymous author from the *New York Times* devotes several paragraphs to a discussion of James and his Harmonial well, framing it as one of a series of curious events that transpired in the chaotic period before Standard Oil intervened, regulating prices and stabilizing the market (it is noteworthy that Standard Oil functioned as the hero here, in the wake of the 1911 breakup and mass public disapproval). James in this article is guided by a figure called "Mountain Bear," who was ostensibly a Seneca leader. Journalist Herbert Asbury, who wrote a series of "informal histories" of subjects ranging from gambling to the San Francisco and New York underworlds, discussed James and his "friendly spirit of a Seneca Indian chief" at some length in *The Golden Flood: An Informal History of America's First Oil Field* (1942). Asbury added a few flourishes to the then-familiar story of James and his placement of the Harmonial well; he wrote that when James was possessed by his spirit-guide, "a stream of Indian talk pour[ed] from his lips" and he "went skipping across one of Porter's fields like a jack rabbit, bounding high into the air at every jump."[78] According to Asbury, "As long as James remained in the Pleasantville field his control was infallible, but when he attempted to operate in other parts of the region the Indian chief appeared to have lost his bearings."[79] Nevertheless, Asbury does concede that James became wealthy in the oil regions, despite the fallibility of his spirits.

Shortly after the publication of Asbury's history, popular histories transformed the identity of James's spirit from a Seneca chief to Lalah, described as a beautiful Indian maiden, a romanticized figure who served to further distance James from the practical, scientific oil industry of the present day. Harry Botsford included an entire chapter on James (whom he called Abram James) in his 1946 book *The Valley of Oil*, although the chapter was titled "Lalah, the Spirit Guide." The story of Lalah, which would be repeated by later writers, appears first in Botsford's work, a sensationalized vision of the early oil industry that contrasted sharply with the contemporary moment:

> It is a far cry from these modern pioneers, who have planned and applied scientific methods to the secondary recovery of oil from fields that have had the appearance a few decades ago of being depleted, to the days of Abram James. Lalah, the spirit guide, has been replaced, as has the old rule-of-thumb, by engineering, hydraulics, and the slide rule. Today's oil men prefer to deal with tangibles. If another Abram James should appear on the scene, even if the spirit guide were to be streamlined and jet-propelled, it is most unlikely that the oil industry would accept the proffered guidance.
>
> Today's oil men are practical folk; if they have anything to do with spirits they want them straight from the bottle."[80]

Here Botsford conflated James and "Lalah" with the nineteenth-century oil industry as a whole so as to contrast the "modern" and "scientific" methods (engineering, hydraulics, and the highly advanced slide rule) of his own era with the premodern methods

used by James (spiritual communication with romanticized preindustrial Indian figures). These representations of Indian spirits are less about advancing a kind of frontier logic, as in the earlier accounts, and more about distinguishing James from the
"practical folk" of the 1940s.

James's increasing marginalization reflected broader trends in narratives produced
by and about the oil industry, in which various forms of psychometrics were replaced
by geology. According to Frehner, a 1940 advertisement by the Union Oil Company
of California that appeared in *Fortune Magazine* celebrated the triumph of science
over the doodlebug and his black box.[81] Like the doodlebug, James and his psychometric methods were distinguished from both practical and scientific methods of
oil location, recast as folklore, and, in some cases, dismissed entirely as a fraud. Paul
Giddens, preeminent historian of the early oil industry, acknowledged the success of
the Harmonial but also claimed that "no one put any faith in the highly supernatural
narrative of James."[82] James appeared in Hildegarde Dolson's *The Great Oilderado*
(1959) as a "rabid-eyed spiritualist" named James Abram in a chapter entitled "Spies
and Robbers," an example of the corrupt outsiders who were drawn to the oil fields
(Coal Oil Johnny here is framed as "one of the few natives who forgot himself").[83]
Mody Boatright's *Folklore of the Oil Industry* (1963) featured James along with several
other "oil locators whose methods were professedly occult" in a work that also
includes Paul Bunyan and tall tales of the oil region.[84] Claiming that "in many respects
the oil witch bears the same relation to petroleum geology and geophysics that the
alchemist bears to chemistry," Boatright recounted the story of James and his Indian
guide (who in this account locate oil on the Potter rather than Porter farm) as part of
a larger argument for the "importance of folk-belief in the history of the oil industry."[85] While arguing for the importance of James and his fellow "occult" practitioners, Boatright's study nonetheless served to amplify the distinction between James's
psychometrics and the practical, scientific efforts of other oil operators.

Although the practice of divination has received more attention in late twentieth-
and twenty-first-century accounts of the early oil industry, James and his role in the
Pleasantville excitement have virtually disappeared from the historical record.[86] Owen's
magisterial *Trek of the Oil Finders*, published as part of the semicentennial of the founding of the American Association of Petroleum Geologists, chronicled the "methods,
men, and the industrial mechanism involved in the oil finding effort" and included
brief discussions of doodlebugs, dowsing, and divining, but does not mention James
at all in the more than fifteen hundred pages that it devotes to the topic of oil finding.[87]
Challenging narratives produced by companies such as Union Oil, Frehner's *Finding
Oil* (2011) argued that "a number of different constituencies contributed to the body
of knowledge that large corporations used for finding oil," constituencies that included

doodlebugs and clairvoyants.[88] Although he does not mention James or his Harmonial Well, Frehner's work invites a reconsideration of figures such as James, who listed his occupation as "oil operator" in U.S. Census records and was viewed by contemporaries as an embodiment of both the practical and spiritual dimensions of oil location.

Although nineteenth-century accounts recognized the ways in which spiritualists such as James, Shufeldt, and others were an integral part of the industrial landscape, the contributions of James and the other "occasional," "incidental," and "practical" oilmen were marginalized in historical accounts written by and about the oil industry during the first decades of the twentieth century, a period Owen connects with the establishment of "the professional practice of petroleum geology."[89] Industry, it seemed, no longer need to connect itself with Americans' spiritual lives or offer a governing morality for its actions, and the "practical" oilman was supplanted by the "professional" geologist. Recovering Abraham James and reintroducing him and "practical" spiritualism to accounts of the early oil industry continue the work of challenging teleological narratives by and about the oil industry in which the "scientific" and professional triumphed over the varied practices of amateurs and charlatans. James and his Harmonial wells remind us of the practical applications of spiritualism and the psychometric aspects of the oil industry and force us to rethink widely held distinctions between the spiritual and scientific.

CONCLUSION

It was by no means a foregone conclusion in the 1860s that the United States would become the oil culture that it is today. In the years following Drake's "discovery" of oil, investors and operators were not entirely certain of its value. They largely acted on blind faith and sheer force of will, hoping to develop a reliable system for placing wells and to demonstrate the utility of crude oil. Practical spiritualism, which took mediums and spirits out of the parlor and into the oil fields, was deeply involved in marketing the oil industry to the American people. It provided not only established networks (publishers, lecture circuits and lyceums, and so forth), but also a rhetoric of correspondence between the physical and spiritual worlds that allowed the excitement and pursuant dangers of drilling and speculation to be recast in terms of "discovery" and divine revelation, languages of frontier exploration that resonated with the nineteenth-century American national psyche.[90] Abraham James, constructed as a liminal figure who traversed the physical and the metaphysical, was deployed in nineteenth-century print discourse to legitimate both practical spiritualism and the oil industry. While later historians recast him as an anomaly in order to celebrate the modernity and technological know-how of their own historical moment, James reveals dualities and tensions that continue to characterize the oil industry writ large and U.S. oil culture. Commenting on the catastrophic events in the Gulf, a *New York Times* article published on December 25, 2010, notes that "for all of the Horizon's *engineering wizardry* [emphasis added], it was tangling with powerful and highly unpredictable geological forces."[91] After more than 140 years, the oil industry still pits its wizards against the forces of nature and geology, staking its success on the combined efforts of science and psychometrics.

NOTES

A significant debt of gratitude is owed to Kathy Cooper, Reid Mahlon Whitaker, and the library staff of Carleton College for providing me with a copy of J. M. Peebles's *The Practical of Spiritualism*; James Green, Connie King, and the staff of the Library Company of Philadelphia; Susan Beates and the staff of the Drake Well Museum Library; the staff of the Charles L. Suhr Library at Clarion University, Venango Campus; and Douglas Shephard of the Barker Historical Museum in Fredonia, New York, for helping me track down references to James, vitapathy, and Fredonia. Thank you also to Susan Maher, Carolyn Sigler, Craig Stroupe, the English and Writing Studies Departments at UMD; to Abram Anders and Evan Brier for their generous feedback on an early version of this essay; to Ron Nagy, Historian of the Lily Dale Assembly; and to Samuel Zuck and Clarence Pelaghi for suggesting primary sources related to Pennsylvania's oil history.

1. J. M. Peebles, *The Practical of Spiritualism: A Biographical Sketch of Abraham James. Historic Description of his Oil-Well Discoveries in Pleasantville, Pa., Through Spirit Direction* (Chicago: Horton & Leonard, 1868), 36. Copy available in the Lawrence McKinley Gould Library at Carleton College.

2. Ibid., 77, 37. The *Oxford English Dictionary* defines psychometry thus: "The supposed practice of obtaining information about an object's history, or about people or events with which it has been associated, purely by touching it or through close proximity to it." For more on the various uses of this term, see "psychometry, n," *OED Online*, Oxford University Press.

3. Accounts of the number of professed spiritualists in America vary. For more, see Whitney Cross, *The Burned-Over District: The Social and Intellectual History of Western New York, 1800–1850* (Ithaca: Cornell University Press, 1950), 349; Sheri Weinstein, "Technologies of Vision: Spiritualism and Science in Nineteenth-Century America," in *Spectral America: Phantoms and the National Imagination*, ed. Jeffrey Andrew Weinstock (Madison: University of Wisconsin Press, 2004), 125; and Molly McGarry, *Ghosts of Futures Past: Spiritualism and the Cultural Politics of Nineteenth-Century America* (Berkeley: University of California Press, 2008), 3. According to Robert Fuller, by the 1870s spiritualism boasted as many as eleven million followers, while Barbara Goldsmith offers a more conservative estimate, putting the number of believers around two million in 1870. See Robert Fuller, *Mesmerism and the American Cure of Souls* (Philadelphia: University of Pennsylvania Press, 1982), 95; and Barbara Goldsmith, *Other Powers: The Age of Suffrage, Spiritualism, and the Scandalous Victoria Woodhull* (New York: Alfred A. Knopf, 1998), 78.

4. Parker quoted in McGarry, *Ghosts of Futures Past*, 3.

5. Pond freshets were an early means of transporting oil from the oil fields to Oil City, where it could be transported by boat to Pittsburgh. The tributaries of Oil Creek were dammed and

barrels of oil loaded onto flat-bottomed boats. A rider would indicate when each dam should be released, and the resulting rise in water levels would push the boats downstream. An average of fifteen thousand to twenty thousand barrels would be transported in a single pond freshet, and collisions and other accidents were common occurrences. On May 31, 1864, between twenty thousand and thirty thousand barrels of oil were lost near Oil City. For more, see J. T. Henry, *The Early and Later History of Petroleum: With Authentic Facts in Regards to its Development in Western Pennsylvania, the Parkers' and Butler County Oil Fields* (Philadelphia: Jas. B. Rodgers Co., 1873), 287; and Charles A. Babcock, *Venango County, Pennsylvania: Her Pioneers and Her People, Embracing a General History of the Region*, 2 vols. (Chicago: J. H. Beers & Co., 1919), 1:304.

6. Harmonial philosophy asserted, among other things, that the material world (i.e., Nature) was a reflection of and existed in harmony with the spiritual realm. "Davis's version of Swedenborg's correspondence guaranteed that the existence and activity of heavenly spheres were replicated in the worldly arrangements of an American earth." Catharine Albanese, "On the Matter of Spirit: Andrew Jackson Davis and the Marriage of God and Nature," *Journal of the American Academy of Religion* 60, no. 1 (1992): 6.

7. Discussions of the oil industry and other forms of industry in the mid-nineteenth century were informed by American understandings of classical republicanism, which juxtaposed agrarian virtue with the greed and corruption of manufacturing and commerce. This discursive tradition often operated in conjunction with evangelical Protestant warnings about the single-minded pursuit of wealth and conservative fears about how the newly rich might misuse their money and upset established social norms. A figure such as the infamous John W. Steele (aka "Coal Oil Johnny") was offered as a cautionary tale of how individuals, families, and communities could be destabilized by the economic promise of the oil fields. For more on Steele as a cautionary tale, see Roger M. Olien and Diana Davids Olien, *Oil and Ideology: The Cultural Creation of the American Petroleum Industry* (Chapel Hill: University of North Carolina Press, 2000), 23–24.

8. The term "oil wizards," which is also evoked in the title of this essay, appears in *History of Venango County, Pennsylvania, and Incidentally of Petroleum, Together with Accounts of Early Settlement and Progress of Each Township, Borough, and Village, with Personal and Biographical Sketches of the Early Settlers, Representative Men, Family Records, etc. By An Able Corps of Historians*, ed. J. H. Newton (Columbus: J. A. Caldwell, 1879), 404.

9. While most accounts refer to him as Abraham James, he is also called Abram James, James Abram, William F. James, and "Crazy James."

10. Brian Black notes that two major journals, *Merchant's Magazine* and the *Journal of the Franklin Institute*, began what he calls "a decade-long search for uses of the abundant petroleum." For more on the search for uses of crude oil, see Brian Black, *Petrolia: The Landscape of America's First Oil Boom* (Baltimore: Johns Hopkins University Press, 2000), 34–36 (35).

11. "Early Days in Oil Creek. Death of the Discoverer of the Pleasantville Oil Fields," *New York Times* (29 November 1884), 3.

12. An 1891 letter from Abraham James to John Bunyan Campbell, the founder of vitapathy, is reprinted as "Letter from a Former Graduate" in Campbell's *Spirit Vitapathy: A Religious Scientific System of Health and Life for Body and Soul, with All-Healing Spirit Power, as Employed by Jesus, the Christ, His Apostles, and Others, That Cures and Saves All Who Receive It* (Cincinnati: H. Watkin, 1891), 299–300. In this letter, James writes that he is "extensively engaged in the development of a new and prolific oil field in the State of Kentucky" (299). Vitapathy, which

seems to have been closely linked with spiritualism, was defined elsewhere in Bunyan's book as a "religious system of health" (11).

13. An "Abrham James" who appears the 1880 Census as born in 1828 (not 1827 as listed in the 1900 Census), living in Fredonia, New York, in 1880, and married to Martha Anne James, listed his occupation as "oil operator." Abrham James, age 52, "United States Census, 1880," Census Records, FHL microfilm 1254553, *United States Bureau of Census*, National Archives, Washington, D.C., *FamilySearch*, April 8, 2011; Abraham James, n.d., "United States Census, 1900," Census Records, FHL microfilm 1241238, *United States Bureau of Census*, National Archives, Washington, D.C., *FamilySearch*, April 8, 2011.

14. On James's Quaker background, see George Shufeldt, *Petroleum Near Chicago. Its Existence a Spiritual Revelation* (Chicago: Chicago Evening Journal Print, 1864), 7; and *History of the Chicago Artesian Well* (Chicago: Religio-Philosophical Publishing Association, 1865), 15–16. Copies are available at the Library Company of Philadelphia. See also Peebles, *The Practical of Spiritualism*, 7–9, who notes that James's parents broke ranks with Quaker orthodoxy and maintained a spirit of religious freedom; and "Early Days in Oil Creek," 3. Shakers (also known as "Shaking Quakers") and Quaker mystics, who proclaimed a direct experience of the divine, profoundly influenced American spiritualism.

15. Peebles, preface, *The Practical of Spiritualism*, n.p.

16. Peebles, *The Practical of Spiritualism*, 7–8.

17. Ibid., 22.

18. The lyrics of the song "Famous Oil Firms" is quoted in Olien and Olien, *Oil and Ideology*, 25.

19. The one exception is Harry Botsford, *The Valley of Oil* (New York: Hastings, 1946), who described James as "a fine figure of a man" who was "inclined to be stout" (243). In contrast to other accounts, Botsford describes James as having a full beard, dark hair, and dark eyes.

20. Shufeldt, *Petroleum Near Chicago*, 7.

21. Shufeldt, *History of the Chicago Artesian Well*, 16.

22. Peebles, *The Practical of Spiritualism*, 25.

23. For an account of a conversation that James had with his deceased mother on the subject of his clairvoyance, see ibid., 21.

24. McGarry, *Ghosts of Futures Past*, 28.

25. Shufeldt, *Petroleum Near Chicago*, 7.

26. Shufeldt, *History of the Chicago Artesian Well*, 16.

27. Peebles, *The Practical of Spiritualism*, 17. For more on James's education and work as a school teacher, see Peebles, 14.

28. Ibid., 18.

29. Ibid., 19.

30. Bret A. Carroll, *Spiritualism in Antebellum America* (Bloomington: Indiana University Press, 1997), 26.

31. Skeptics of James's spiritual powers would claim that he knew more about geology than he let on. "Early Days in Oil Creek," 3.

32. Peebles, *The Practical of Spiritualism*, 38.

33. "Early Days in Oil Creek," 3.

34. Mary Louise Pratt, *Imperial Eyes: Travel Writing and Transculturation* (New York: Routledge, 1992), 9.

35. The 1891 letter to Bunyan is the only piece of writing attributed to James that I have been able to locate.

36. For more on the limestone, Joliet Marble, and surface indications, see Shufeldt, *Petroleum Near Chicago*, 7–8.

37. Chicago Avenue was renamed Artesian Avenue to commemorate the placement of the well.

38. A. T. Andreas, *History of Chicago: From the Earliest Period to the Present Time*, 3 vols. (Chicago: A. T. Andreas, 1884), 1:165.

39. Ann Braude, *Radical Spirits: Spiritualism and Women's Rights in the Nineteenth Century*, 2nd ed. (Bloomington: Indiana University Press, 2001), 165.

40. McGarry, *Ghosts of Futures Past*, 47, describes the *Religio-Philosophical Journal's* support of women's rights.

41. "The Artesian Well of Chicago, and the Spirits," *Spiritual Magazine*, January 1, 1866, 5.

42. William Cleveland, *The Religion of Modern Spiritualism and Its Phenomena. Compared with the Christian Religion and Its Miracles* (Cincinnati: The Truth of Light Publishing Co., 1896), 149.

43. Shufeldt, *Petroleum Near Chicago*, 9

44. Ibid., 6–7.

45. Ibid., 10.

46. Ibid., 10–11.

47. Adaline Buffum, *History of the Spirits' Oil Well, Alias Artesian Well Near Chicago. The Facts as the Spirits Gave Them. The Artesian Well No Demonstration of Spirit Power* (Chicago: Religio–Philosophical Publishing Association, 1866), 6–7. Copy available at the Library Company of Philadelphia.

48. Ibid., 8.

49. Ibid., 7.

50. Ibid., 11.

51. Ibid., 15.

52. Peebles, *The Practical of Spiritualism*, 29.

53. Ibid., 30.

54. See, for example, George Shufeldt, "Letter no. 1," *Scientific American*, vol. 14, September 1867, 163; Ambrose, "Chicago Revisited," *The New York Evangelist* 36, no. 46 (November 16, 1865), 1; D.W., "Subterranean Rivers," *The Round Table. A Saturday Review of Politics, Finance, Literature, Society* (April 3, 1869), 219; J. M. Ellis, "The Great West," *Friends' Intelligencer* 23, no. 20 (July 21, 1866), 300, 314–15. The above sources were accessed online through *American Periodicals Series*. For an account published in London, see Benjamin Coleman, "Passing Events—The Spread of Spiritualism," *Spiritualist Magazine*, vol. 4 (London: James Burns, 1869), 68–72.

55. "Religious Notices," *New York Times*, February 23, 1867, 6.

56. Andrew Cone and Walter R. Johns noted that "several good wells were obtained that had been located, as asserted, by parties who profess to be guided by revelations made by spirits or spiritualistic manifestations." Andrew Cone and Walter R. Johns, *Petrolia: A Brief History of the Pennsylvania Petroleum Region, Its Development, Growth, Resources, Etc., From 1859 to 1869* (New York: Appleton and Company, 1870), 122.

57. For more on the professionalization of oil geology, see Brian Frehner, *Finding Oil: The Nature of Petroleum Geography, 1859–1920* (Lincoln: University of Nebraska Press, 2011). He

has noted that "the discipline of petroleum geology coalesced in approximately the second decade of the twentieth century" (22).

58. Black, *Petrolia,* 180–88.

59. Quoted in P. C. Boyle, ed., *The Derrick's Hand-Book of Petroleum: A Complete Chronological and Statistical Review of Petroleum Developments,* 2 vols. (Oil City: The Derrick Publishing Company, 1898), 1:88.

60. Peebles, *The Practical of Spiritualism,* 35.

61. *The Derrick's Handbook,* 98.

62. Quoted in Peebles, *The Practical of Spiritualism,* 64.

63. Olien and Olien, *Oil and Ideology,* 24.

64. Edgar Wesley Owen, *Trek of the Oil Finders: A History of Exploration for Petroleum* (Tulsa: The American Association of Petroleum Geologists, 1975), 1577.

65. Frehner, *Finding Oil,* 21.

66. Black, *Petrolia,* 46.

67. Frehner, *Finding Oil,* 27.

68. Black, *Petrolia,* 46.

69. George Shufeldt, "Spiritual Phenomena. Abram James—Man and Medium," *Buchanan's Journal of Man* 1, no. 2 (March 1887), online, *Project Gutenburg,* April 17, 2010; Emma Hardinge, *Modern American Spiritualism* (New York: The Author, 1870), 517, 528–30. Spiritualists also extended their influence across the Atlantic; excerpts from J. T. Trowbridge, "A Carpet-Bagger in Pennsylvania," *Atlantic Monthly* 24, no. 140 (1869): 739, were reprinted as "Readings in the *Atlantic Monthly.* Spiritualism Everywhere," *Spiritual Magazine* 4 (June 1871; London: James Burns, 1871), 242–43.

70. William Wright, *The Oil Regions of Pennsylvania* (New York: Harper & Brothers, 1865), 62–63.

71. Trowbridge, "A Carpet-Bagger in Pennsylvania," 739.

72. One 1866 account by Rev. S. J. M. Eaton, framed as a "popular description of the oil region," dismisses those who listen to the advice of "lying spirits" without naming James specifically. Rev. S. J. M. Eaton, *Petroleum: A History of the Oil Regions of Venango County, Pennsylvania* (Philadelphia: J. P. Skelly & Co., 1866), iii, 88.

73. *History of Venango County,* 404.

74. James's Harmonial no. 1 was a victim of its own success. So many wells were sunk in its vicinity that its overall production was greatly decreased. *The Derrick's Hand-Book,* 108.

75. Cone and Johns, *Petrolia,* 467; Coleman, "Passing Events," 71–72.

76. *The Derrick's Hand-Book,* 113; Henry, *The Early and Later History of Petroleum,* 216; *History of Venango County,* 404; "Early Days in Oil Creek," 3.

77. Gary S. McKinney, *Oil on the Brain: The Discovery of Oil and the Excitement of the Boom in Northwestern Pennsylvania,* 3rd ed. (2003; Chicora, Pa.: Mechling Bookbindery, 2008), 273.

78. Asbury, *The Golden Flood: An Informal History of America's First Oil Field* (New York: Alfred A. Knopf, 1942), 270. Twenty-first-century audiences are perhaps most famous with one of Asbury's other "informal histories," *The Gangs of New York: An Informal History of the Underworld* (New York: Alfred A. Knopf, 1927), which informed the 2002 Martin Scorsese film.

79. Asbury, *The Golden Flood,* 270.

80. Botsford, *The Valley of Oil,* 249.

81. Frehner, *Finding Oil*, 21. He wrote, "Although doodlebugs prospered until about 1898, the ad says that they disappeared shortly afterward because Union eradicated them as the first company to establish a department dedicated to studying the field of geology and hiring practitioners to find oil" (21).

82. Paul Giddens, *The Early Days of Oil: A Pictorial History of the Beginnings of the Industry in Pennsylvania* (Princeton: Princeton University Press, 1948), 89.

83. Hildegarde Dolson, *The Great Oilderado: The Gaudy and Turbulent Years of the First Oil Rush: Pennsylvania: 1859–1880* (New York: Random House, 1959), 239–40.

84. Mody C. Boatright, *Folklore of the Oil Industry* (Dallas: Southern Methodist University, 1963), 23.

85. Ibid., 4, 3.

86. For more on divination and the early oil industry, see Owen, *Trek of the Oil Finders*, xii, 9, and 239; Black, *Petrolia*, 46–48; or Frehner, *Finding Oil*, 21–35. James is not mentioned in these three sources.

87. Owen, preface, *Trek of the Oil Finders*, ix.

88. Frehner, *Finding Oil*, 22.

89. Owen, *Trek of the Oil Finders*, 157.

90. James was by no means the last spiritualist to work in the oil fields. When the focus of the oil industry moved to Texas, a new generation struggled with the seemingly arbitrary nature of drilling wells. In 1919, Edgar Cayce gave his first oil reading for Day Matt Thrash of the Sam Davis Petroleum Company and would continue to offer readings for a number of Texas oilmen in the 1920s. As late as 2003, efforts were still being made to find Cayce's "Mother Pool" in Texas. For more, see Sidney D. Kirkpatrick, *Edgar Cayce: An American Prophet* (New York: Riverhead Books, 2000), 205–36. For other discussions of twentieth-century "seers," see Boatright, *Folklore in the Oil Industry*, 24–33.

91. David Barstow, David Rohde, and Stephanie Saul, "Deep Water Horizon's Final Hours," *New York Times*, December 25, 2010, *New York Times Online*.

3

Picturing a Crude Past

Primitivism, Public Art, and
Corporate Oil Promotion in the United States

ROSS BARRETT

Not long after the Deepwater Horizon blowout of April 20, 2010, the *Miami Herald* published an article that considered the disaster's place in oil history. Moving toward an argument against a federal moratorium on offshore drilling, the story offered a brief historical sketch of the U.S. oil industry that emphasized the dependence of Gulf State economies on deep-ocean extraction. While setting the disaster within the specific history of offshore drilling, the *Herald* story invoked another "context" for the affair, noting: "Oil and humanity have been linked since the dawn of civilization. In ancient times, oil . . . was collected and used to make ointments, medicines, building materials, adhesives and lamp fuel."[1] Locating the roots of the oil industry in the originary moments of "civilization," the *Herald* article introduced a maximally expansive perspective for the evaluation of the Deepwater Horizon disaster, a perspective that implicitly refigured the blowout as one moment in an eternal sequence of human interactions with oil.

In its efforts to archaize oil, the *Herald* story drew on a cultural discourse long employed to assess and recode the crises of the modern petroleum industry, a set of narratives and representational tropes that first appeared in corporate oil promotions of the early twentieth century. In the midst of a 1901 dedicatory speech in the Pennsylvania oil region, Congressman John Dalzell offered an early articulation of the central themes of this discourse:

So long as men have known anything they have known of the existence of petroleum. . . .
It has been known in all places and by all races; by Egyptian, Persian, Greek, and Roman;
by European and Asiatic and by dwellers in the islands of the sea. . . . It served in the
construction of the ark in which Noah and the progenitors of our race successfully rode
the waters of the deluge. It burned in the sacred lamps of the Temple of Jupiter, and
made part of the wrappings in which the Egyptians sought to preserve the bodies of
their dead.[2]

Imagining petroleum production as a timeless practice transcending the vicissitudes of
sociopolitical history, Dalzell's speech drew inspiration from its ceremonial object: the
Drake Memorial in Titusville, Pennsylvania (Figure 3.1, 1899–1901). Commissioned
by Standard Oil, the Drake Memorial shaped a complex allegory of oil extraction that
rehabilitated the public image of the corporation and the broader oil industry in a
moment of intensifying crisis. Conceived as a symbolic response to the period's intensifying concerns about the sustainability and social effects of oil capitalism, the
monument worked to reinvest the turbulent petroleum industry with deep historicity.
The Drake Memorial, I will argue, organized an image of the petroleum business that

FIGURE 3.1. Charles Brigham and Charles Henry Niehaus, Drake Memorial, 1899–1901,
Titusville, Pennsylvania. Photograph by the author.

encouraged visitors and national audiences (reached through an extensive publicity campaign) to take the long view on oil: to adopt a sweeping historical perspective that reconceived the fugitive resource as a timelessly abundant element and the boom-and-bust oil industry as one phase in an age-old and steadfast venture. In so doing, the memorial set the terms for a cultural discourse that would inform promotional imaginings of the oil industry throughout the twentieth century, and (as suggested by the *Miami Herald* article) that continues to find voice in contemporary assessments of oil.

This essay reconstructs the Drake Memorial project, examining its plastic forms and press notices as components of the monument's promotional meaning. I focus on the looming bronze figure at the center of the memorial—Charles Henry Niehaus's bronze *The Driller* (Figure 3.2, 1900–1901)—as the symbolic core of the monument's reimagining of the oil industry. Nude, heavily muscled, and crouching over a barren rock shelf, *The Driller* articulates a primitivist vision of petroleum extraction that addressed and reframed the material conditions of oil capitalism that deeply unsettled turn-of-the century Americans. This archaizing vision pursued several symbolic objectives. By realigning oil with deep historicity, *The Driller* answered period concerns about the industry's sustainability. In conjuring a primordial past for oil production, moreover, the sculpture recoded the atavistic associations that circled around the industry in the moment. Returning extractive industry to its imagined roots, *The Driller* refigured the violent and unpredictable process of oil production as an elemental contest between body and nature. This fantastic vision resonated with contemporaneous antimodernist arguments that decried the strictures of bourgeois society and urged Americans (and especially white men) to rediscover primal energies and corporeal capacities smothered by "civilization." Tapping into these popular primitivist fantasies, *The Driller* recast the modern oil industry as a field of savage exertion that held the promise of personal and national renewal.

Alhough keyed to the specific concerns facing the turn-of-the century oil industry, the Drake Memorial established the foundations for a cultural discourse—I will call it "petro-primitivism"—that would reappear in later promotional displays and gradually filter into the rhetoric of advertising and publicity. To suggest the enduring power of petro-primitivist themes, I will briefly consider two displays mounted by U.S. oil companies: the Sinclair Oil exhibit at the Chicago World's Fair of 1933–34 and Sun Oil's *Oil Serves America* exhibit at the Franklin Institute in Philadelphia (1953–54). Undertaken in moments of conflict and crisis, these projects employed the image of oil's primordial past to reframe the chaotic cycles of oil capitalism, counter conservationism, and reassert the viability of domestic oil in the face of industrial globalization. By examining the primitivist arguments advanced by these exhibits and the earlier Drake Memorial, I aim to offer a new account of the promotional initiatives of the

FIGURE 3.2. Charles Henry Niehaus, *The Driller*, 1899–1901, bronze. Photograph by the author.

early petroleum industry that explores the intersections between the traditional arts and corporate publicity and that reestablishes the important role that representations of petroleum's deep past played in accommodating twentieth-century Americans to modern oil capitalism.[3]

Bleak Visions: Popular Accounts of Petrolia and the Early Oil Industry

Efforts to raise a monument to Edwin Drake (1819–1880) began not long after the operator struck a large petroleum pool on the banks of Oil Creek, a tributary of the Alleghany River in northwest Pennsylvania, on August 27, 1859.[4] Although Drake would subsequently recede from the industry that grew up around his well, the oilman enjoyed iconic status among the small producers of the Pennsylvania oil region. Embracing Drake as a founding father of "Petrolia" (as the region was quickly dubbed) and an icon of independence in an era of corporatization, Pennsylvania oilmen began calling for a monument to his memory in the late 1860s. The most ambitious attempt to realize this goal unfolded in 1881–82, when the businessmen of Titusville (the nerve center of the oil region) began soliciting subscriptions for a sculptural monument within the city's new oil exchange. After making preliminary arrangements with Washington, D.C., sculptor Vinnie Ream for a monumental figure, however, the commission failed to raise enough money to complete the project.[5]

After several subsequent efforts went unrealized, the memorial project was taken up in 1899 by Standard Oil executive Henry H. Rogers, who pledged $50,000 and named Calvin Payne, general manager of the Trust's pipeline subsidiary, as the director of the undertaking.[6] Payne hired Boston architect Charles Brigham to design a granite memorial for a triangular lot in Woodville Cemetery, a picturesque burying ground on the western edge of Titusville that catered to the city's elite.[7] Brigham settled on a design that featured a curving granite exedra organized around a central sculptural niche and capped by two shallow-relief panels, and selected models submitted by New York sculptor Charles Henry Niehaus (1855–1935) to fill the monument's sculptural spaces. Niehaus was a popular choice among municipal and corporate patrons of the moment. The artist worked in a classicized sculptural language that was accessible to popular audiences and readily adapted his creative ambitions to suit the demands of his clients. This flexibility served Niehaus well over the years, allowing the artist to win commissions for sculptures of a range of subjects from Abraham Lincoln to Nathaniel Bedford Forrest.[8]

After Niehaus completed a plaster model of *The Driller*, the sculpture was cast in bronze by the John Williams Foundry in New York and set on its base in August 1900; the artist finished maquettes for the allegorical figures of *Memory* (at the exedra's left edge) and *Grief* (at the right) soon after, which the New England Granite Company executed in 1901.[9] The completed Drake Memorial was dedicated on October 5, 1901, with an elaborate ceremony that included several speeches, a brass-band performance, and a procession of schoolchildren.[10] Widely reported in the oil region and national press, the ceremony culminated an extensive publicity campaign that carefully tracked the memorial's development. This campaign ensured that the Drake Memorial spoke to two audiences: even as it intervened in the civic landscape of the oil region, the heavily promoted monument organized a cultural spectacle that reached a national viewership.

Although it would later fall into obscurity, the Drake Memorial was the centerpiece of a sweeping promotional initiative undertaken by Standard Oil at the turn of the century. Breaking from its previous policy of public taciturnity, the Trust sponsored exposition displays, published books and periodicals that advanced its views, and hired an advertising agency to place company-friendly articles in the national press.[11] This campaign had acquired increasing urgency by the last decade of the nineteenth century. Brian Black has demonstrated that the rapid expansion of oil production in Pennsylvania generated a wave of popular representations of the region that emphasized the danger, social chaos, and environmental ruin of the early industry.[12] As Standard Oil attained control over production and refining in the oil region, these dystopian visions of Petrolia were increasingly employed to attack the Trust. Representing the first major cultural attempts to make sense of the oil business, these journalistic, artistic, and reformist accounts called into question the sustainability of oil and the compatibility of the petroleum industry with period notions of technological modernity.

The petroleum industry was associated with disorder and ruin almost from its inception. The only regulatory framework guiding early oil production was the common-law doctrine of the rule of capture, which granted drillers the right to exhaust underground deposits regardless of surface property demarcations.[13] By disconnecting subterranean oil pools from property boundaries, the rule of capture removed any incentive for traditional land stewardship and instead ensured a flush-field model of production in which drillers sought to drain pools as quickly as possible. East Coast reporters covering the oil boom marveled at the despoliation produced by this system. An 1864 article in *Harper's Magazine*, for example, imagined Petrolia as a dystopian inversion of the pastoral ideal, noting that "the whole aspect is as unattractive as anyone with a prejudice for cleanliness, a nose for sweet smells, and a taste for the green of country landscapes can well imagine. Everything you see is black ... [and] saturated with waste petroleum."[14] Echoing other period accounts, the *Harper's* article figured Petrolia as a "black" landscape of industrial dreck, an abject space that offended refined aesthetic tastes and short-circuited normative sensory experiences of the natural landscape.

Other observers noted and decried the radically unstable character of oil-region development. The boomtown Pithole frequently served as a metonym for the region's disposable mode of settlement in these accounts. Just eight months after oil was discovered on sparsely populated farmland near Pithole Creek in January 1865, Pithole boasted fifteen thousand residents; by 1868, local oil pools had been exhausted and the instant city was abandoned. Noting the absence of traditional communal hubs (schools, churches, etc.) and the prevalence of illicit entertainments within Pithole, visitors construed the boomtown as a flimsily built and morally debased encampment headed for ruin. An 1866 *Beadle's* article thus predicted that in the near future "Pithole will have run its race. It may never become a Pompeii or Herculaneum, buried in the body of the earth, but it will expire when oil ceases to flow in its midst."[15] The eastern press regularly revisited the fallen boomtown in the following decades, appropriating Pithole's "forlorn and shattered" ruins as an illustration of the oil region's alarmingly evanescent forms of settlement.[16]

The chaos of the early oil industry also inspired several painters to produce dystopian scenes of Petrolia. In 1861, the Philadelphia artist James Hamilton composed an otherworldly scene of that year's Rouse Well fire (Figure 3.3), which killed nineteen men and raged out of control for three days.[17] *Burning Oil Well at Night* rendered oil as a horrific force that eradicated the picturesque landscape: verdant trees are remade as fiery skeletons, the night sky is replaced by clouds of red energy, and a central plume of flame refuses the sort of deep prospect viewers associated with optimism and progress.[18] Pittsburgh artist David Gilmour Blythe (1815–1865) shaped a similarly dark vision of the industry. As Sarah Burns, Rina Youngner, and Bryan Wolf have shown, *Prospecting* (Figure 3.4) satirizes the speculative turmoil and environmental devastation unleashed by the oil boom.[19] Blythe's scene, I would suggest, also offers a bleak account of oil's social impact in the region. The composition of *Prospecting* invokes the transitions remaking Petrolia: scattered debris from the region's agricultural past (wagon wheel, yoke, etc.) in the foreground creek gives way to modern derricks and refineries in the distance. And yet the filthy waterway also seems to complicate any notion of progress in the oil fields. Littered with bones and inhabited by antediluvian creatures (the turtle at left), the oil-fouled creek appears as a primordial marl pit that ensnares the petroleum operator. This murky morass locates a regressive impulse within the oil boom, suggesting that the processes that remade Petrolia as a field of ultramodern financial and industrial endeavor also entailed a return to base primitivity.

Imagining the oil region as a space of commercial disorder, environmental despoliation, and social devolution, *Prospecting* and *Burning Oil Well at Night* amplified claims

FIGURE 3.3. James Hamilton, *Burning Oil Well at Night, Near Rouseville, Pennsylvania,* c. 1861, oil on canvas. Smithsonian American Art Museum, Museum Purchase.

FIGURE 3.4. David Gilmour Blythe, *Prospecting*, 1861–63, oil on canvas. Private collection. Photograph courtesy of Berry-Hill Galleries.

that were made in other popular accounts of the early industry. By aligning oil technology with catastrophe and oil-region industrialization with debasement, Blythe and Hamilton's paintings stressed the incompatibility of oil capitalism with the dominant narratives of modern industrial society. On the one hand, *Burning Oil Well at Night* depicts a subject (oil's violent debilitation of extractive technology) that clashed dramatically with the body of feeling that Thomas Hughes has called "technological enthusiasm": late nineteenth-century Americans' burgeoning confidence in the capacity of industrial technology to remake the natural world.[20] On the other hand, Blythe's dark scene transgressed deeply rooted doctrines of material progress, which framed U.S. historical development as a linear process of improvement fueled by technological innovation and unrestrained market exchange.[21] Elaborating on contemporary media responses to the oil boom, Hamilton and Blythe framed the oil industry as an alarming deviation from, and even an obstacle to, technological modernity and national progress.

This bleak image in turn informed the arguments of later scientists and reformers who questioned the sustainability and social effects of the monopolized oil industry. Having secured control over the Pennsylvania fields in the 1870s, the Standard Oil Trust achieved dominance over domestic petroleum refining, transportation, and marketing the following decade.[22] The ruthless business practices that created this dominant position, however, sparked resistance from local producers and antimonopoly reformers who increasingly aligned Standard Oil with a phenomenon that Michael Watts has called "evacuative despoliation." Working to theorize the material conditions of advanced oil capitalism, Watts has proposed that the contemporary petroleum industry is structured by "a peculiar sort of double movement," wherein "wealth literally flows out" of oil-producing regions even as their terrain is socially, politically, and ecologically devastated.[23] Critics of Standard Oil detected similar processes of distant enrichment and local ruination taking shape around the Trust. As early as 1879, an antimonopoly *New York Daily Graphic* cartoon (Figure 3.5) imagined Standard as a remote entity whose local applications of power (in the form of a crushing fist) produced socioeconomic disorder and dangerous calamity (the smoking oil tank at right) within the oil region. Later critics extended this line of critique. Henry Demarest Lloyd's *Wealth Against Commonwealth* (1894), for example, forcefully tied the monopoly's enrichment to oil-region impoverishment. To dramatize the effects of the Trust's dominance, the text thus juxtaposed "the spectacle of a few men at the centre of things in offices rich with plate glass and velvet plush" with the "strange spectacle . . . of widespread ruin" throughout Petrolia.[24] Spokesmen for independent oil-region companies emphasized similar themes in their own attacks on Standard Oil. A 1901 article in the antimonopoly *Petroleum Gazette* thus noted that the "[oil] region has contributed to other great fortunes, but somehow or other comparatively little of

FIGURE 3.5. "The Iron Hand," *New York Daily Graphic*, March 12, 1879, wood engraving.

the wealth has found its way back."[25] An 1899 article in the same journal made a more pointed argument, noting, "If the Standard . . . desired to erect a monument . . . a fitting course for it would be to make one heap of the remnants of the millions of dollars worth of oil refineries it has put out of business here."[26]

Unsettled by the expansion and centralization of the oil industry, late nineteenth-century artists, journalists, and reformers worked out a grim cultural image for the petroleum business. Linking oil extraction to disorder and decline, observers of the first oil boom questioned the compatibility of petrocapitalism with the fundamental ideologies of industrial modernity. Drawing on these dark accounts, reformers and rivals framed the ascendant Standard Oil Trust as an engine of upheaval and desolation. Both cultural discourses undermined confidence in the long-term prospects of the industry; these doubts were further stoked by new scientific arguments stressing the productive limitations of the Pennsylvania oil fields. As Roger and Diana Olien have shown, geologists Pete Lesley and John F. Carll began to warn of the imminence of petroleum depletion in the 1880s and 1890s.[27] In 1888, for example, Carll noted that "we have no reasonable ground for expecting . . . that these oil fields will continue to supply the world with cheap light."[28] An 1889 *American Geologist* article dramatically affirmed this prognostication, noting that in the oil region: "Exhaustion stares the visitor in the face everywhere. . . . The quiet old towns that sprang into sudden energy during the oil-craze have sunk back to their former quietness and lethargy. . . . Corner lots are at a discount, and often the ruins of derrick and engine house alone mark the spots where once a 'city' stood."[29] Drawing on the familiar trope of ruination, this passage reimagines the Pennsylvania oil fields as a desolate wasteland that made the unsustainable nature of oil production palpably visible.

OIL HISTORY IN BRONZE AND STONE:
SCULPTING THE "PRIMITIVE IDEA OF MAN"

Undertaken even as these critiques proliferated, the Drake Memorial attempted to rehabilitate the public image of Standard Oil and rebuild popular confidence in the oil industry. To do so, the monument organized a spectacle that reconnected oil with traditional forms of civic community, realigned the dynamic industry with stability and permanence, and reimagined oil as a vehicle for social and national reinvigoration. The classicized architectural forms and heavy stone masses of the Drake Memorial initiate this symbolic project: balancing a vertical sculptural niche with the sweeping curve of the granite exedra, the monument gives palpable form to the themes of harmonious order, solidity, and permanence. Arrayed around the swirling bronze, the exedra's classical forms seem to counter and contain the dynamic energies of the central bronze, invoking a suggestive tension between fixity and fluidity that is (as we will see) central to the promotional program of the memorial. By situating a bench at the base of the exedra, moreover, designer Charles Brigham created an amphitheatrical space that tied the memorial to the traditions of civic architecture. Stressing this connection, promoters presented the exedra as a didactic space where visitors might gather for contemplative experiences. In his dedication speech, for example, Calvin Payne declared, "This monument should be an object lesson. . . . It should be an incentive to succeed in what you are called upon to do."[30] Echoing other backers, Payne framed the monument as a civic space where visitors might rededicate themselves to the cultivation of traditional virtues (diligence, ambition, duty, etc.) and the reforging of communal bonds.[31]

At the heart of this space was the allegorical figure of *The Driller* (Figure 3.2), which advanced several "object lesson[s]" about the oil industry. Taken together, the medium, pose, and narrative irresolution of Niehaus's sculpture worked to recode extraction as a process whereby the explosive material of oil was powerfully rechanneled into an endlessly productive flow. Period reviews outlined the mechanisms of this complicated symbolic project for their audiences. In noting the "tremendous energy in suspense shown by the bronze symbolical figure," for example, a *New York Times* review encouraged readers to address the sculpture as a construct holding in tension the themes of power and duration, energy and stasis.[32] Various aspects of Niehaus's figure in turn seem to bear out these tropes, particularly its medium. As a malleable metal carefully manipulated in a molten state and then allowed to solidify in a cast, bronze already carried connotations of fluidity and stasis. Niehaus's handling of surface texture suggestively emphasized the dual status of the medium: incised marks and lumps of congealed metal appear on the figure and its base, palpably invoking the malleability of the metal and encouraging the visitor to read *The Driller* as a controlled manipulation of fluid material.

The productive performance of the gleaming sculpture seems to realign oil extraction with these themes, framing oil production as a carefully controlled application of technological power to dynamic natural matter. Squatting over his drill, the driller models a willful command over the natural landscape that visualizes the complete harnessing of the dynamic material of oil through extractive labor. Drawing on the idealized forms of the ancient *Belvedere Torso* (first century BCE, Vatican Museum, Rome), the driller's heavily muscled and coiling torso dramatically affirms the theme of corporeal potentiality. These invocations of productive power are counterbalanced with signs of rational control. Read together, the figure's focused gaze, clenched jaw, and firmly gripping fist signal an inner discipline guiding the extractive process.

Even as it visualizes the power and rationality of modern oil extraction, the laborer's productive performance remains open-ended. Holding his hammer aloft at the final point of a backward stroke, *The Driller* appears poised in a moment of suggestive stasis. Reviews encouraged audiences to understand this pause as replete with potentiality; a writer for the *Oil City Derrick* noted, for example, that Niehaus had rendered "the exact point wherein the greatest potential of power is secured."[33] Considered together, the forms of the monument's niche might be read as a visualization of the eventual realization of that power: rising vertically from a curving stone foundation, the black-and-gold figure visually invokes the inky plume of an oil "gusher" bursting from the earth. And yet this conclusion remains visible only in abstract and transient fashion. Poised on the verge of an explosive "strike," *The Driller* remains locked enduringly in a fruitful moment of productive potential, a moment in which an extractive conclusion is both immediately foreseeable and infinitely deferred. The sculpture, that is, visualizes a permanent transitional state of productive capability that both invokes the inexorable completion of industrialized oil extraction and endlessly delays that final moment. In organizing this fantasy of unending capacity, I would suggest, *The Driller* advanced an effective response to period arguments about the imminent exhaustion of the oil region and the ultimate transience of all oil deposits.

Embodying productive power, rational control, and infinite potentiality, *The Driller* offered a glittering spectacle of "energy in suspense"—a spectacle of dynamic power rationally contained that was calculated to dispel lingering concerns about the stability and long-term viability of the oil industry. This vision of oil futurity takes shape, however, within an archaizing symbolic frame: rather than picturing the mechanized techniques of the modern petroleum industry, *The Driller* describes an imagined scenario of primitive oil production. In organizing this primordial spectacle, I will argue, the sculpture attempted to recode the cultural image of the oil industry in two significant ways. By conjuring oil's ancient past, *The Driller* buttressed the tempestuous petroleum business with the weight of deep history. And, by reconceiving oil extraction as an exercise of primal energies, the sculpture aligned the industry with an emerging set of antimodern arguments that connected visceral struggle with personal and national renewal.

Eschewing the collective and technologically mediated processes employed by period oilmen, *The Driller* reduced extraction to a spartan struggle between the individual body and the natural landscape. Both terms of this binary appear in their sparest form. Invocations of the landscape are limited to the sculpture's base, which reduces the primitive oil field to three elements—the stump on which the figure rests, a gnarled cluster of oak leaves concealing the driller's genitalia, and a rock shelf that serves as the sculpture's foundation. The heaving body above toils with simple tools and without protective equipment of any kind. While certain elements (including the figure's parted hair and shaven face) maintain a subtle connection to the sculpture's turn-of-the-century milieu, the drastically simplified forms of *The Driller* project the pictured scenario into a remote past. The primitive character of the unfolding struggle is heightened by the figure's "crude" physiognomy (low brow, thick nose, broad cheeks) and awkward earthbound pose. While resting much of his weight on his left knee, the laborer balances on the ball of his right foot; the tension in the right calf and flexed position of the foot suggest that the laborer will use the right leg to propel his hammer downward. Despite the seemingly elegant character of the figure's resting pose, these cues suggest that the impending stroke will take the form of a violent full-body lunge—an ungainly action that reconfirms the crudity of the driller's labor.

Taking note of these cues, period reviews underscored the primordiality of the sculpted scenario. An *Oil City Derrick* report thus argued that Niehaus's figure embodied "the primitive idea of man, wresting from the earth her secrets and gifts, with . . . energy, determination, and vigor."[34] Extending this interpretation, a *New York Times* review declared that "the monument will . . . recall first of all the various laborers in these oily fields—the Indians who collected oil to mix with their paints and used to set fire to the scum of petroleum on lakes and rivers on days of high festival—the settlers who used the oil as liniment and for lighting purposes, and the great army of workmen who have been laboring in the oil fields since 1859."[35] This sweeping passage figures *The Driller* as a "monument" to the "various laborers" of an ever-receding past of oil production, a past that encompasses the extractive practices of nineteenth-century "workmen," Anglo-American colonial "settlers," and "Indians" before them. Oil extraction appears here as a transhistorical practice rooted in the distant past and linking successive phases of civilization. Echoing other reviews, the *Times* article found encoded within the primitive nude an alternative perspective on the history of oil that reframed the turn-of-the-century industry as one fleeting phase in an age-old temporal arc.[36]

In conjuring oil's primordial past and expansive history, the Drake Memorial extended an impulse that appeared throughout Standard Oil's promotional ventures in the moment. The Trust's exposition displays, for example, took various measures to visualize oil's timeless roots. Standard Oil's classicized display at the Columbian Exposition thus included two lamp-bearing sculptural nudes that tied the company to the ancient past.[37] Pro-Trust writers explored these archaizing themes in even more strident terms. John McLaurin's oil history *Sketches in Crude Oil* (1896) accordingly noted that "petroleum . . . flourished 'ere Noah's flood had space to dry.'"[38] *The Derrick's Hand-Book to Petroleum* (1900) likewise informed its readers that "petroleum . . . has been known and used by mankind in various parts of the world from time immemorial."[39] Like the Trust's exposition displays, these accounts had some historical basis. Although oil was little used in the ancient Western world, archaeological and documentary evidence suggests that ancient Near East civilizations used petroleum to produce lighting fuels, bituminous sealants, and incendiary weapons as early as 3000 BCE.[40] As the above texts suggest, however, pro-Trust boosters used references to ancient petroleum practices alongside biblical passages and creative historical fabrication to construct sweeping narratives that reimagined oil as an eternal object of universal "human" endeavor. Advanced even as reformers questioned oil's future viability, these narratives refigured the turn-of-the-century industry as a modern extension of an "immemorial" field of activity—a field that, having already endured for millennia, would continue to flourish into the distant future.

The primordial struggle described by *The Driller* condensed these promotional claims into a striking vision of oil's deep historicity and unbounded futurity. In its efforts to conjure oil's primordial past, moreover, the sculpture organized a spectacle of rapacious masculinity that aligned the petroleum industry with new primitivist discourses emerging in the moment. Gail Bederman and T. J. Jackson Lears have shown that traditional constructs of self-restrained bourgeois masculinity were dramatically undermined at the end of the nineteenth century by the corporate restructuring of the economy, growing working-class political agency, and the new activism of middle-class women.[41] In an effort to consolidate a new manhood that might shore up gender and class hierarchies, Bederman and Lears argue, reformers urged bourgeois men to cultivate a "strenuous" mode of masculinity characterized by vigorous physicality, imperialist militancy, and aggressive sexuality. Anthony Rotundo and John Pettegrew have shown that the spokesmen of strenuousness frequently espoused primitivist ideals, urging men to rediscover primal energies purportedly suppressed by the "feminizing" forces of civilization.[42] By tapping long-hidden reserves of animalistic vitality, reformers argued, American men could reinvigorate themselves and renew the nation's flagging energies.

Composed even as these arguments proliferated, *The Driller* gave striking visual form to the precepts of masculinist primitivism. In so doing, the sculpture contributed to a broader primitivist discourse of work that focused on the industrial laborer's body. Melissa Dabakis and Edward Slavishak have demonstrated that corporate boosters regularly employed artistic and journalistic representations of the working body to refigure industry as a realm of masculine independence and primal vitality.[43] Paralleling these contemporaneous projects, *The Driller* reframed the oil-field worker as a dynamic nude engaged in an autonomous performance of strenuous labor. The crouching pose of Niehaus's laborer further imbues this performance with primitivist resonances. Squatting over his drill and poised to thrust forward, the laborer appears ready to direct a sexualized act of aggression toward both the drill and the monument visitor, who confronts the elevated sculpture at the level of the driller's genitalia. Noting the physique, suggestive pose, and crude labor act of *The Driller*, reviewers framed the sculpture as a vision of masculine potency, describing the primordial worker as a "splendid figure of male physical power," a "virile" hero, and a predator "quivering with virile power."[44] These reviews frequently employed penetrative rhetoric that made the sexualized connotations of the worker's drilling labor explicit. Critics thus described the laborer as tapping nature's "secrets and gifts" or delving into "the magnificent riches . . . hidden in [Mother Nature's] bosom."[45] In characterizing the driller as an explorer of hidden resources, these reviews simultaneously tied the figure to the strenuous man's rediscovery of long-concealed primal energies, a process frequently described in extractive terms by period reformers.

Even as it invoked a distant past for the petroleum industry and dramatically redrew the arc of oil history, then, *The Driller* reframed oil extraction as a primitive practice of heroic struggle that offered access to reinvigorating vital forces. In pursuing these various impulses, the sculpture countered critical accounts that tied the oil industry to disorder, transience, and social regression. By resituating the modern oil business within a broad and unending history of human interactions with oil, the sculpture answered concerns about the fleetingness of petroleum and the industrial system that developed around it. And, by drawing creatively on the themes of primitivism, the sculpture acknowledged and recoded the atavistic associations that swirled around the oil industry at the turn of the century, converting these devolutionary connotations into signs of oil's regenerative potential for citizenry and nation. Trumpeted in a variety of periodicals across the country, the Drake Memorial encouraged period Americans to reconceive oil as a timeless natural element, a vehicle of social stability and national renewal, and a reliable platform for limitless industrial expansion.

The Cultural Afterlife of Petro-Primitivism

As it contended with the specific concerns of its moment, the Drake Memorial established the key terms for a promotional discourse that would inform corporate oil publicity throughout the twentieth century. Although subsequent promotional projects would largely abandon the masculinist constructs that spoke to anxious turn-of-the-century audiences, corporate oil publicity continued to meet reformist critiques, regulatory initiatives, and economic crises with invocations of oil's immeasurably distant past. Two later public displays offer glimpses at the ways that twentieth-century oil companies adapted the tropes of petro-primitivism.

The Sinclair Oil pavilion at the Chicago World's Fair of 1933–34, first, revolved around a dramatic invocation of the primordial past: a spectacular exhibit of seven animatronic dinosaurs (Figure 3.6) clashing in a rocky Mesozoic landscape.[46] Guided by informational placards and souvenir newspapers distributed onsite, visitors wound their way around a series of outdoor tableaux featuring groupings of sculpted reptiles. This sequence ended in the "grotto" (Figure 3.7), a promotional gallery that celebrated the products and operations of Sinclair Oil. In part, the spectacular displays of the Sinclair pavilion can be understood as a dramatic extension of the corporation's ongoing promotional appropriation of the dinosaur. After the company selected the brontosaurus as an official logo in 1930, Sinclair launched an advertising campaign that tied the superior quality of the company's oil products to their ancient age and employed the dinosaur as an immediately legible emblem of primordiality.[47] Elaborating on the company's advertisements, the exposition pavilion presented its mechanized dinosaurs as dramatic emblems of the ancientness, and by extension the superior caliber, of Sinclair reserves.[48]

I would suggest, however, that the pavilion's prehistoric spectacle also worked to symbolically bolster the position of Sinclair Oil in a context of intensifying crisis. By the time the Chicago fair opened in summer 1933, Sinclair had become implicated in new conflicts over the sustainability of oil. These struggles were fueled by the chaotic oil boom that unfolded in East Texas in the early 1930s. Heedless overproduction in the

FIGURE 3.6. Photograph of Sinclair Dinosaur Exhibit, from *Millions See Weird Sinclair Dinosaurs* (New York: Sinclair Refining Company, 1933).

FIGURE 3.7. Photograph of the "Grotto," Sinclair Dinosaur Exhibit, from *Millions See Weird Sinclair Dinosaurs* (New York: Sinclair Refining Company, 1933).

Texan fields quickly glutted crude markets, inspiring new conservationist arguments emphasizing the nonrenewable character of petroleum and predicting the imminent exhaustion of domestic oil.[49] Echoing arguments voiced elsewhere, Interior Secretary Harold Ickes warned that unless overproduction and waste were "checked, this most valuable and necessary natural resource ... must inevitably soon reach practical exhaustion."[50] Buffeted by a groundswell of conservationist and anticorporate feeling, East Texas oil companies worked to obstruct local and federal regulatory initiatives. Chief among these was Sinclair Oil, which had come to rely on cheap Texan crude as a means of expanding its operations.[51] The company had long led efforts to discredit conservationist arguments; in a 1921 speech, for example, chief executive Harry Sinclair declared, "There is plenty of petroleum and always will be. Exhaustion of the world's supply is a bugaboo."[52] Sinclair Oil again spoke out against conservationist measures debated by state and federal authorities in the 1930s, with the hope of preserving its supply line to the Texan fields.[53]

Read against this backdrop, the dinosaur exhibit can be understood as an effort to address the regulatory initiatives that undermined Sinclair's market position and to bolster the company's tenuous public image. As dramatic embodiments of the deep past, the pavilion's dinosaurs palpably visualized the eternality of oil for fairgoers, answering new concerns about the resource's transience (concerns that threatened Sinclair's market position) with signs of its immeasurable timelessness. The robotic

creatures' periodic performances dramatically elaborated on this theme. Animated by
electric motors and equipped with loudspeakers that projected simulated cries, the
dinosaurs ceaselessly reinvigorated the prehistoric past for the onlooker, shaping a
mechanized spectacle that allegorized the unending viability of ancient oil deposits.
The displays of the "grotto" (Figure 3.7), finally, visualized Sinclair's role as a key catalyst of oil's infinite futurity. Juxtaposing a model of the company's brontosaurus logo
with photographic panels describing the operations that made Sinclair "America's
fastest-growing oil company," the hall framed the company as a conduit by which the
ultra-archaic material of oil was productively channeled into a golden future of economic growth.[54]

Twenty years after the Sinclair exhibit, the Sun Oil Company unveiled its promotional exhibit *Oil Serves America* (1953–c. 1962) at the Franklin Institute in Philadelphia.[55] Comprising twelve separate displays arranged in a spiraling sequence, the
exhibit explained the sources of oil, described modern extraction technologies, and
celebrated Sun Oil's innovations in refining (chiefly its role in developing fluid catalytic
cracking). The exhibit's parade of industrial achievements began, however, with a
dramatic invocation of oil's past: Morris Berd's mural *The History of Oil and Natural*

FIGURE 3.8. Photograph of Morris Berd, *History of Oil and Natural Gas*, 1953, from *Oil
Serves America* (Philadelphia: The Company, 1960).

Gas (Figure 3.8, now unlocated).[56] Painted in a cheery cartoonish style on a semicircular wall, the mural presented visitors with a panoramic composition that traced oil's development across time and space. Shifting from scenes of the "early uses of petroleum obtained by primitive peoples" (at right) to a vignette of cutting-edge extraction and refining (at left), Berd's painting framed the exhibit's displays with an expansive vision of oil history.[57]

In so doing, *The History of Oil and Natural Gas* firmly situated that history within a North American context. Although the mural included a vignette of ancient Mediterranean oil use (the ship-caulking scene at upper right), its various components emphasize developments unfolding within the continent. Set at the center of the mural, Drake's inaugural derrick functions as the organizing hub of the composition. The "modern" vignette at left pictures bits of cutting-edge infrastructure developed by U.S. companies, including a steel derrick, a fractionating tower (at the far left, essential to catalytic cracking), and a Hortonsphere tank (for the storage of liquid petroleum gas). Given emphasis by its large size and visual proximity to the visitor, the "primitive" grouping at right further affirmed the "American" character of the pictured history. Picturing a group of Native Americans engaged in archaic practices of oil extraction and consumption, the vignette firmly situated oil's past within the confines of the continent. Read together, the three main vignettes of *The History of Oil and Natural Gas* foreground the specific history of American oil, and in so doing imagine that broad history as an inexorable and linear process of technological advancement.[58]

This historical spectacle, I would suggest, perfectly answered the symbolic needs of its corporate patron in the early 1950s. Having built its operations on Texan and Albertan crude, Sun Oil remained a staunch defender of North American oil production long after its major competitors had begun to concentrate their efforts overseas.[59] This position allowed Sun to escape the legal scrutiny and public hostility directed at the corporate members of Aramco, which the Federal Trade Commission investigated from 1949 to 1952 as a potential violation of antitrust regulations.[60] Sun's reliance on domestic oil nevertheless weakened its market position in these years: already squeezed by competitors who bought cheap Middle Eastern oil, the company was further undercut by Texas authorities' 1952 decision to impose multiyear controls on the state's oil production.[61]

Considered in this context, *The History of Oil and Natural Gas* can be understood as a symbolic attempt to reinvigorate Sun Oil's identity as a domestic producer. By reworking oil history as an American phenomenon, the mural affirmed the company's commitment to domestic oil and its divergence from the unpopular multinationals. In tracing the expansive history of American oil, moreover, Berd's painting advanced a complex response to the regulatory initiatives that constrained Sun Oil in the moment. By conjuring a primitive past for American oil, the mural emphasized the already lengthy life of domestic oil pools. In rendering oil history as a sequential process of technological advancement, moreover, the painting visualized a claim for domestic oil's futurity long made by Sun Oil officials. Called to testify before a Senate Committee investigating wartime oil shortages, chief executive J. Edgar Pew declared in 1945 that "this country possesses hydrocarbon resources sufficient to meet this Nation's requirements for oil products for generations to come."[62] To support this bold assertion, the executive cited a succession of technological breakthroughs that had consistently extended the productivity of proven fields and enabled drilling in previously inaccessible sites. Noting that "oil has always been discovered and produced when and as needed," Pew reimagined the history of domestic oil as an enduringly sustainable progression toward ever-greater productivity facilitated by persistent technological innovation and minimal state regulation.[63] Undertaken in a moment in which this vision appeared increasingly tenuous, *The History of Oil and Natural Gas* advanced a dramatic restatement of the company's investments in domestic oil. Appearing at the entrance to the *Oil Serves America* exhibit, the mural situated the displays of Sun Oil's operations that followed in a deeply rooted and infinitely extendable history of domestic oil production—a history that countered the pressures of oil's globalization with a fantasy of timeless American oil.

Taken together, *Oil Serves America* and the Sinclair pavilion offer a suggestive glimpse at the rich cultural life of petro-primitivism in the century after the dedication of the Drake Memorial. Faced with supply crises, market spasms, and environmental disasters, individual companies and industry groups continually employed images of oil's primitive origins to redraw the public face of oil capitalism. Indeed, a wide variety of promotional endeavors focused on oil's distant past, including advertisements, children's toy giveaways (such as ARCO's "Noah's Ark" play sets of the 1970s), exposition displays (including Sinclair Oil's Dinoland at the 1964 New York World's Fair), and educational films (such as *Oil for Tomorrow*, 1944, and *As Old as the Hills*, 1950). Extending tropes crystallized by the Drake Memorial, these projects countered concerns about the violent instability of oil with visions of primordiality meant to suggest the permanence and infinite futurity of petrocapitalism. Fluid and readily adaptable, the tropes of petro-primitivism have encouraged generations of Americans to embrace oil—a resource long understood to be dirty, dangerous, and fleeting—as an entirely natural and unassailable component of everyday life in the United States. In so doing, these archaizing imaginings of oil contributed to a much broader field of symbolic practices and cultural representations that have worked to forestall efforts to see beyond oil capitalism and conceive alternative energy economies.

Notes

1. Mark Washburn and Maria Davis, "Answering Oil Questions," *Miami Herald*, June 13, 2010, http://www.miamiherald.com/2010/06/13/1677599/answering-oil-questions-under standing.html.

2. "Honoring Col. E. L. Drake," *The Petroleum Gazette* (October 6, 1901), 2.

3. In tackling the early cultural projects of Standard Oil, this essay aims to complement the excellent work done by Ulrich Keller, John Ott, Roland Marchand, and other scholars who have addressed Standard's later patronage of modernist photography and painting. See Ulrich Keller, *The Highway as Habitat: A Roy Stryker Documentation, 1943–1955* (Santa Barbara: University Art Museum, 1986); John Ott, "Landscapes of Consumption: Auto Tourism and Visual Culture in California, 1920–1940," in *Reading California: Art, Image, and Identity, 1900–2000*, ed. Stephanie Barron, Sheri Bernstein, and Ilene Susan Fort (Berkeley: University of California Press, 2000), 51–68; Roland Marchand, *Creating the Corporate Soul: The Rise of Public Relations and Corporate Imagery in American Big Business* (Berkeley: University of California Press, 1998).

4. On Drake, see Harold F. Williamson and Arnold R. Daum, *The American Petroleum Industry: The Age of Illumination* (Chicago: Northwestern University Press, 1959), 1–135; Brian Black, *Petrolia: The Landscape of America's First Oil Boom* (Baltimore: Johns Hopkins University Press, 2000), 1–59.

5. See "Monument to Colonel Drake," *Titusville Morning Herald*, February 13, 1872, 3; "Colonel E. L. Drake," *Titusville Morning Herald*, March 19, 1880, 4; M. N. Allen, "In a Reminiscent Vein," *Titusville Morning Herald*, October 21, 1899, 3; E. C. Bell, *History of Petroleum* (Titusville, Pa.: The Bugle, 1900), 132–39. On Ream's interest in the project, see D. J. Morrell, Johnstown, Pa., to Vinnie Ream, Washington, D.C., August 26, 1881, Papers of Vinnie Ream Hoxie, box 2, Correspondence 1874–89, folder 8, Library of Congress.

6. Letters from Calvin Payne, Oil City, Pa., to Laura Drake, Germantown, Pa., August 7, 1901; September 25, 1901; August 11, 1902, Charles B. Stenger Collection, Drake Well Museum Library, Titusville. On Payne, see *The Derrick's Handbook to Petroleum* (Oil City, Pa: Derrick Publishing Company, 1898), 932–34.

7. On Woodlawn Cemetery, see Samuel Bates, *Our County and Its People* (Boston: W. A. Ferguson & Co., 1899), 357–58.

8. See Regina Niehaus, *The Sculpture of Charles Henry Niehaus* (New York: De Vinne Press, 1901); Charles Caffin, *American Masters of Sculpture* (New York: Doubleday, Page & Company, 1903), 117–28.

9. "In Granite and Bronze," *Titusville Morning Herald*, October 5, 1901, 1; "Artists and Their Work," *New York Times*, June 23, 1901, 12.

10. "A Beautiful Memorial," *Oil City Derrick*, October 5, 1901, 1.

11. Harold F. Williamson and Arnold R. Daum, *The American Petroleum Industry: The Age of Energy, 1899–1950* (Chicago: Northwestern University Press, 1963), 8–9.

12. Black, *Petrolia,* 60–82.

13. Ibid., 38–41.

14. "After Petroleum," *Harper's New Monthly Magazine* (December 1864), 59.

15. "Three Days Among the Oil Wells," *Beadle's Monthly* (March 1866), 242.

16. See "A Carpetbagger in Pennsylvania," *Atlantic Monthly* 23 (1869), 746; and "A Ruined City," *Youth's Companion* (December 25, 1873), 421.

17. On *Burning Oil Well*, see Arlene Jacobowitz, *James Hamilton: American Marine Painter* (Brooklyn, N.Y.: Brooklyn Museum 1966), 42. I thank Heather Semple for her assistance in tracking down information about Hamilton's painting.

18. On the Rouse Well fire, see Williamson and Daum, *The American Petroleum Industry: The Age of Illumination*, 112. On the implications of the distant gaze, see Albert Boime, *The Magisterial Gaze: Manifest Destiny and American Landscape Painting, 1830–1865* (Washington, D.C.: Smithsonian Institution Press, 1991); and Angela Miller, *Empire of the Eye: Landscape Representation and American Cultural Politics, 1825–1875* (Ithaca: Cornell University Press, 1993), 154–65.

19. Sarah Burns, *Painting the Dark Side: Art and the Gothic Imagination in Nineteenth-Century America* (Berkeley: University of California Press, 2004), 72–74; Rina Youngner, *Industry in Art: Pittsburgh 1812 to 1920* (Pittsburgh: University of Pittsburgh Press, 2006), 41–55; Bryan Wolf, "All the World's a Code: Art and Ideology in Nineteenth-Century American Painting," *Art Journal* 44 (1984), 334–36.

20. Thomas Hughes, *American Genesis: A Century of Invention and Technological Enthusiasm* (Chicago: University of Chicago Press, 2004).

21. On the ideology of progress, see T. J. Jackson Lears, *No Place of Grace: Antimodernism and the Transformation of American Culture* (Chicago: University of Chicago Press, 1994), 4–26.

22. Williamson and Daum, *The American Petroleum Industry: The Age of Illumination*, 372–429.

23. Michael Watts, "Petro-Violence: Community, Extraction, and Political Ecology of a Mythic Commodity," in *Violent Environments,* ed. Nancy Paluso and Michael Watts (Ithaca: Cornell University Press, 2001), 205.

24. Henry Demarest Lloyd, *Wealth Against Commonwealth* (New York: Harper and Brothers, 1894), 42, 45. Perhaps the text's most spectacular passage imagines the Trust's corporate headquarters as a "heart of a machine . . ." counting "out a gold dollar for every drop of blood that used to run through the living breasts of the men who divined, projected, accomplished, and lost." See Lloyd, 117.

25. "For Andrew Carnegie," *The Petroleum Gazette* 9 (December 1901), 16.

26. "Drake Monuments," *The Petroleum Gazette* 4 (November 1899), 8. See also untitled article, *The Petroleum Gazette* 3, no. 11 (February 1899), 1.

27. The Oliens' helpfully incisive account of early conservationism appears within a text otherwise devoted to a curiously defensive and elaborate apologia for Standard Oil. See Roger M. Olien and Diana Davids Olien, *Oil and Ideology: The Cultural Creation of the American Petroleum Industry* (Chapel Hill: University of North Carolina Press, 2000), 120–27.

28. John F. Carll, cited in Harold Williamson and Arnold Daum, *The American Petroleum Industry: The Age of Illumination*, 376.

29. E. W. Claypoole, "The Future of Natural Gas," *The American Geologist* 1 (January 1888), 33.

30. "The Day's Proceedings," *Titusville Morning Herald*, October 5, 1901, 1.

31. See, for example, "A Beautiful Memorial," *Oil City Derrick*, October 5, 1901, 1.

32. "Monument for Titusville," *New York Times*, March 2, 1901, 8.

33. "A Work of Art," *Oil City Derrick*, October 5, 1901, 1.

34. Ibid. See also "In Granite and Bronze," *Titusville Morning Herald*, October 5, 1901, 1.

35. "Monument for Titusville," *New York Times*, March 2, 1901, 8.

36. See also "A Shaft for E. L. Drake," *Washington Evening Times*, October 7, 1901, 3.

37. For an illustration of Standard Oil's display at the Columbian Exposition, see Hubert Bancroft, *The Book of the Fair* (Chicago: Bancroft Company, 1893), 506. The Trust's exhibits at the Paris Exposition of 1900 and the Pan-American Exposition of 1901 similarly invoked petroleum's past; both included painted murals depicting "the discovery of oil." See *Report of the Commissioner-General for the United States to the International Universal Exposition* 4 (Washington, D.C.: Government Printing Office, 1901), 120; *The Exhibit of the Standard Oil Company of New York at the Pan American Exposition* (New York: Standard Oil, 1901), http://www.panam1901 .org/mines/standard_oil.htm.

38. John McLaurin, *Sketches in Crude Oil* (Harrisburg, Pa.: The Author, 1896), 1.

39. *The Derrick's Hand-Book of Petroleum* (Oil City, Pa.: The Derrick Publishing Company, 1898–1900), 5.

40. See Robert James Forbes, *Bitumen and Petroleum in Antiquity* (Leiden: E. J. Brill, 1936), 1–42; Forbes, *Studies in Early Petroleum History* (Leiden: E. J. Brill, 1958).

41. Gail Bederman, *Manliness and Civilization: A Cultural History of Gender and Race in the United States, 1880–1917* (Chicago: University of Chicago Press, 1995), 12–17; Lears, *No Place of Grace*, 103–17.

42. Anthony Rotondo, *American Manhood: Transformations in Masculinity from the Revolution to the Modern Era* (New York: Basic Books, 1994), 227–32; John Pettegrew, *Brutes in Suits: Male Sensibility in America, 1890–1920* (Baltimore: Johns Hopkins University Press, 2007), 1–20.

43. Edward Slavishak, *Bodies of Work: Civic Display and Labor in Industrial Pittsburgh* (Durham: Duke University Press, 2008), 73–95; Melissa Dabakis, *Visualizing Labor in American Sculpture: Monuments, Manliness, and the Work Ethic, 1880–1935* (Cambridge: Cambridge University Press, 1999), 82–104.

44. Charles de Kay, "Essays on Sculptors," *New York Times*, December 26, 1903; "Monument for Titusville," *New York Times*, March 2, 1901, 8; "Memorial to Edwin L. Drake," *Titusville Morning Herald*, October 5, 1901.

45. "In Granite and Bronze" and "Memorial to Edwin L. Drake," *Titusville Morning Herald*, October 5, 1901.

46. On the Sinclair Exhibit, see Cheryl Ganz, *The 1933 Chicago World's Fair: A Century of Progress* (Urbana: University of Illinois, 2008), 74; *Official Guidebook of the Fair* (Chicago: A Century of Progress, 1933), 105; *Millions See Weird Sinclair Dinosaurs* (New York: Sinclair Refining Company, 1933) [souvenir newspaper].

47. On Sinclair's appropriation of the dinosaur, see W. J. T. Mitchell, *The Last Dinosaur Book: The Life and Times of a Cultural Icon* (Chicago: University of Chicago Press, 1998): passim.

48. A souvenir paper distributed onsite thus noted that while "nothing we could use would graphically portray tremendous age as well as do these dinosaurs," the company's "oils were formed *before* the dinosaurs came into being [original emphasis]." See *Millions See Weird Sinclair Dinosaurs*, 3.

49. Olien and Olien, *Oil and Ideology*, 185–208; Norman Nordhauser, *The Quest for Stability: Domestic Oil Regulation, 1917–1935* (New York: Garland Publishing, 1979), 74–115.

50. Harold Ickes, "The Crisis in Oil: A Huge National Problem," *New York Times*, June 11, 1933.

51. Olien and Olien, *Oil and Ideology*, 191.

52. Reprinted in Olien, *Oil and Ideology*, 137.

53. Nordhauser, *The Quest for Stability*, 82.

54. *Millions See Weird Sinclair Dinosaurs*, 6.

55. The latest reference I have found to the exhibit is a 1962 *Life* article; see "Franklin Institute," *Life*, April 13, 1962, 12.

56. On Morris Berd, see "Morris Berd," 93, Artist," *Philadelphia Inquirer*, October 8, 2007; *Who's Who in American Art* (New York: R. R. Bowker, 1984), 67; "Four Seasons: Morris Berd," *American Artist* 44 (February 1980), 52–55.

57. *Oil Serves America: A Presentation by Sun Oil Company* (Philadelphia: The Company, 1960), 5.

58. Period reviews reaffirmed the nationalist emphasis of the display; see, for example, "Educational Oil Exhibit," *World Petroleum* 24 (1953): 174.

59. Arthur M. Johnson, *The Challenge of Change: The Sun Oil Company, 1945–1977* (Columbus: Ohio State University, 1983), 57–58, 99–106.

60. Daniel Yergin, *The Prize: The Epic Quest for Oil, Money, and Power* (New York: Simon and Schuster, 1991), 472–75.

61. Johnson, *The Challenge of Change*, 101.

62. *Investigation of Petroleum Sources: New Sources of Petroleum in the United States* (Washington, D.C.: Government Printing Office, 1946), 4.

63. Ibid., 25; see also 15–21, 31–35.

4

A Short History of Oil Cultures; or, The Marriage of Catastrophe and Exuberance

FREDERICK BUELL

Vaclav Smil begins *Energy in World History* with a daring proposition. He considers Leslie White's assertion that the link between energy and culture is the first important law of cultural development. "Other things being equal," White writes, "the degree of cultural development varies directly as the amount of energy per capita per hour harnessed and put to work." Smil then cites the further claim by Ronald Cox that "a refinement in cultural mechanisms has occurred with every refinement of energy flux coupling."[1] Smil's book, he then says, is an attempt to evaluate these assertions.

Only at the end of his survey of energy history does Smil return to the subject. His conclusion is plain. "The amount of energy at a society's disposal puts clear limits on the overall scope of action" but does little more than that. Still more pointedly, Smil goes on to assert that "timeless literature, painting, sculpture, architecture, and music show no correlation with advances in energy consumption."[2] Case closed.

Yet today, oil presents society with a large portfolio of dread problems—rapid global warming that threatens lives, lifestyles, and ecosystems; an expanding number of serious, world-altering globalized environmental crises all related to fossil-fuel-fueled population and economic growth; increasing geopolitical instability, conflict, and terrorism related to control of oil supplies or affecting the production/distribution of oil; and a possibly imminent failure of supply—peak oil—that would wreck the world's economic and social systems. All of these crises have led to new, widespread awareness of just how completely oil has become essential to all aspects of humans' way of life, from agriculture to health care, transportation to consumer goods. Oil has become an obsessive point of reference in and clear determinant over the daily lives of many, either victimizing them directly and cruelly as with Shell in Nigeria, or Texaco in Ecuador, or making them increasingly feel that their developed-world normalities are a shaky house of cards. Indeed, it has become impossible not to feel that oil at least partially determines cultural production and reproduction on many levels. Nowadays, energy is more than a constraint; it (especially oil) remains an essential (and, to many, *the* essential) prop underneath humanity's material and symbolic cultures.

Yet no effective response to the huge conceptual gulf between energy and culture that Smil found has been made. Is asking how oil inflects culture like asking how the weather (or, worse, how air, or worse still, how oxygen) affects it? Clearly, without weather, air, or oxygen, no culture would exist. But can one say with any specificity that any of these is a cultural determinant? Jonathan Bate and others have made connections between weather and culture; indeed, links between air and culture would engage pollution studies (which, in turn, would engage a small niche in literary/artistic tradition and theory). But these movements are peripheral at best—or nonexistent, as in the case of oxygen.[3] And unlike most of today's theory-inspired advances in cultural study that have focused on race, colonialism, gender, class, sexuality, and, most recently, environment, the study of oil does not uncover a large trove of important old literature, even though it does feature a growing body of contemporary art, literature, and popular cultural work. But what oil does have, unlike oxygen, weather, and air, is a reasonably well-elaborated and defined human history, one with a complex set of filiations, fissures, ruptures, and breaks. And oil's possible collapse, as imagined today, provides both motivation and a heuristic for asking many interesting questions about oil's relationships to culture, *both* in the past and present. We need to ask what we start finding when we cease living in oil as if it were our oxygen and look back on its histories—material, technological, social, and cultural—from the standpoint of today's startled awareness of the fragility of the system "Colonel" E. L. Drake and John D. Rockefeller built. Perhaps the gap between energy and culture can be credibly bridged and made available to the traffic of a new field of study.

William Catton's book *Overshoot: The Ecological Basis of Revolutionary Change* takes the first step in building this necessary bridge. Modern westerners and their immediate ancestors, Catton declares, "have lived through an age of exuberant growth, overshooting permanent carrying capacity [of the earth] without knowing what we were doing." This historically novel exuberance came, Catton argues, from two sources: "(a) discovery of a second hemisphere, and (b) development of ways to exploit the planet's energy savings deposits, the fossil fuels."[4] The first method, which

Catton calls "takeover," was simply "behaving as all creatures do. Each living species has won for itself a place in the web of life by adapting more effectively than some alternative form." European colonization, which took over land and developed its eco-system resources more completely than the hunter gatherers it displaced, multiplied Europe's per capita resources by five times. Far less "natural" and more determinative was the second method, which Catton calls "drawdown." This involved "digging up energy that had been stored underground millions of years ago" and then "drawing down a finite reservoir of the remains of prehistoric organisms."[5]

Catton's inflexible, single-step dialectical narrative (ending in disaster) limits his ability to say much about the specifics of fossil-fuel culture. Nonetheless, it does allow him to make a few important macro-observations about it. Colonialism and then, more important, fossil-fuel-energy use allowed "quite a marked rise in prosperity *and* . . . a phenomenal acceleration of population increase."[6] These, in turn, helped create in the West an important cultural attitude: a faith in progress so strong that "the idea that mankind could encounter hardships that simply will not go away" was not just unlikely but in fact "unthinkable."[7] Fossil-fuel culture can be, in short, described as an "age of exuberance"—an age that is also, given the dwindling finitude of the resources it increasingly makes social life dependent on, haunted by catastrophe.[8]

A far more sophisticated theoretical lens is required to see the welter of smaller shapes in this larger history. Again there is an excellent place to start: Jean-Claude Debeir, Jean-Paul Deléage, and Daniel Hémery's *In the Servitude of Power*. Unlike Smil, Debeir, Deléage, and Hémery do not just chronicle a history of energy-related technical advances, but find a way to theorize that process to reveal a much more finely grained social history of energy than ever before. All of this will allow me to move to a still finer resolution, and to extend the process into culture as well as history.

For Debeir, Deléage, and Hémery, energy materializes as energy only with the development of technologies they call "converters"—which include everything from sails to atomic reactors. Only thus does a resource or environmental process become, in fact, "energy." Further, these converters do not exist singularly; they emerge and develop as parts of "converter chains," ones that run throughout society. The Neolithic revolution in food energetics, for example, did not occur only with the domestication of animals and plants. A whole chain of converters materialized: "The deployment of new capacities for large-scale harvesting, transporting, and storage (silos for cereals, drying of fish, for example) and the diversification of culinary preparation methods (grinding grain, pottery for cooking)" were equally necessary. But with converter chains, a third theoretical entity also appears: converters and converter chains are always a part of a society, and the three together materialize as an "energy system." This is a system that "includes, on the one hand, the ecological and technological characteristics of the chains (evolution of sources, converters, and their efficiency) and, on the other hand, the social structures for the appropriation and management of these sources and converters." In an energy system, simple energy determinism does not exist. For example, the "first converter of thermal energy into mechanical energy," the steam engine fed by coal, was not what "produced the factory system by replacing human labor, but quite the opposite": it was "the factory system that made possible the use of steam engines," something that then had the "effect, if not the goal, of establishing the domination of capital over labor."[9] Causation is not simple; a whole environmental, technical, and social system ultimately bootstraps itself into existence. This system is "a determination [that] is itself determined: it is the result of the interplay of economic, demographic, psychological, intellectual, social and political parameters operating in the various human societies."[10]

Debeir, Déleage, and Hémery then use this framework to historicize energy. History becomes a succession of distinct energy systems. In considering oil history and ultimately culture, then, we need to consider the previous energy system that it disrupted and transformed: we need to orient oil in relation to the energy system it emerged within, the system Deleir and colleagues call "coal capitalism." Coal capitalism deployed the steam engine, humanity's first converter capable of turning thermal into mechanical power; coal, thus converted, extended itself far beyond its extensive precapitalist uses (for heating and medieval industry once firewood became scarce), transforming the previous medieval energy system into the more modern coal-capitalist one. Importantly, however, the new coal capitalism was not just the latest in a series of energy systems; it "signaled a radical break with all previous energy systems known to humanity. With it, the primacy of biological energies ended and that of fossil energies was established."[11]

Coal capitalism was thus unique among previous systems in being the first truly exuberant one. Debeir, Deléage, and Hémery (along with many other writers on fossil fuels) regularly describe it as liberatory. For example, Debeir, Deléage, and Hémery repeatedly claim that coal capitalism freed "societies from the restrictive relationship to nature imposed upon land-based production, a liberation which came about thanks to the ever-growing use of energy"; "it enabled the European economies to by-pass the natural limitations of organic energy."[12] Steam engines placed in coal mines—all of which in England had been pioneered during medieval times—pumped away water that would flood them, allowing them to go deeper and become more productive. Improved steam engines in ships and railroads made the coal's energy more portable than ever before, thereby freeing English industrialism and empire from geographic limits. Coal refined into coke removed another "critical organic constraint on the growth of the industrial economy"; the limit imposed on the iron-making process by charcoal fell away, thereby liberating the manufacture of machinery (including steam engines).[13] The factory system itself was liberated from an organic constraint—geography in this case—as steam power replaced water power: factories no longer had to be placed on one of the rapidly dwindling number of sites on the banks of usable rivers but could be put anywhere. All these "liberations" paid off. As Barbara Freese puts it, they were crucial to Britain's rise as an industrial and colonial power. "By the time London held the first World's Fair in 1851," Freese writes, "Britain was hailed as the workshop of the world, and its markets and its empire reached global scale."[14]

Liberation from "nature" released "mechanical power," decisively changing both the discourse of nature and that of machinery. Eighteenth-century characterizations of nature as lawful and orderly and their persistent imaging of that order as a clockwork mechanism clearly accommodated Enlightenment enthusiasm for "improvement": a delicate mechanism could be perfected. Now, however, coal-fueled mechanical power—embodied eventually in huge locomotives—rumbled into town, took over the machine metaphor, and promised open-ended progress. Steam engines, engines of motion and change, replaced clocks as the paradigm of machinery. A contributor to a journal edited by Charles Dickens evoked a specter that, though frightful, took the story's protagonist into the depths of a coal mine to teach him that coal was placed on earth so that "man may hereafter live, not merely a savage life, but one civilized and refined, with the sense of a soul within. . . . Thus upward, and thus onward ever."[15] Similarly, Leo Marx, in his classic study *The Machine in the Garden*, notes how, in the United States, writers in the leading magazines "adduce the power of machines (steam engines, factories, railroads, and, after 1844, the telegraph) as the conclusive sanction for faith in the unceasing progress of mankind." In both high cultural and popular discourse, Marx concludes, "the fable of Prometheus [was] invoked on all sides."[16]

But if exuberance ran high, the growth of coal capitalism also produced the opposite: Britain, the workshop of the world, became also (most famously in the views of Karl Marx and Friedrich Engels) the workhouse of the world, even as it sought to globalize that condition by becoming the world's preeminent colonial power. Initially, Romantic Prometheanism opposed this new mechanical power, demonizing the machine; at the same time, however, it offered its own augmentation of power on another level, as it transformed "nature" from clockwork regularity into a dynamic organic/organicist force, one operating both in nature and the human imagination.[17] Subsequent literary naturalism, however, gave a decisive victory to the demonic power of machinery over its organic/imaginative competitor; think of the sheep destroyed by the steam engine in Frank Norris's *The Octopus*. More significant, naturalism represented how the liberation of human society from organic constraints ironically ended up creating a variety of machine-made organic nightmares, from Dickensian

miasmic environments to Dickensian oppression of the poor. In the process, coal
capitalism developed (appropriately, given its mode of extraction) a sinister cultural
geography of depths and instructive descents. The narrator of Rebecca Harding
Davis's "Life in the Iron Mills" tells her readers at the start, "This is what I want you
to do. I want you to hide your disgust, take no heed to your clean clothes, and come
right down with me,—here, into the thickest of the fog and mud and foul effluvia. I
want you to hear this story. There is a secret down here, in this nightmare fog."[18] The
fossil-fueled fires of hell were brought close to hand, "down" in the factory district.

Thus historicized, exuberance is no longer just surplus energy creating optimism,
and its catastrophe is not hapless dependency on what is running out. Exuberance
and catastrophe materialized as historically specific forms of capitalist triumph and
oppression, of environmental domination and destruction, and of human liberation
and psychic and bodily oppression. All of these versions of the two motifs were,
moreover, embedded in the materiality of coal itself, be it Promethean coal that gave
humanity its new modes of and uses for fire, or stygian coal that re-created the ancient
fiery nether region as polluted industrial district and city. With these reflections,
clearly, we have moved energy history into cultural history.[19]

Oil entered this history and began reshaping it in two phases: first, as part of what
I will call "the culture of extraction," and second, as a key part of a new culture for a
new energy system, which I will call "oil–electric–coal capitalism." In its first phase, oil
(formerly used as a medicine) appeared quickly and exuberantly as a remarkable, new
energy source within a bootstrapped system of extraction, refining, transportation,
and marketing.[20] Oil, in this phase, also had a role in creating what I call "extraction
culture," a specific formation that is still alive today. In its second phase, oil proceeded
to thoroughly reshape coal capitalism and do so culturally as well as technologically,
expanding dynamically not just into new industries but also into new areas of cultural
life. The new system integrated industry with society and culture more completely
than ever before, even as it erased or sublimated most of the highly visible evils of the
pervious era of stygian coal capitalism.

First, oil-extraction culture. The opening of this era in the United States began when
Drake struck oil in Titusville. This was, Ida Tarbell wrote, the "signal for a rush such
as the country had not seen since the gold rush of '49."[21] It was a triumph of wildcatting, speculation, development, pollution, booms, and crashes, a moment of legendary exuberance in American history. Unlike coal mining, which was a capital-intensive
operation with a large labor force working underground, often in appalling conditions, oil in Pennsylvania promised immense reward for little investment and less
hard labor. So much for the workhouse of the world. Oil, tapped, rose to the surface

by itself—albeit sometimes calamitously—to reward the efforts of a few daring and lucky men. Thus oil's geography of depth differed greatly from coal's. People did not have to go underground to get it; they stayed on the surface to tap it, already pressurized and ready to go.[22]

But the oil boom was no mere gold rush. It was not a one-shot, extract-and-run proposition. It established a new industry and brought wealth and power to the United States. As such, Tarbell saw oil extraction as signaling a resurgence of the old epic-heroic ideology of democratic, self-reliant, community- and nation-building individualism. Oil extraction "used men of imagination who dared to risk all they had on the adventure of seeking oil . . . used capital wherever it could be found . . . used the promoter and the speculator . . . called on the chemist to evaluate the products and had set him up a laboratory to enlarge and improve them . . . [and] called on the engineer to apply all known mechanical devices." Evoking this epic-scale mobilization of talents in nationalist, Whitman-like prose, Tarbell concluded: "The way that all these varied activities fell in line, promptly and automatically organizing themselves, is one of the most illuminating exhibits the history of our industry affords, of how things came about under a self-directed, democratic, individualistic system: the degree to which men who act on 'the instant need of things' naturally supplement each other—pull together."[23]

Although Tarbell, writing this as an introduction to Paul Giddens's 1938 *The Birth of the Oil Industry*, also foregrounds the excesses of speculators, the sometimes spectacular environmental and human disasters brought on by rapid growth (Pithole went from seven to fifteen thousand people in just a few weeks), she dismisses these as peripheral to the epic of oil individualism. "Men did not wait to ask if they might go into the Oil Region," Tarbell wrote, "they went. They did not ask how to put down a well: they quickly took the processes which other men had developed for other purposes and adapted them to their purpose. . . . It was a triumph of individualism."[24] Thus coal's backbreaking labor in extraction became the thrill of creation; coal's widening of social castes became individualist opportunity; and the gloom of impoverished cities and dismantled, wrecked environments seemed to lift.

But more interesting still, with oil extraction, catastrophe did not simply remain on the periphery of exuberance. It became, for Giddens and even Tarbell, an integral part of the exuberance of oil, not as with coal, its squalid nemesis. Enthusiastically describing one such catastrophe, Giddens writes how a well at the lower end of Oil Creek sent up a large gusher—3,000 barrels per hour. About 150 people gathered to watch, when "a sheet of fire, as sudden as lightning, burst forth . . . [and] instantly, an acre of ground with two wells, oil vats, a barn, and over 100 barrels of oil were ablaze. . . . The well continued to spout oil high into the air, which fell to the ground, igniting as soon as it fell and adding dense smoke and sheets of flame to the horrors of the scene."[25] Most of the onlookers became "human torches and frantically tried to escape from the fiery furnace." Epic catastrophe came with epic actions. This tone prevails even in writing about slower, seamier aspects of oil damage. Huge volumes of oil poured out into rivers and onto the ground due to the failure or absence of containers; oil-river transport featured the exciting release of "freshets" downstream to float the barges—an event that ended often in wreckage that blackened the streams; boomtowns like Pithole famously lacked all sanitation ("The whole place smells like a camp of soldiers when they have the diarrhoea"; fights, drunkenness, and "garroting [were] almost common"); and so much oil spilled from teamsters' wagons onto the already muddy roads that they became a "perpetual paste, which destroyed the capillary glands and hair of the horses," many of which died along the way, so that "hundreds of dead horses could be seen along the banks of Oil Creek."[26] Add to this wildly fluctuating oil prices and boom and then bust land prices and it becomes almost impossible to separate out catastrophe from exuberance and vice versa. Indeed, the two were mutually reinforcing in Giddens's and even Tarbell's prose.

Things changed quickly, Tarbell's *History of Standard Oil* makes clear, as Rockefeller transformed extraction culture into a vertically integrated monopoly that stifled this resurgence of American individualism and frontier spirit. Oil, once systematized, began transforming social life—sending out tentacles into people's private lifeworlds to change them in what seemed to many (but not all) in exuberantly positive ways. Unlike coal capitalism, oil did not remain culturally inscribed as mostly an affair of production machinery for industry and commercial transport. "Give the poor man his cheap light, gentlemen," Rockefeller famously told his colleagues, and the ancient organic constraint of darkness was gone; the lives of the poor were "lightened."[27] Huge machinery now shrank in size and scattered about the factory floor, and then drove in the form of new Fords out the door as parts of a new consumer culture, ones that even the working class could enjoy. Old constraints on both physical and social mobility for even the working class were suddenly relieved. Everyman seemed to have now individual access to real power: oil concentrated into one gallon of energy "equal to the amount in almost five kilograms of the best coal"—itself the equivalent of fifty "well-fed human slaves toiling all day."[28]

Urban environments also began to lose the customary organic miasma caused by coal; pollution abated significantly at industrial sites and in cities. Oil-electric industrial production was materially and culturally refigured as clean, efficient, and modern (think, for example, of Henry Ford's Rouge River Plant and the artist Charles Sheeler's images of it). At the same time, oil-electric capitalism exported coal's miasma as far away as it could. The hellish depths were resited as backward, stagnant, unpleasant spots outside the system. Ironically, in retrospect, even cars were hailed as a great sanitary improvement, replacing the thousands of animals that had daily deposited millions of tons of waste in the streets—and therefore also the atmosphere, as dried dung particles were swept into the air. In consequence, cultural geography changed again: people more and more valorized living within the clean, new apparatus of oil-electric production-consumption, not apart from it.

In doing all this, oil had a partner: electricity.[29] What oil did, electricity furthered, taking over the role of light-giver from oil, increasing cleanliness, mobility, and speed with electric motors for factories, trains, and appliances. Together, oil and electricity wrapped people within their many infrastructures—roads, pipelines, telephone lines, power cables—even as it began doing something else of great cultural importance: reaching into and restructuring peoples' private worlds, identities, bodies, thoughts, sense of geography, emotions.[30] Perhaps the most important product of oil-electric capitalism was modern consumerism. Half-concealed, half-fetishized, oil-electric infrastructures extruded numerous cultural infrastructures (converters), which modern people, including modern artists and writers, chose as preferred dwelling places.

In this transformation, the extraction era's exuberance modulated into the exuberance of a new dynamic system that sought stability in change. The oil industry pioneered that goal; oil historians Harold Williamson, Ralph Andreano, Arnold Daum, and Gilbert Klose discuss how attempts to stabilize the boom and bust oil industry appeared first in Oklahoma in 1914, and then nationally, as industry and government, impelled by fear of scarcity, came tensely together to manage oil during World War I. These efforts continued after the war, resulting, by World War II, in a dynamically growing system "far from perfect" but nonetheless "the basic, essential structure" necessary for "attempts to meet old and new difficulties" even today.[31] If oil, first illuminant and then automobile fuel, was essential to the construction of the new system, it also, in its third major use, as a lubricant, may be seen metaphorically as equally essential to the dynamic stability and stable dynamism of oil–electric–coal capitalism.

Upton Sinclair's novel *Oil!* chronicles one aspect of this immense social and cultural change. We meet its father and son protagonists as "Dad" (J. Arnold Ross, already a multimillionaire "big operator" in the oil business) takes his son, "Bunny," for a high-speed drive along a California highway. Dad appears to his son, Bunny, as a figure of epic proportions: accessing an "engine full of power" by the mere pressure of "the ball of [his] . . . foot" and rocketing down roads "twisting, turning, tilting inward on the outside curves, tilting outward on the inside curves, [the road having been engineered] so that you were always balanced, always safe." Dad was a man of money who had commanded the magic necessary to create all this. He "said the word . . . and surveyors and engineers had come, and diggers, by the thousand, swarming Mexicans and Indians, bronze of skin, armed with picks and shovels; and great steam shovels with long hanging lobster-claws of steel. . . . All these had come, and for a year or two they had toiled, and yard by yard they had unrolled the magic ribbon. . . . Never since the world began had there been men of power equal to this."[32] Although the novel goes on to expose this system as predatory and corrupt, Dad is nonetheless far from the big capitalists Sinclair depicted in his earlier novel, *The Jungle*, a novel that drama-tized as few American texts have the hellish underworld of coal capitalism. Dad never quite loses completely his new oil-era, Tarbell-like appeal as an epic individualist and adventurer remarkable for "the ingenuity by which [he] . . . overcame Nature's obsta-cles."[33] He is also a loving father who never lets his son's radical, anti-oil-corporation politics interrupt their close relationship. Dad is, in short, positioned in between: in between Tarbell's democratic extractor epic and a system in the process of forming its top-down, vertically integrated combinations. He drives at high speed, yet he does this on a road engineered for both speed and safety.

If Dad is favorably depicted, so are the physical operations of his industry, which have none of the coal-capitalist miasma that infused every aspect of *The Jungle*. At the site of one of Dad's new wells, Bunny thinks:

> It was all nice and clean and new, and Dad would let you climb, and you could see the view, clear over the houses and trees, to the blue waters of the Pacific—gee, it was great! And then came the fleet of motor trucks, thundering in just at sunset, dusty and travel-stained, but full of "pep." . . . [The men] went to it with a will; for they were working under the eye of the "old man," the master of the pay-roll and their destinies. They respected this "old man," because he knew his business, and nobody could fool him. Also, they liked him, because he combined a proper amount of kindliness with his stern-ness; he was simple and unpretentious.[34]

Alhough clearly portraying Bunny as naive, Sinclair shares Bunny's excitement about the ingenuity involved in oil extraction, as Sinclair's subsequent fascinated descrip-tion of the intricacies of drilling shows. Depicting the industry, Sinclair once again channels some of Tarbell's exuberance, which in turn channeled a previous era of U.S. national ideology.

This exuberant portrayal of oil drilling is not, however, solely retrospective. It also faces forward. Sinclair shows how, incorporated into the oil-electric system, exube-rance takes on key new forms. In the new energy system, men have "pep" and Dad is a "real guy" who has "'the stuff,' barrels of it."[35] Dad is, in short, an enlivened, positive, capable, always energetic machine himself—one that is fueled by oil. Dad thus is part of a long line of figures styled and self-styled as "modern." That identification, along with the new energetics that is one of its chief signs, exuberantly marks off these individuals, together with the larger U.S. oil-electric capitalist energy system, as part of a new and, for some, exuberantly better world.

In this new era, American exceptionalism leaves the frontier and invests itself in the modernity of the United States, and the gap between it and the world outside modernity becomes reinscribed as a gulf between advanced and developing or backward places. This new societal exceptionalism promotes a new notion of individualism, which in turn becomes a new place for oil-electric cultural invention. In popular and also high cultural discourse, people's bodies and psyches are refigured as oil-electric-energized systems, and avant-garde artists become the experts who most aggressively convert these energetics into new styles, new aesthetics, new poetics.[36]

I will let Dad stand as a sufficient early example of a new kind of bioenergetics, pep, produced by oil. His foot connected him to engine power that augmented him, even as his charisma as a "big operator" yet a "real guy" gave him attractive force over his men. As it was with Dad, so it was with many. Slang was a fertile seedbed for their invention. People started (bodily and psychically) to "rev up" and "step on the gas." Sometimes they operated on all their cylinders and stopped, when necessary, to refuel. Electricity, oil's partner in the new energy system, provided a seedbed for even more fertile invention: as David Nye, in *Electrifying America*, puts it, electricity became "a metaphor for mental power, psychological energy, and sexual attraction," and it "merged with new therapeutic conceptions of the psyche and the self." Examples include "'She really got a charge out of seeing you,' or 'He's gone on a vacation to recharge his batteries. . . . An 'energetic' person was 'a human dynamo,' a powerful performance was 'electrifying.'"[37] The kinetics in all of these examples are so pronounced, catastrophe is not simply banished or geographically relocated to a hell; as in extraction culture, it is fused with exuberance. Thus, people also crash, undergo crackups; they blow a fuse; they burn out. But unlike extraction culture, this fusion—as modernist art and aesthetic invention reveal—is complex and polyvalent, anything but simple.

Sinclair Lewis's title character Babbitt, for example, "whose god was Modern Appliances," embodied his ego in his Dutch Colonial home in Floral Heights and his automobile, which he drives and parks in "a virile adventure masterfully executed."[38] He commutes to work in Zenith, a city transformed, so that new "clean towers . . . thrust" "old . . . factories with stingy and sooted windows, wooden tenements colored like mud" from the business center. Further, he smugly sees himself as filled with new energy, as "capable, an official, a man to contrive, to direct, to get things done."[39]

Exuberant in his views of himself and his world, Babbitt is, however, Lewis makes abundantly clear, psychologically, socially, and aesthetically a catastrophe—an emblem of the stupidity and vulgarity that the new modern energies are in fact bringing about. These are qualities Babbitt has mostly not because he partakes too fully of modern energetics, but because he partakes too little: he is, in short, a dim bulb.

At the other extreme end of the spectrum of modern energetics is Hart Crane, who styled himself as "quite fit to become a suitable *Pindar* for the dawn of the machine age, so called." Crane's stylistic innovations sought to "absorb" the machine into poetry, and he pursued it by cultivating "an extraordinary capacity for surrender, at least temporarily, to the sensations of urban life," to the end of internalizing the "power of machinery" so completely that it might become "like the unconscious nervous responses of our bodies, its connotations emanat[ing] from within."[40] From this stunningly romantic surrender, Crane writes poetry that creates, more than anything else in existence, the kinetic tactile and kinesthetic effects felt by bodies and psyches impelled by oil-electric powered machinery into motion—by elevators, airplanes, trains, and subways. Packing sensations like the sudden, stomach-churning initial drop of an elevator or subway in descent into his always dynamically forward-rushing verse, Crane also incorporates into his style the new perceptual kinetics explored in oil-electric-powered film, its capacity for representing dynamic sudden motion in shifts of scene and perspective and in cut and zoomed shots. This ecstasy of motion is, however, nearly as catastrophic as it is exuberant. Kinetic catastrophe is the subject of "Kitty Hawk," and even his poems' authentic ecstasies, like "Atlantis," are wedded to the surmounting of almost equal extremes of despair, as in "The Tunnel."

This same argument could be developed in regard to Ezra Pound, early T. S. Eliot, and Ernest Hemingway. They too self-consciously invented modern, expert-created, and widely advertised styles, styles that could perhaps be thought of as aesthetic converters, that formed a key part of their project to rescue literature and thereby civilization in a time of acute crisis. Representing a new kind of alienation and social fragmentation—a nightmare side of the modernity brought in when oil-electric-capitalism banished coal to the peripheries—they also explored what seemed like qualitatively new modes of mind and perception that aestheticized those experiences, again as a new kind of energetics.[41] With Hemingway and Eliot, that meant a new energetics of hyperconsciousness—for example, the light in Hemingway's well-known short story "A Clean Well-Lighted Place"—that aesthetically haunted and mesmerized even as it paralyzed, rather than powered, bodies and psyches. In the midst of their portrayal of cultural and existential catastrophe, a clean catastrophe, not a coal-miasmic one, exuberance subtly accompanies even the most desolate depictions, thanks to the entrancing and self-consciously transformative novelty of the writers' styles.

Perhaps the most clear-cut example of modern catastrophic-exuberant energetics comes, however, from the new oil-electric technology of film. Arguing that film represents not just a new medium but a change in the very "way in which human perception is organized," Walter Benjamin relates the jerky motion of (early) film to the new Fordist system of manufacture, arguing that it is embodied visibly in the assembly line.[42] In *Modern Times*, Charlie Chaplin (a favorite of Benjamin's) simultaneously embodies and disrupts this new energetics, creating, with astonishing comic grace, a body that both channels and subverts the assembly line's motion—which is, of course, also the motion of his medium, the mechanism of film itself. What were the cultures of coal have become now the aesthetics, even the poetics, of oil.

If modern oil–electric–coal capitalism sought both dynamism and stability, it was never more than precariously achieved. In World War I, oil exuberance was wedded all too clearly to oil catastrophe in a high-profile marriage of absolute opposites. Oil powered destructive new machinery (tanks, airplanes, trucks, diesel submarines), was used in making destructive weapons (TNT and even mustard gas), and fueled a refitted British navy, superior to Germany's, which remained tied to coal. In contrast, it was what saved the Allies and won the war, according to some influential voices: as Daniel Yergin writes, in his history of oil, *The Prize*, at a celebratory dinner ten days after the armistice, Lord Curzon uttered the famous words: "The Allied cause had floated to victory upon a wave of oil."[43] Oil helped kill millions. Oil led to victory. Immediately after World War I, as noted above, modernist exuberance was accompanied by the attempt to structure oil and society into a dynamic yet stable system.

In the context of World War II, the same description still fits: again the allies floated to victory on a sea of oil, and again war was followed by an attempt to stabilize. Once again, one finds a period of postwar exuberance, as the 1950s and 1960s saw a new expansion of consumer society. But this exuberance marked not just simple continuance; it accompanied a reinvention of and a new phase in the oil-electric energy system, as oil extruded a crucial new set of converter chains. The petrochemical industry, development of which started after World War I, but which blossomed only after World War II, created a huge new array of products to add to its consumer repertoire. As wartime petrochemistry was reworked into the chemical equivalent of plowshares, oil, chemically metamorphosed, became central to many new productions, from plastics to pharmaceuticals, print inks to pesticides. It changed into what people dressed in, evacuated into, viewed, and even ate, not just what they put into their power machinery. Oil thus now reappeared as an agent of chemical *and* social metamorphosis. Bodies became literally oily, in what they ate, and in the cosmetics and clothes they put on; pharmaceuticals began doing the same thing for minds.

On the heels of this exuberance came a much more insistent form of catastrophe. In 1964, Rachel Carson's *Silent Spring* made chemical metamorphosis seem the start of an apocalypse. With the transformation of fear of nuclear destruction into fear of environmental self-destruction that came with the 1970s environmental crisis, an apocalypse that involved oil, in many ways beyond Carson's carcinogenic and ecocidal toxics, seemed likely. People died in New York and London from the pall of fossil-fuel air pollution. Global warming made an early appearance on the popular stage with the film *Soylent Green* (1973), even as the oil crises of the 1970s added the threat of economic chaos to environmental meltdown. But then came exuberance again, with what seemed like no transition. Ronald Reagan, arguably, was swept to power on unhappiness with oil scarcity, an unhappiness that was quickly salved by the release of a new sea of oil, one that floated his new conservatism to new victories—the most significant of which was the collapse of the USSR, a collapse that can be linked to a resulting plunge in oil prices.

These rapid oscillations between oil exuberance and catastrophe, I would argue, signal the arrival of a new cultural regime—one that we dwell in today. This new regime involves a fusion of the two motifs and links them in a mutually reinforcing symbiosis that recalls early extraction culture. But now the fusion takes place not against the background of celebratory nationalism, or modernist neoexceptionalism, but a combination of multiplied scenarios for global apocalypse and theoretical advances toward antifoundationalism, the breaching of apparently secure cultural boundaries, and the embrace of disequilibrium and emergence. Stability seems to be completely gone—gone simultaneously in a runaway dynamism of exuberance and catastrophe.

On the exuberant side, the dynamic growth of new industries (computers, genetics, robotics, and nanotechnology) has been accompanied by a new, exuberant rhetoric that rejects the very notion of stability and equilibrium and that celebrates risk and even imminent catastrophe as part of this new dynamism.[44] Important also are exuberant versions of postmodern theory, celebrating human supersession of nature and evolution, and the breaching of boundaries between the human and the technological.[45] Simultaneously, psychic and bodily energetics have been taken to a catastrophically exuberant extreme in fictions like William Gibson's *Neuromancer* (1984), in which psyches wired into cyberspace experience qualitatively new and addictive kinds of out-of-body acceleration. Equally, new catastrophic-exuberant fantasies of postevolutionary metamorphosis and hybridization have been fetishized by writers from Bruce Sterling in *Schismatrix* (1986) to China Miéville in *Perdido Street Station* (2003).

On the catastrophic side, myriad environmental, technological, economic, and geopolitical crisis scenarios have now become key reference points for U.S. culture's construction of normality. Oil is central or significant in many of these crisis scenarios, even as worries about it have become a key part of today's norms.[46] More and more people today feel they dwell in what Ulrich Beck calls "Risk Society." U.S. popular culture (in blockbuster films and video games especially) now exuberantly sets its high-tech exciting narratives in postapocalyptic milieus. Many of these films, like the James Cameron's *Terminator* films (1984, 1991), *Children of Men* (2006), or *I Am Legend* (2007), are at best only very indirectly related to oil, but they do the oily cultural work of injecting exuberance into catastrophe in postapocalyptic settings. More directly engaging oil are films like *The Day After Tomorrow* (2004), a film that works to make global warming thrilling, and Cameron's *Avatar* (2009), which wrests a visually stunning utopian vision from energy woes.

A few more serious texts, however, attempt to unravel this fusion of catastrophe and exuberance. Octavia Butler's *The Parable of the Sower* (1984) and *Xenogenesis Trilogy* (1987–89), Cormac McCarthy's *The Road* (2006), and Kazuo Ishiguro's *Never Let Me Go* (2005) all present meltdowns and narratives of painful, slow, on-foot struggle that resist the exuberance that is today so persistently inscribed in postapocalyptic space. A small, more recent wing of such writing is now devoted to specifically post-oil fictions, including Sarah Hall's *The Carhullan Army* (2007), James Howard Kunstler's *World Made by Hand* (2008), and Andreas Eschbach's *Ausgebrannt* (2009),[47] fictions which, in that order, focus attention on the question, possibility, and even possible character of postexuberant societies.[48] In these texts, most notably, fantasies of postphysical acceleration and quick-time metamorphosis are stifled.[49] What the significance of these cultural attempts to resist the contemporary postapocalyptic fusion of catastrophe and exuberance might be is of course not yet clear. But what is clear is that the old faith in stability is gone. Oil's power, complexity, and serious woes are not only transparent to people today as never before, they are also themselves a hot cultural commodity in oil capitalism. In the process, the old traditions of exuberance and catastrophe, embedded in the earliest oil literature, have taken on extreme new forms.

NOTES

1. Vaclav Smil, *Energy in World History* (Boulder: Westview Press, 1994), 2.

2. Ibid., 252.

3. See the chapter entitled "Major Weather" in Jonathan Bate, *The Song of the Earth* (Cambridge: Harvard University Press, 2000). The closest thing I know to oxygen history is Peter D. Ward's *Out of Thin Air: Dinosaurs, Birds, and Earth's Ancient Atmosphere* (Washington, D.C.: Joseph Henry Press, 2006). It is a history calibrated in million-year intervals that speculatively relates the evolution of larger brains in early hominids to rising oxygen levels on earth and forecasts further change in 250 million years, when oxygen levels might drop. These speculations make me doubt that oxygen history will become an important theme in cultural history anytime soon.

4. William R. Catton Jr., *Overshoot: The Ecological Basis of Revolutionary Change* (Urbana: University of Illinois Press, 1982), 5–6.

5. Ibid., 28–29.

6. Ibid., 30.

7. Ibid., 6.

8. "Catastrophe" and "exuberance" are Catton's terms, but they need far more sensitive and complex descriptions than he gives them—and also need to be far more variable, specific, and context dependent. Consistently, however, the two terms interpenetrate, albeit in different fashions. For example, the term "exuberance" properly suggests a certain precariousness and even a measure of bad faith; it represents a departure from a sturdy sense of likelihood and normality. Even when used robustly, then, it is always shadowed by what fossil-fuel discourse persistently structures as its opposite partner—catastrophe. The two terms also vary for different times, places, issues, discourses, and speakers.

9. Jean-Claude Debeir, Jean-Paul Deléage, and Daniel Hémery, *In the Servitude of Power: Energy and Civilization through the Ages*, trans. John Barzman (London: Zed Books, 1991), 7.

10. Ibid., 13. A determination that is itself determined is, of course, very different from the determinisms that are regularly used to inspire or dismiss work on culture and technology, environment, and biology.

11. Ibid., 87. In looking at this break and the era that follows, one must acknowledge that both "exuberance" and "catastrophism" are cultural concomitants not just of fossil-fuel development but also of the larger acceleration of demographic–technological–economic–social growth that the combination of fossil fuels and capitalism inaugurated. In this complex, capitalism temporally preceded fossil-fuel development, but fossil-fuel exploitation soon became arguably as fundamental.

12. Ibid., 91, 99.

13. Barbara Freese, *Coal: A Human History* (New York: Penguin Books, 2003), 66.

14. Ibid., 69. The well-known domestic effects of the new coal capitalism were supplemented by coal-facilitated reorganization in the colonies. To note one concrete example: by 1826, the steam-powered gunship *Diana* (called the "fire devil") entered Burmese waters, easily destroying local opposition. More important, the *Indus*, in 1837, sailed up into Indian rivers, and, in 1841, the *Nemesis* did the same in China. About this process, historian Daniel Hedrick comments: "We cannot claim that technological innovation caused imperialism, nor that imperialist motives led to technological innovation. Rather the means and the motives stimulated each other in a relationship of positive mutual feedback." Daniel Hedrick, *The Tools of Empire: Technology and European Imperialism in the Nineteenth Century* (New York: Oxford University Press, 1981), 54.

15. Freese, *Coal*, 11.

16. Leo Marx, *The Machine in the Garden: Technology and the Pastoral Ideal in America* (New York: Oxford University Press, 1964), 192–93.

17. Mary Shelley's *Frankenstein* is an excellent (and extreme) attempt to represent and measure the stresses of this double endeavor: a destructively powerful, yet tenderly, poetically sensitive and intelligent monster is assembled mechanically out of scavenged biological parts and then galvanized (doubtless by electricity, thought by many to be the *élan vital*) into life.

18. Rebecca Harding Davis, *Life in the Iron Mills and Other Stories* (New York: The Feminist Press at CUNY, [1861] 1993), 13.

19. The Promethean myth, of course, was woven into old cultural and techno-cultural traditions; its fusion with coal came only with the invention of the steam engine. Coal's stygian features, however, are part of an old tradition of coal as a pollutant, one that begins well before the Industrial Revolution, in Medieval and Renaissance accounts of the appalling conditions in the English mines and of massive air pollution events. Fossil fuels, moreover, lit Milton's hell, and perhaps were also implicated in its brimstone, as English coal had a very perceptible, high sulfur content, and fossil fuels were lively features of depictions of hell all the way back to early Christian sources. For a general discussion, see Freese, *Coal*, 14–42; in Milton's *Paradise Lost*, see book 1, lines 725–29; for early Christian depictions of hell, see book 8, lines 100–106 of the Christian Sybillines in *New Testament Apocrypha*, vol. 2, ed. Wilhelm Schneemelcher, trans. Robert McLoed Wilson (Nashville: James Clarke & Co., 1992).

20. In fact, the (logical) order in which I have listed these converters is misleading. Before the development of extraction techniques came experiments with refining oil and the development of lamps suited for its use as a luminant. Also before extraction, capital accumulation began and marketing was pioneered, two other crucial parts of the oil converter chain. And together with extraction, storage and transportation converters had to be immediately developed—and go through many phases, as teamsters hauling carts with barrels yielded to railroad tankers and then to pipelines.

21. Ida Tarbell, Introduction to Paul H. Giddens, *The Birth of the Oil Industry* (New York: Macmillan, 1938), xix.

22. Oil geography suggests fascinating homologies with psychoanalytic theory and modern cultural practice, from therapy to poetry and art. The subject lies, unfortunately, beyond the reach of this essay. The new cultural geography of the later oil system is a separate but also important and interesting topic; see note 29.

23. Tarbell, Introduction, xxxvii–xxxviii.

24. Ibid., xxxix.

25. Giddens, *The Birth of the Oil Industry*, 76–77.

26. Ibid., 137, 139, 102.

27. Sonia Shah, *Crude: The Story of Oil* (New York: Seven Stories Press, 2004), 6. The new oil-flavored exuberance was distinctive in yet another way. No longer a Promethean intervention from above, or agent of capitalist oppression creating underworlds, energy became fused with widespread social desire. Indeed, it and the invention it stimulated and fetishized became an important attractor of peoples' imaginations and fantasies. Henry Adams's concept of history as a response to *attractive*, not compulsive, forces, and his use of energy production (the dynamo) as a central symbol for these was one response. More concrete was another change noted by Debeir, Déleage, and Hémery. By the twentieth century, energy production "reversed the demand-supply relation [a scarcity of supply relative to demand] which characterized early industrialization" (as it indeed had all previous energy systems). Now "energy production acquired unprecedented elasticity," and it "anticipated demand" and even "generated new needs." *In the Servitude of Power*, 108.

28. Shah, *Crude*, 3.

29. Oil was a new energy source, materialized as such by the growth of complex sets of converter chains; electricity was, however, simply a converter, sometimes connected to oil, sometimes to coal. But both allowed the miasmas of the coal era to be situated farther and farther away (culturally and geographically) from not just the well-to-do, but the growing middle classes.

Early observers, like Henry Adams in *The Education of Henry Adams*, were well aware of this. In his famous celebration of the dynamo, Adams writes that, clean and quiet, "it would not wake the baby lying close to its frame." Adams meaningfully explains why this is the case, noting that the dynamo utilized an "ingenious channel for conveying somewhere the heat latent in a few tons of poor coal hidden in a dirty engine house carefully kept out of sight." Henry Adams, *The Education of Henry Adams* (Boston: Houghton Mifflin, [1918] 1961), 380. Jill Jonnes also emphasizes how important oil's and electricity's ability to distance or erase coal was to the very idea of modernity. Writing about the dynamo in the 1893 World's Columbian Exposition and Fair in Chicago, Jonnes notes that, installed for the "White City's magnificent lighting displays, [it] was powered by one great 2000-horsepower Allis Chalmers engine, as well as numerous 1000-horspower engines, all fueled with oil (supplied by Standard Oil) rather than coal." The reason was that the display was meant to symbolize an ideal modern world displacing/replacing the miseries of actual Chicago: "The White City would have no smoky pall." Jill Jonnes, *Empires of Light: Edison, Tesla, Westinghouse, and the Race to Electrify the World* (New York: Random House, 2003), 261. Theodore Dreiser made the same point in writing about "a certain oil refinery," a highly polluting oil facility that was banished to the hinterland of Bayonne. Theodore Dreiser, "A Certain Oil Refinery," *American Earth: Environmental Writing Since Thoreau*, ed. Bill McKibben (New York: Literary Classics of the United States, 2008), 188–91.

30. Jacques Ellul has pithily (if androcentrically) characterized this key modern transformation as a move from "man and the machine" to "man in the machine." Jacques Ellul, *The Technological Society* (New York: Vintage, 1964), 6.

31. Harold Williamson, Ralph Andreano, Arnold Daum, and Gilbert Klose, *The American Petroleum Industry: The Age of Energy, 1899–1959* (Evanston: Northwestern University Press, 1963), 565–66.

32. Upton Sinclair, *Oil!* (New York: Penguin, [1926] 2006), 6.

33. Ibid., 76–77.

34. Ibid., 59.

35. Ibid., 61.

36. True, this development is not wholly novel: Walt Whitman, in his remarkable poem "To a Locomotive in Winter," enthusiastically converted a steam engine into a new energetics for American bodies, psyches, and art. He also did the same with electricity in "I Sing the Body Electric," absorbing a widespread romantic discourse of electricity and bodies, as Paul Gilmore discusses at length. Paul Gilmore, *Aesthetic Materialism: Electricity and American Romanticism* (Stanford: Stanford University Press, 2009), 143–76. Oil-electricity's revision and great expansion of both these discourses subsequently did much to constitute "the modern."

37. David E. Nye, *Electrifying America: Social Meanings of a New Technology, 1880–1940* (Cambridge: MIT Press, 1992), 155. Nye's conclusion was that "electricity was not merely one more commodity; rather, it played a central role in the creation of a twentieth-century sensibility. Electricity seemed linked to the structure of social reality; it seemed both to underlie physical and psychic health and to guarantee economic progress." Ibid., 156.

38. Sinclair Lewis, *Babbitt* (New York: Oxford University Press, [1922] 2010), 15.

39. Ibid., 3, 6.

40. Hart Crane, *The Complete Poems and Selected Letters and Prose of Hart Crane*, ed. Brom Weber (Garden City: Doubleday, 1966), 262.

41. Alienation may be seen, I believe, as the oil-era replacement for, and the descendant of, the exploitation and environmental immiseration of the coal-capitalist working class. Modernist alienation is clean, not miasmic; individualized, not collective; higher up on the class ladder than coal-misery; and an affliction of the refined consciousness, not the degraded laboring body.

42. Walter Benjamin, "The Work of Art in the Age of Its Technological Reproducibility," in *The Work of Art in the Age of Its Technological Reproducibility and Other Writings on Media,* ed. Michael W. Jennings, Brigid Doherty, and Thomas Y. Levin (Cambridge: Harvard University Press, 2008), 23. See also Thomas Levin's introduction to the section on film in ibid., 315–22.

43. Daniel Yergin, *The Prize: The Epic Quest for Oil, Money, and Power* (New York: Free Press, 1991), 183. Curzon's rhetoric (and the tone of Yergin's title and book) is perhaps a bit exaggerated, the former being a tribute to the wartime contribution of the American oil industry, and the latter clearly indebted to the (extraction-era) discourse of the epic of oil. Still, oil's contribution to World War I was great, and by World War II, Curzon's comment would apply without qualification.

44. On the new exuberance, see Kevin Kelly, *Out of Control: The New Biology of Machines, Social Systems, and the Economic World* (New York: Basic Books, 1995); Alvin Toffler, *The Third Wave* (New York: Bantam Books, 1984); and Ilya Prigogine and Isabelle Stengers, *Order out of Chaos: Man's New Dialogue with Nature* (New York: Bantam Books, 1984). On its involvement with risk, see Julian L. Simon, *The Ultimate Resource 2* (Princeton: Princeton University Press, 1966); and Naomi Klein, *The Shock Doctrine: The Rise of Disaster Capitalism* (New York: Picador, 2008). See also the discussion of risk and the new exuberance in Frederick Buell, *From Apocalypse to Way of Life* (New York: Routledge, 2005), 177–246.

45. See, for example, Fredric Jameson, *Postmodernism; or, The Cultural Logic of Late Capitalism* (Durham: Duke University Press, 1991); and Donna Haraway, *Simians, Cyborgs, and Women: The Reinvention of Nature* (New York: Routledge, 1990).

46. This is the central argument of my *From Apocalypse to Way of Life.*

47. I include Eschbach's fiction in the list because, while it is by a German writer and has not yet been translated, it is substantially set in and influenced by reflection on U.S. culture. Its deviation from the Anglophone postapocalyptic mode is very refreshing because it explores different postcatastrophic post-oil futures for different societies.

48. A slightly different, but very interesting, example of resistance to fused catastrophe and exuberance is Kim Stanley Robinson's global warming trilogy (*Forty Signs of Rain* [2004], *Fifty Degrees Below* [2005], and *Sixty Days and Counting* [2007]). A speculative fiction and alternative history of the present, the trilogy shoehorns attempts to deal with the first, catastrophically large disruptions of global climate into a realistic fiction of mixed subgenres. Partly Washington novels of political and scientific-political intrigue, partly suspense novels dealing with internal spying, partly romances, and partly novels of the education and growth of a large cast of interesting and likable good people dealing with domestic and personal issues, the trilogy not only confines exuberance and catastrophe within these different frames, but also manages to end in a strikingly complex fashion. On the one hand, it concludes nonexuberantly, as catastrophic climate disruptions (dramatized quite vividly) will certainly continue. On the other hand, it also concludes noncatastrophically, as the crisis is now in the hands of a good president and staff, elected in a narrow defeat of the scientifically illiterate far-right candidate [Bush] even as characters' romantic and familial problems happily resolve.

49. That such fantasies are directly and/or indirectly related not only to today's culture, one dependent on oil, but to oil in its contemporary material and technologically reworked forms is, I think, clearly arguable. Today's postbiological acceleration is clearly a descendant of futurist versions of modern automotive speed, and apocalypses that have characters trudging along disused highways pushing shopping carts play both on automobile and oil-midwifed twentieth-century consumer culture. And quick-time metamophoses, while inspired by the baby steps genetic engineering has actually taken, play on the postwar reshaping of motive energy into metamorphic energy. Motive energy literally became metamorphic with the rise of post–World War II petrochemistry and its transformation of oil into so many different forms. In a different sense, motive energy also became metamorphic with the more recent cultural fascination with robotics. In fact, and far more in fantasy, today's robotics has transformed the instrumental mobile machinery of modernity (for example, automobile culture) into a wide variety of lively postbiological cyborg life forms (from malign terminators operating in a militarized postapocalypse to Spielbergian A.I.'s, active in a Disneyfied postapocalypse).

PART II

Oil's Golden Age

Literature, Film, and Propaganda

5

Essential Driving and Vital Cars

American Automobile Culture in World War II

SARAH FROHARDT-LANE

On December 8, 1941, as part of the same campaign that included the bombing of Pearl Harbor, Japan seized control of rubber plantations in the Pacific, cutting off over 90 percent of the U.S.'s rubber supply.[1] Thus as the United States entered World War II, the country faced a severe rubber shortage. This scarcity, as well as shortages of gasoline and spare parts at various points in the war, created a significant transportation dilemma for the U.S. home front. The federal government's response was two-pronged: it regulated Americans' use of limited resources—rationing rubber and gasoline and lowering speed limits, for example—and it undertook a widespread propaganda campaign to encourage compliance with these regulations.

At first glance, propaganda that implored Americans to alter their driving habits during World War II appears to undercut the importance of driving to Americans' daily lives. For example, one poster demanded, "Have you REALLY tried to save gas by getting into a car club?" above a drawing of a wounded soldier with furrowed brow and blood streaming from underneath his helmet. Another, atop a picture of an oil tanker "torpedoed off the Atlantic Coast" demanded rhetorically, "Should brave men die so you can drive?" And most famously, a 1942 poster urging drivers to "join a car-sharing club today" proclaimed, "When you ride alone, you ride with Hitler."[2] In these and other government messages, excessive or improper driving was tantamount to treason and conservation of scarce materials became patriotic.

"Promoting a Culture of Driving: Rationing, Car Sharing, and Propaganda in World War II" by Sarah Frohardt-Lane. *Journal of American Studies* 46, special issue 2 (May 2012): 337–55. Copyright 2012 Cambridge University Press; reprinted with permission.

But government propaganda and corporate advertisements encouraging drivers to change their habits also established the notion that Americans could make key contributions to the war through proper driving, not by giving up driving. As this essay argues, rather than dampening Americans' affection for the automobile, wartime propaganda reinforced the notion that automobiles were essential to the war effort and portrayed driving as integral to the American way of life. In this way, the war helped establish driving as both patriotic and constituent of citizenship. Yet this citizenship was not available to all; posters, pamphlets, and magazine and newspaper advertisements portrayed the American driver almost exclusively as white and most often as male. Such depictions implied that the mobility and independence that driving afforded were the sole domain of white American men.

This essay first situates itself in current scholarship and outlines the rubber scarcity that prompted rubber and gasoline rationing. It explores how government propaganda and advertising on the subject of wartime transportation measures portrayed driving as essential to Americans while simultaneously denigrating alternative modes of transportation, including public transit and walking. The proper role of women in the transportation crisis became the subject of much discussion in advertisements and government press releases, demonstrating continued uneasiness about female drivers. Finally, despite being inundated with messages to comply with wartime transportation guidelines and regulations, many Americans chose not to comply. How residents in Detroit, Michigan, resisted the imposition of gasoline rationing and asserted their need to drive in spite of the war is the subject of the last section.

STUDIES OF THE AUTOMOBILE AND OF WARTIME PROPAGANDA

One measure of the new directions in scholarship on automobiles is the growth of "automobility studies." In this field, Cotten Seiler's recent work has used the term "automobility" to capture the "multiple, heterogeneous, intersecting components" of the automobile's impact.[3] John Urry has employed "automobility" to encompass the vast "system" of material objects, culture, consumption, mobility, and environmental aspects that make up the automobile's dominance.[4] By using the term "automobility" in this work, I draw on Seiler's and Urry's multiple layers to include material, sociocultural, and environmental elements of the car.

Increasingly, automobility studies have taken into consideration racialized and gendered divisions in Americans' relationships with cars and driving. Cotten Seiler's insightful essay, "So That We as a Race Might Have Something to Travel By," looks at African American drivers in the mid-twentieth century.[5] As Seiler shows, cars offered opportunity for blacks to experience the freedom of mobility, but black drivers were particularly susceptible to discrimination. Thomas Sugrue, Paul Gilroy, and George Lipsitz each explore whites' prejudice against black drivers and the threat black drivers' autonomy posed in the white imagination.[6] As this essay illustrates, World War II propaganda contributed to an image of driving as an exclusively white domain, at the same time that it emphasized the mobility and freedom that driving offered. Wartime events furthered the racialization of driving as white and public transit as black, creating an increasing divide in status and desirability of public and private forms of transportation.[7]

Virginia Scharff, Margaret Walsh, and Deborah Clarke, among others, have studied female drivers and changing cultural norms with regard to women's driving.[8] Although many histories of the car have focused on male drivers, these authors demonstrate that women drove cars from their first inception in the United States, and advertising drew upon female beauty to promote particular models of cars for male and female drivers. Walsh argues that cars were integral to women's labor in the twentieth century, and ultimately women's widespread use of the automobile has made it "a sex neutral vehicle."[9] This work complicates the relation of women to driving and automobiles by examining the multiple and contradictory messages about female drivers that government propaganda and car advertisements conveyed during World War II. While advertisements increasingly featured solo female drivers near the end of the war, female drivers were particularly targeted in appeals to restrict driving in order to save rubber and gasoline, and were overrepresented in commentary on the benefits of foregoing automobile driving as part of the war effort.

Despite the growth of automobility studies, the war years have not figured prominently in histories of the automobile. In fact the majority of such works focus on the widespread proliferation of automobile ownership in the 1920s and on the development of infrastructure to support increased automobility beginning in the 1950s. To the extent that World War II enters into these histories, it is a momentary stagnation or interruption in the story of an increasingly motorized society.[10] World War II, in this line of reasoning, interrupted Americans' "love affair" with the automobile; gasoline rationing temporarily curtailed car driving and suspended car culture before both surged in the postwar era.[11] As Ted Steinberg has written, World War II "put a dent in American auto culture" and created "a moment of ecological and social possibility" in which Americans might have turned away from increasing reliance on automobile-based transportation.[12]

By contrast, wartime propaganda and advertising have received extensive scholarly attention.[13] During World War II, the Office of War Information (OWI) was responsible for coordinating much of the U.S.'s propaganda. At first, President Franklin Roosevelt was reticent to create an agency that would have much control over information about the war—in light of criticism of the heavy-handed work of George Creel in the Committee on Public Information during World War I—and eager to contrast American efforts with those of Nazi Minister of Propaganda Joseph Goebbels.[14]

After three incarnations of a propaganda office floundered in part for want of a clear mandate, Roosevelt created the OWI in June 1942 under the direction of journalist Elmer Davis.[15]

The OWI was tasked with disseminating information about the war and drumming up support for home-front campaigns such as those that urged conservation of materials. Attempting to shape Americans' behaviors and simultaneously convey "just the facts" of the war proved contradictory and divided OWI's staff, as Allan Winkler and Sydney Weinberg have shown.[16] In terms of transportation, the OWI sought to convince Americans that the answer to the transportation shortage was to reduce all "unnecessary" travel and maximize use of private automobiles and public transit systems for all "essential" trips. Its strategies for obtaining compliance centered on convincing American drivers that they must make their cars last for the entirety of the war because their cars were necessary to victory, and because there would be no new cars and few spare parts.

A rich body of scholarship has shown how corporate advertising during World War II encouraged the public to see consumption as integral to winning the war and central to Americans' self-definition. Advertisers built demand for future goods to be available after the war and encouraged Americans to think of it as a sacrifice to go without these products. It became patriotic to deny consumption during the war, at the same time that Americans' identities as consumers solidified.[17] As Donnagal Goldthwaite Young has argued, "sacrifice and consumption" went hand in hand in wartime advertisements. And in the transportation industries in particular, "by articulating a vision of abundance that would be experienced by all once their engines, rubber, steel, fan belts, and bolts won the war, corporations succeeded in defining a postwar American way of life in terms of material goods."[18] Thus by suggesting that the war was being fought to secure an idealized, abundant society built on free enterprise, advertisements conflated a potential outcome with the war's purpose. The automobile industry was no exception, and automobile companies flooded newspapers and magazines with advertisements that kept their names in front of the public and associated their products with the war effort. Automobiles were not for sale to the general public during the war, but as we shall see, car companies, service stations, and parts manufacturers found ample opportunity to promote their businesses in relation to the war. At the same time, the OWI's campaigns stressed the importance of the private automobile to wartime transportation and the successful completion of the war.

RUBBER AND THE WARTIME TRANSPORTATION CRISIS

From tires on tanks and airplanes to gas masks, rubber was essential to combat operations during World War II. At the same time, rubber was a key component to the smooth operation of American civilian life. In particular, during peacetime it took approximately 250,000 tons of crude rubber to keep the country's 27 million private cars running for a year.[19] But as estimates in the fall of 1942 revealed, even if no rubber were set aside to replace tires on private cars, the United States would have an overall deficit of 211,000 tons of crude rubber by the beginning of 1944 due to military needs for rubber.[20] Public transit systems that used buses also were dependent on rubber to operate, and at the dawn of World War II many mass transit systems had recently converted their streetcar lines to bus lines.[21] Thus from the beginning of the U.S.'s military involvement in World War II, scarcity of rubber was a significant military problem with widespread consequences for civilian transportation as well.

To address the rubber shortage, the Office of Price Administration (OPA) began rationing rubber in January 1942, making rubber the first commodity to be rationed.[22] When it became clear that rubber rationing had failed to sufficiently reduce rubber consumption, the Office of Defense Transportation (ODT) imposed a speed limit of thirty-five miles per hour for the duration of the war.[23] In December, the OPA implemented nationwide gasoline rationing to more effectively curtail the wear of rubber tires by making it more difficult for drivers to use their vehicles.[24] Gasoline rationing spawned widespread opposition and disbelief in its necessity, prompting the OWI to undertake a massive campaign to convince Americans of the gravity of the transportation problem and the need to comply with wartime regulations.

Selling Automobility without Selling Cars

Although popular histories of World War II routinely laud automobile companies for patriotically halting civilian production and converting to war production, car manufacturers actually delayed conversion in order to sell more cars. Some plants began to make aviation equipment in mid-1940, in expectation that the United States might enter the war, but it was not until February 1942 that civilian automobile production ceased.[25] According to economist John Kenneth Galbraith, who was an assistant to OPA director Leon Henderson, heads of the major automobile manufacturers were unwilling to discontinue car production completely despite strong pressure from the government coupled with the U.S. entry in the war in December 1941. It took a secret "deal," proffered by Henderson, to halt auto manufacturing and force full conversion to the war effort. Henderson agreed to allow continued car production into early 1942 and promised not to claim that the car companies were being uncooperative.[26]

When it became clear that Henderson's public explanation would not be convincing, OPA staff sat in on multiple sessions in which he practiced making it sound believable. In time for a press conference announcing the cessation of civilian car production, Henderson found language to make his case convincing and cover up the fact that he had bowed to the auto executives' wishes. According to Galbraith, Henderson was both imaginative and eloquent. Staff cars were needed for the military; passenger cars were needed for defense workers; the components would otherwise be piled up in the factories during the duration. The weeks until the order went into effect would be used to plan the production of tanks and military vehicles and ensure that Detroit would truly be an arsenal of democracy.[27]

Galbraith's account suggested that top executives at Ford and General Motors were able to ensure that the automobile industry secured a reputation for full cooperation and sacrifice as part of the war effort, while actually consuming limited supplies in order to turn greater profits until they were forced to stop. Despite an extreme rubber shortage for military and civilian needs, the federal government acquiesced to automobile manufacturers' desire simultaneously to continue using scarce resources for their own benefit and to appear patriotic and sacrificing for the war effort. This was a harbinger of what was to come, and of the extent to which notions of citizenship and patriotism would infuse the rhetoric of automobility during the war.

The OWI was instrumental in portraying the automobile as fundamentally important to the war effort. With entry into the war, it was clear that there were insufficient tires, spare parts, and gasoline to support the country's overreliance on individual automobiles as the primary means of transit. But the same material shortages limited the possibility of greatly expanding public transit systems to absorb much former automobile traffic. To address the wartime "transportation crisis," the ODT focused on preserving the function of Americans' private automobiles, maximizing use of the existing public transportation systems, and reducing overall travel needs.[28]

As the propaganda wing for the federal government, the OWI created information campaigns to support these transportation directives. With a limited supply of tires available through rationing, and no new civilian automobiles for consumers from February 1942 until after the war ended, government officials were very concerned that drivers would wear out their tires and lose the functionality of their automobiles. As a result, the OWI sought to convince the public that it was imperative to keep their cars functional by changing their driving habits. In making their case for the need to change driving habits to keep cars operational, several of the OWI's information campaigns underscored the importance of cars to the war effort and even cast driving as patriotic, despite the resource shortages.

OWI press releases, radio shorts, and posters fundamentally promoted the automobile as an essential part of the war effort and spoke to the American public primarily as a nation of drivers. As a March 1943 press release put it, "The privately owned automobile is no longer the exclusive concern of its owner. The car owner has in fact become the custodian of a vital unit in America's wartime transportation system."[29]

This portrayal elevated the importance of automobiles: no longer was a car solely within the owner's purview, it was now "vital" to America. Emphasizing the importance of maintaining one's automobile to "America's wartime transportation system" portrayed automobiles as a weapon in the fight, and drivers as contributing to the war's successful execution.

Similarly, the OWI's guidelines for radio broadcasts suggested local stations emphasize that the country was counting on Americans to keep their cars working. As one memo on carpooling stressed, "All car operators have in their possession a valuable unit in their nation's transportation plan . . . an automobile."[30] In this way, the rhetoric surrounding driving offered car owners an opportunity to assist their country by keeping their cars in operation. Indeed, in September 1944 the OWI proclaimed: "All conservation measures—mileage rationing, group riding, tire conservation, automobile maintenance—have been directed toward one single objective: assurance that the private automobile will, according to ODT, 'continue to perform its indispensable function as an integral unit of the nation's transportation facilities.'"[31] Such language emphasized the automobile as central to American mobility, and that all attempts to control Americans' driving habits during the war were for the larger goal of preserving automobiles as the primary form of transportation.

Private companies involved in the making and service of automobiles expanded upon government propaganda that linked proper car maintenance with helping win World War II. Car dealers, filler (gas) and service stations, tire manufacturers, oil companies, and other large and small endeavors advertised their products as helping Americans defeat the Axis. Just before halting car production, companies such as Chevrolet tapped into the rhetoric of conservation and advertised the dealer's expertise in helping drivers "conserve" their cars. An ad published in *Collier's* in early February 1942 sought to channel drivers' concern about maintaining their tires into business for the company's service stations.[32] It touted twelve "conservation services" that Chevrolet dealers offered, while describing "Chevrolet owners" as "patriotic" and "forward-looking." The ad encouraged "all motorists" to have their cars serviced because "a mobile nation is a strong nation." In this line of thought, the driving public contributed to America's strength and to the successful execution of the war.

Advertisements spread name recognition for Chevrolet and other brands of car-related products, depicted these companies' (white male) workers as heroic, and created an image that connected service at particular businesses with patriotic duty. For example, a Chevrolet ad in the *Saturday Evening Post* in May 1943 proclaimed, "Your Chevrolet Dealer serves for victory, day after day, in many, many ways. Conserves the lives of the vital cars and trucks which are serving industry—serving agriculture—

serving all America. Preserves the greatest motorized transportation system which is helping America to win the war. See him for service today."[33] By describing "vital cars" and "the greatest motorized transportation system" as contributing to the war effort, the advertisement underscored automobiles' importance and desirability, even at a time when new cars could not be purchased.

Results of a survey conducted by the OWI in December 1943 demonstrated that many drivers had been exposed to messages about the need to preserve one's automobile and take care of its individual parts to ensure it ran properly. Survey researchers asked participants if they were familiar with the government's "tire conservation campaign," and if so what the campaign asked drivers to do. When thinking of tire conservation and actions they were encouraged to take, slightly more drivers recalled the importance of having their tires inspected and inflated than recalled the need for reducing driving or driving more slowly.[34] Thus keeping one's car fully functional seemed to have stuck in drivers' minds more than limiting or altering one's driving in any fundamental way.

Because rubber for civilian use was very scarce and many workers drove to their jobs, the government stressed car drivers' opportunities to aid the war effort by reducing their tire use. In heavily promoting representations of drivers who had their tires checked regularly as patriotic, this tire conservation propaganda conflated driving and American citizenship. Suggesting it was Americans' responsibility to take proper care of their tires, these messages portrayed a nation comprising drivers and ignored the millions of Americans who did not own a car.

AUTOMOBILES AS PREFERABLE TO
ALTERNATIVE MEANS OF TRANSPORTATION

Government appeals for workers to form carpools targeted only drivers. The carpooling arrangement that the OPA backed most vigorously was "car-sharing," in which four or five drivers took turns driving the others to work in their car, with each person responsible for driving approximately one week per month. Radio announcements and posters asked drivers to do their part by joining a car-sharing club with other motorists, and noted that there was no need to pick up nondrivers. As an OWI memo in the summer of 1944 assured drivers, "Car pooling does not mean 'thumbing' a ride- or the casual picking up of people waiting for buses or street cars. The kind that counts is the organized car pool whose members rotate the use of cars, ride together, leave some cars at home that would otherwise be burning up gasoline, wearing out precious rubber and parts."[35] In this situation, owning a car, and subsequently regulating one's use of it, was key to helping the war effort. Only those who already drove cars could be patriotic by joining a carpool; those who did not drive could not contribute by riding with another person, even though doing so might free up much-needed space on public transit vehicles.

Government propaganda promoting car-sharing programs encouraged Americans to identify driving with freedom and to see car sharing as the best way to preserve this freedom during the war. A press release in July 1944 explained, "The driver who drives alone just because he likes his freedom probably doesn't realize that by saving his car, gasoline, and rubber he can help insure a lot more freedom for himself later on."[36] Such an appeal framed carpooling as a key measure to ensure drivers could continue to use their cars when they needed them during the war, and would not lose this freedom of mobility at a later date. The pitch for carpooling was premised on the idea that it was a temporary and tolerable inconvenience in return for continued use of one's automobile.

Portrayals of carpooling that explained that "neighbors" should arrange carpools implicitly reassured the driving public—represented as all white—that they had total control over who entered their cars. They discounted the prospect of taking strangers "waiting for buses or streetcars" and offered instead an image of coworkers and neighbors joining together to cut down on their use of rubber.[37] For whites, taking neighbors and coworkers more likely brought to mind visions of sharing one's car with other whites than did the vision of picking up intending passengers at a bus stop.

Casting public transportation in a more negative light than carpooling was one method that government agencies used to promote car-sharing programs. Describing a film he hoped would be produced to encourage car sharing, the OPA's Malcolm Lund explained, "The selling job should be entirely positive. It should be pointed out that car sharing actually works no hardship on anyone, that it is infinitely more desirable than waiting indefinitely while crowded buses pass you up, or fortunately finding a square foot or two on which to stand in a crowded bus."[38] Thus even in addressing the inadequacies of individual automobile use, and promoting restricted driving, the government disparaged public transit as unpleasant and crowded. Adding to the riders' actual experiences, this public characterization of streetcars and buses as undesirable reinforced the importance of cars and driving to the American way of life and undercut acceptance of alternative modes of transit.

But government propaganda went even further in depicting the good citizen as a driver and undermining public transit. With public vehicles carrying on average 53 percent more passengers during the war than in 1941, and limited equipment to accommodate the surge in ridership, the OWI sought to dissuade drivers from taking public transit.[39] Given the vast number of automobiles in the country compared to the number of public transit carriers, the OWI explained in a 1943 press release that "the man who climbs aboard a public conveyance, when he is able to use his own car for essential transportation, is rendering a disservice to his country."[40] In this depiction it was unpatriotic for car owners to take the bus. As an undated pamphlet to promote car sharing laid out, "It is [the car owner's] duty to use his car in a Car Sharing Club and *stay off the street cars and buses*. Our public transportation facilities cannot carry the entire load."[41] In fact, "the individual car owner, who makes his automobile and tires last longer by sharing his car with his neighbors is serving his country's best interest by the very fact that he *does not* become a street car or bus rider."[42]

Government communications about driving asserted that cars were necessary to victory and deplored the possibility of losing the use of a car. This is seen particularly in the joint ODT and War Advertising Council print advertisements on the theme of "Take Care of Your Car."[43] Prefacing a series of these ads, ODT director J. Monroe Johnson explained the premise to heads of private companies, such as service stations, whose sponsorship he solicited. Johnson wrote that these ads "have been prepared on a common sense, practical basis around the question, 'Will they [American drivers] ride . . . or will they walk?' It is of first importance to this nation as well as to every individual and every automotive service dealer in this country that the answer be, 'THEY WILL RIDE!'" The memo emphasized the basic premise that driving was the primary mode of transport for Americans, even despite resource shortages that reduced the availability of rubber and gasoline. The question "Will they ride . . . or will they walk" is a rhetorical one, meant to underscore the importance of maintaining one's car at all costs in order to avoid having to walk.

Ads within the portfolio ODT sent out spoke directly to drivers and urged them "unless you'd rather walk than ride, continue to treat that car with loving care." Walking was the unpleasant consequence of inadequate attention to one's car, hitting home the importance of keeping the car properly serviced in order to avoid having to walk. As one ad pronounced, "WARNING TO ALL CALL OWNERS: Take care of your car and all its parts OR YOU'LL WALK. Sounds drastic, doesn't it?" In another, a man learns that if he does not care for his car he will have to walk: "Walk! Did you say walk? To work . . . every day. That's several miles. I can't walk that far." Upon this realization, he vows to take better care of his vehicle to avoid walking. "Car, you're going to get some of the best service you've ever had. And believe me, you'll get it regular from now on."[44]

Some ads went a step further to depict the negative effects on a man who was forced to walk. One, with the caption "He doesn't get around much anymore!" featured a picture of a man soaking his legs in a bucket of hot water. The text below the caption read, "He's got a headache, too! For he's the man who said, 'New cars are just around the corner!' Today he's walking instead of riding because he let his 'old' car die on its wheels . . . stopped giving it the proper care to help it carry on."[45] Having to walk, as depicted in the ad, was such a difficult change from driving that the hardship of walking instead of driving could make a man immobile.

A government-industry meeting about ODT's "car conservation campaign" in the spring of 1945 emphasized the importance of cars to the war effort and threatened those who failed to regularly service their cars with having to walk instead. As the meeting's secretary clearly noted, "The primary purpose of this campaign is to bring home to the general car owning public the vital need to keep every privately owned automobile in operating condition right on through to final victory—and on beyond that to the day when new cars can again be purchased. . . . It is in the last analysis a question of 'Look after your car—and all the parts of it—or you'll walk!'" By upholding driving as the preferred method of transportation, to be preserved through careful attention to the needs of one's car, other forms of transportation appeared less desirable.

TARGETING FEMALE DRIVERS

Women were specifically targeted in gasoline conservation propaganda as those most likely to violate driving restrictions and those who should reduce their driving. At times women were expected to embrace walking, while at other points advertisements assured women that they were not expected to give up "essential" driving. A United Motors Service advertisement in February 1943, "Housewife with an 'A' card," made clear that "housewives" were entitled to drive for certain functions. Next to a picture of a white woman behind the wheel, the ad declared, "Better than anyone else, you know the meaning of 'indispensable' driving." The ad provided its own definition of indispensable driving: "It's what gets the children to school, takes you marketing, and accomplishes those essential 'odd errands' that can't be predicted or avoided."[46] The ad's language drew on the federal government's categorization of "essential" and "nonessential" driving but offered an expansive view of "essential" that included trips of convenience to run "odd errands," activities that OWI propaganda explicitly told women to avoid doing unless with a group of riders. By identifying driving as "essential" and "indispensable," this particular ad recognized white women's entitlement to driving while it portrayed driving as critical for the perpetuation of daily family life.

Not all ads supported the notion that women's driving was essential. In fact, a disproportionate number of ads that encouraged drivers to reduce their driving were aimed at women. Propaganda that lauded the benefits of walking for women contrasted sharply with depictions of men in pain from having to walk because they failed to take care of their cars, as described above. While ads in the series "Take Care of your Car . . . Or You'll Walk!" portrayed men who had to walk as suffering, ads and newspaper stories suggested that female drivers in particular could benefit from walking instead of driving. Crafters of this message touted walking as a boon for women's beauty and health, in addition to helping relieve the crisis in transportation. One newspaper ad suggested a silver lining of limited automobile use with the headline "Delightful New Way to Reduce Hips!" The text recommended that women should bicycle and walk whenever possible, and that doing so would increase their femininity. "You'll soon be a finer figure of a woman if you'll just help make gas rationing work!" the ad promised.[47]

Such propaganda reflected broader anxieties about women's roles and the appropriateness of female drivers. The toughest ads seeking Americans' compliance with gas rationing singled out female drivers for criticism. Meant to be the words of a Nazi sailor, one cheered, "ATTA GIRL! WASTE YOUR GAS!... Don't pay any attention to that gas rationing stuff. Nuts to the war! Be a gas chiseller. Grab all the gas you can. What's your name, fraulein? We'll put you down for a Nazi medal!"[48] Another, from the voice of an American on an oil tanker at sea, chastised a female driver for wasting gasoline to attend a "fancy pink tea." After a graphic description of the ways one might die on a tanker, the seaman exclaimed, "Listen, lady, we ain't riskin' our hides just so's people can drive to pink teas. Just remember, every drop of that stuff in your gas tank is blood!"[49] In this and other appeals to obey gasoline rationing, women were depicted as especially frivolous in their use of gasoline and the consequences as particularly dire. Propaganda thus simultaneously recognized women's driving, if not car ownership, as a growing trend but put women's access to automobiles as subordinate to men's.

Suggesting that gas rationing offered a chance for women to benefit from walking—but implying that noncompliance with rationing would force men to walk—reinforced men's claims to driving. It also sent mixed messages about women's relationship to automobiles and driving. Margaret Walsh has argued that World War II created the opportunity for more American women to drive when they took on "war work" outside the home and served in the military.[50] This opportunity, however, was circumscribed by media messages implying women had less of a right to use automobiles when limited resources called for changes in driving habits.

OPPOSITION TO GASOLINE RATIONING

Despite the OWI inundating the public with messages about the severity of the transportation problem, and attempting to convince Americans of the need for gasoline rationing, many Americans opposed gas rationing in the late fall of 1942. After an intensive publicity campaign in November 1942, one in three Americans newly subject to gasoline rationing still did not think rationing was necessary.[51] A minority expressed opposition to some rationing throughout the war, but the announcement of nationwide gasoline rationing engendered a particularly bold negative reaction.[52] Officials were "besieged with protests."[53] Organized resistance to gasoline rationing was strongest in Detroit and Indianapolis in the weeks preceding its December 1 implementation; government reports referred to these cities as "trouble spots" and "centers of opposition."[54]

In Detroit, the automobile capital, opponents to gas rationing framed their protest in terms of the dire consequences for their mobility if rationing were implemented. According to an OWI public opinion report based on surveys conducted just prior to and subsequent to the nationwide gasoline rationing, those opposed in Detroit felt that limiting the amount they could drive would exacerbate the city's "transportation problems."[55] One Detroit public transit rider wrote to Washington to explain his concern. After detailing a recent fatal bus accident blamed on the driver's limited visibility due to overcrowding, Steve Reznik described his difficulty in getting to work on time when he had to wait for a streetcar with room to stop for him. He worried that with gas rationing more Detroiters would crowd the city's public vehicles and cause even further delays. Wrote Reznik, "If gas rationing takes effect in Detroit[,] God help those who will have to depend on the Detroit Street Railway System. It's the worst transportation system in the world. Ask those who ride it. They ride it not because they want to but because they have to. They don't own an automobile."[56] What is particularly striking about Reznik's letter is the fact that as a person without a car he was concerned enough about the proposal of gasoline rationing to write to the OPA in Washington, D.C., voicing his opposition to it. His experience with public transportation had led him to believe that any policies that might increase ridership would be disastrous.

Reznik's letter was forwarded to the ODT, whose director at the time, Joseph Eastman, replied two weeks later. Eastman informed Reznik that the most important thing was to use cars effectively, but that noncar owners could also help out: "The one immediate help which almost every individual can give is to cooperate in the more intensive use of the personal automobile." Acknowledging that Reznik did not own a car, Eastman advised, "Even if you do not own a motor car which you could share with other car owners in rotation, you may find a fellow-employee or a neighbor whose terminal-to-terminal trips morning and evening in his machine come at times that would be reasonably convenient for you."[57] Eastman's emphasis on "more intensive use of the automobile" implied that greater employment of the car was what the country needed; the private auto was a key weapon in the war. His reply to Reznik suggested that the opportunity for noncar owners to be patriotic lay in attaining access to a car and a driver willing to take them. Implicitly, walking, taking public transit, and otherwise avoiding using rubber and gasoline in private cars were not patriotic. In the realm of transportation, noncar owners' ability to connect with a car owner determined their contribution to the war effort.

Resistance to gasoline rationing in Detroit demonstrated that many drivers were opposed to direct limitations on their ability to drive as they pleased, even when the restriction was framed as a temporary measure. More significant, as Reznik's letter shows, even those who did not own cars could oppose gasoline rationing. Both those with cars and those without saw restrictions on driving in Detroit as a serious threat to their well-being. In the view of those who opposed rationing, any limits on automobile use that forced more riders onto overcrowded buses and streetcars were a bad idea.

CONCLUSION

John Kenneth Galbraith famously commented about World War II, "Never in the history of human conflict has there been so much talk of sacrifice and so little sacrifice."[58] Indeed the war did not bring extreme hardship to Americans on the home front, as compared to those in other warring nations. But during the war and in popular parlance since, many Americans have embraced the idea that they made significant sacrifices, and did so gladly.[59] As I have shown here, American drivers were asked to make relatively minor changes to their driving habits during the war, and many refused to do even this little. But the rhetoric of driving and sacrifice during the war had enduring effects in the postwar era and created the perfect opportunity for car-related businesses to promote unlimited driving as the ultimate reward after years of sacrifice.

An October 1945 advertisement for United Motors Service articulated the underlying message of longing for the total freedom to drive at war's end that had underpinned its ads throughout the war. It described a driver who had "been waiting, dreaming of the time when gas rationing would be over and he could head for the open road on a well-earned vacation trip." The ad played on its enduring wartime theme of the need for routine car maintenance to keep it in good condition, but substituted the end goal of "enjoy[ing] many miles of safe, satisfying 'pleasure' driving as your reward" for such service instead of the wartime message that car maintenance was integral to the war effort.[60]

Unlimited use of the car as a "reward" for proper automobile upkeep during the war promoted driving as an activity even more than a form of transportation. Often it was also linked to the joys of being an American, and the ability to enjoy these pleasures after years of wartime restrictions. As an ad in late December 1945 put it, "It's time to take your reward: you took the trouble, spent the money, to keep your car in good shape during the war years when it couldn't be replaced. You've earned the right to take those trips . . . drive where you please."[61] Those with cars were all too ready to do so.

NOTES

1. Daniel Yergin, *The Prize: The Epic Quest for Oil and Money* (New York: Free Press, 2008), 309; Richard Lingeman, *Don't You Know There's a War On?* (New York: G. P. Putnam's Sons, 1970), 235; Mark S. Foster, *A Nation on Wheels: The Automobile Culture in America Since 1945* (Belmont, Calif.: Thomson, Wadsworth, 2003), 32.

2. These posters are reproduced online on the George C. Marshall Library website, http://library.marshallfoundation.org/posters/library/index_posters.php, and the National Archives Powers of Persuasion exhibit website, http://www.archives.gov/exhibits/powers_of_persuasion/use_it_up/images_html/ride_with_hitler.html.

3. Cotten Seiler, *Republic of Drivers: A Cultural History of Automobility in America* (Chicago: University of Chicago Press, 2008), 5.

4. John Urry, "The 'System' of Automobility," *Theory, Culture & Society* 21, nos. 4–5 (2004): 25–26.

5. Cotten Seiler, "So That We as a Race Might Have Something to Travel By," *American Quarterly* 58, no. 4 (2006): 1091–117.

6. Thomas J. Sugrue, "Driving While Black: The Car and Race Relations in Modern America," Automobile in American Life and Society, http://www.autolife.umd.umich.edu/Race/R_Casestudy/R_Casestudy1.htm; Paul Gilroy, "Driving While Black," in *Car Cultures*, ed. Daniel Miller (Oxford: Berg, 2001); George Lipsitz, "'Swing Low, Sweet Cadillac': White Supremacy, Antiblack Racism, and the New Historicism," *American Literary History* 7, no. 4 (1995): 700–25.

7. See Sarah Frohardt-Lane, "Race, Public Transit, and Automobility in World War II Detroit" (PhD diss., University of Illinois, 2011).

8. Virginia Scharff, *Taking the Wheel: Women and the Coming of the Motor Age* (New York: Free Press, 1991); Margaret Walsh, "Gendering Mobility: Women, Work, and Automobility in the United States" *Historian* 93, no. 3 (2008): 376–95; Deborah Clarke, *Driving Women: Fiction and Automobile Culture in Twentieth-Century America* (Baltimore: Johns Hopkins University Press, 2007).

9. Walsh, "Gendering Mobility," 376.

10. See James J. Flink, *The Automobile Age* (Cambridge: MIT Press, 1990); Flink, *The Car Culture* (Cambridge: MIT Press, 1975); Foster, *A Nation on Wheels*; David L. Lewis and Laurence Goldstein, eds., *The Automobile in America* (Ann Arbor: University of Michigan Press, 1983). One exception is John Heitman, *The Automobile in American Life* (Jefferson, N.C.: McFarland and Company, 2009), which devotes a chapter to World War II.

11. See, for example, Yergin, *The Prize*, 391; Paul Casdorph, *Let the Good Times Roll: Life at Home in America During World War II* (New York: Paragon House, 1989), 14; Foster, *A Nation on Wheels*, 32–33.

12. Ted Steinberg, *Down to Earth: Nature's Role in American History* (New York: Oxford University Press, 2002), 213.

13. See, for example, Allan M. Winkler, *The Politics of Propaganda: The Office of War Information, 1942–1945* (New Haven: Yale University Press, 1978); Sydney Weinberg, "What to Tell America: The Writers' Quarrel in the Office of War Information," *Journal of American History* 55, no. 1 (June 1968): 73–89; William L. Bird, *Design for Victory: World War II Posters on the American Home Front* (New York: Princeton Architectural Press, 1998); Gerd Horten, *Radio Goes to War: The Cultural Politics of Propaganda During World War II* (Berkeley: University of California

Press, 2002); Clayton R. Koppes and Gregory D. Black, *Hollywood Goes to War: How Politics, Profits, and Propaganda Shaped World War II Movies* (New York: Free Press, 1987).

14. Winkler, *Politics of Propaganda*, 8–9, 20.

15. Ibid., 31. Before the OWI, Roosevelt had established the Office of Government Reports (September 1939), the Division of Information within the Office of Emergency Management (March 1941), and the Office of Facts and Figures (October 1941).

16. Winkler, *Politics of Propaganda*, 35; and Weinberg, "What to Tell America."

17. See Lizabeth Cohen, *A Consumers' Republic: The Politics of Mass Consumption in Postwar America* (New York: Vintage Books, 2004); Amy Bentley, *Eating for Victory: Food Rationing and the Politics of Domesticity* (Urbana: University of Illinois Press, 1998); Maureen Honey, "The 'Womanpower' Campaign: Advertising and Recruitment Propaganda during World War II," *Frontiers* 6, no. 1 (1981): 50–56; Meg Jacobs, "How about Some Meat? The Office of Price Administration, Consumption Politics, and State Building from the Bottom Up, 1941–1946," *Journal of American History* 84, no. 3 (December 1997), 910–41; Mark H. Leff, "Home-Front Mobilization in World War II: American Political Images of Civic Responsibility," in *Regional Conflicts and Conflict Resolution*, ed. Roger E. Kanet (Urbana: Program in Arms Control, Disarmament, and International Security, 1995): 277–97; Leff, "The Politics of Sacrifice on the American Home Front in World War II," *Journal of American History* 77, no. 4 (1991); Charles McGovern, *Sold American* (Chapel Hill: University of North Carolina Press, 2006); James T. Sparrow, "'Buying Our Boys Back': The Mass Foundations of Fiscal Citizenship in World War II, *Journal of Policy History* 20, no. 2 (April 2008): 263–86; Dannagal Goldthwaite Young, "Sacrifice, Consumption, and the American Way of Life," *Communication Review* 8 (2005): 27–52.

18. Young, "Sacrifice, Consumption, and the American Way of Life," 47.

19. Bernard M. Baruch, ed., *Report of the Rubber Survey Committee*, Washington, D.C., 1942, 34.

20. Ibid.

21. On the General Motors buyout of streetcars across the United States in order to replace them with General Motors–produced buses, see Bradford Snell, "American Ground Transport," in *Crisis in American Institutions*, ed. Jerome H. Skolnick and Elliott Currie (Boston: Little, Brown, 1985); Jim Klein and Martha Olson, *Taken for a Ride* (HoHoKus, N.J.: New Day Films, 1996); Kenneth T. Jackson, *Crabgrass Frontier: The Suburbanization of the United States* (Oxford: Oxford University Press, 1985); Edwin Black, *Internal Combustion: How Corporations and Governments Addicted the World to Oil and Derailed the Alternatives* (New York: St. Martin's Press, 2006); Jane Holtz Kay, *Asphalt Nation* (New York: Crown Publishers, 1997); Jack Doyle, *Taken for a Ride: Detroit's Big Three and the Politics of Pollution* (New York: Four Walls Eight Windows, 2000); Jerry Kloby, *Inequality Power and Development: The Task of Political Sociology* (Atlantic Highlands, N.J.: Humanities Press, 1997).

22. Cohen, *A Consumers' Republic*, 65.

23. Bradley Flamm, "Putting the Brakes On 'Non-essential' Travel: 1940s Wartime Mobility, Prosperity, and the U.S. Office of Defense," *Journal of Transport History* 27, no. 1 (March 2006): 79.

24. While nationwide gasoline rationing was implemented because of the rubber shortage, on the East Coast gasoline rationing began in March 1942 because of the success of German submarines in sinking oil tankers off the Atlantic coast, causing temporary gasoline shortages in the region. "Nonessential" driving, informally known as pleasure driving, was also temporarily banned on the East Coast because of the gasoline shortage.

25. John Rae, *The American Automobile: A Brief History* (Chicago: University of Chicago Press, 1965), 149.

26. John Kenneth Galbraith, *A Life in Our Times* (Boston: Houghton Mifflin, 1981), 157.

27. Ibid., 157–58.

28. Flamm, "Putting the Brakes On."

29. "Information Campaign for the Car Sharing Program," File Government Campaign Materials: Car Sharing- Mileage-Tires-Transportation, Record Group 208, Entry 141, box 751, National Archives at College Park, College Park, Md.

30. Ellipsis in original. Office of Price Administration, Department of Information, "The Facts about the Need for Car Pooling," 3, File "Gas Rationing and Car Sharing, Office of War Information," Record Group 208, Entry 69, box 221, National Archives at College Park, Md.

31. OWI P Release, File "Gasoline Situation Box 1042," Entry 198 RG 208 OWI NC-148; September 13, 1944.

32. Chevrolet Motor Division, General Motors Corporation, "Chevrolet Car Conservation Plan," February 7, 1942, *Collier's*, 39, Reel 10, D'Arcy Collection, University of Illinois Communications Library.

33. Chevrolet Motor Division, General Motors Corporation, "Your Chevrolet Dealer," May 29, 1943, *Saturday Evening Post*, 33, Reel 10, D'Arcy Collection, University of Illinois Communications Library.

34. Office of War Information, Bureau of Special Services, Surveys Division, *Special Memorandum No. 102: Car Owners Look at the Tire Situation*, January 28, 1944.

35. Office of War Information Domestic Radio Bureau Fact Sheet No. 256, "New Car-Pooling Regulations," July 24, 1944, File "Gas Rationing & Car Sharing, Office of War Information," Record Group 208, Entry 69, box 221, National Archives at College Park, Md.

36. Ibid.

37. "The Facts about the Need for Car Pooling," 2, File "Gas Rationing and Car Sharing," Office of War Information, Record Group 208, Entry 69, box 221, National Archives at College Park, Md.

38. Malcolm Lund to William H. Wells, September 9, 1943, File "Gas Rationing & Car Sharing," Office of War Information, Record Group 208, Entry 69, box 221, National Archives at College Park, Md.

39. U.S. Office of Defense Transportation, *Civilian War Transport: A Record of the Control of Domestic Traffic Operations, 1941–1946* (Washington, D.C., 1948), 297.

40. Office of War Information, "U.S. Government Transportation Campaigns- to Save Rubber, and to Conserve America's Truck, Bus, Railroad, and Local Transit Facilities," March 1943, File "Government Campaign Materials: Car Sharing-Mileage-Tires-Transportation," Record Group 208, Entry 141, box 751, National Archives at College Park, Md.

41. "Information Campaign for the Car Sharing Program," File "Government Campaign Materials: Car Sharing- Mileage-Tires-Transportation," Record Group 208, Entry 141, box 751, National Archives at College Park, Md.

42. Ibid. Underlining in original.

43. File "Gasoline Program and Conservation," Office of War Information, Record Group 208, Entry 66, OWI Records of Deputy Director Maurice Hanson, November 1943–October 1945, box 1, National Archives at College Park, Md.

44. "Go 'Way Ghost!," File "Gasoline Program and Conservation," Office of War Information, Record Group 208, Entry 66, OWI Records of Deputy Director Maurice Hanson, November 1943–October 1945, box 1, National Archives at College Park, Md.

45. "He doesn't get around much anymore!," File "Gasoline Program and Conservation," Office of War Information, Record Group 208, Entry 66, OWI Records of Deputy Director Maurice Hanson, November 1943–October 1945, box 1, National Archives at College Park, Md.

46. United Motors Service, "Housewife with an 'A' card," February 20, 1943, *Saturday Evening Post*, 40, Reel 10, D'Arcy Collection, University of Illinois Communications Library.

47. "Delightful New Way to Reduce Hips," File "Gasoline Rationing," Office of War Information, Record Group 208, Entry 40, box 144, National Archives at College Park, Md.

48. File "Programs 9-1, Gas and Rubber 1942–44," Office of War Information, Record Group 208, Entry 1, Records of the Office of the Director, 1942–45, box 6, National Archives at College Park, Md.

49. "Look, Lady—No Sneakin' Sub Can Keep Us From Deliverin' The Gas!," File "Gasoline Rationing," Office of War Information, Record Group 208, Entry 40, box 144, National Archives at College Park, Md.

50. Walsh, "Gendering Mobility," 384.

51. Before the campaign, only 48 percent of Americans in these areas thought gasoline rationing necessary. Office of War Information, Surveys Division, Memorandum No. 42, "Public Opinion after Two Weeks of Nationwide Gasoline Rationing," January 5, 1943.

52. In September 1943, when asked what problem they would like to discuss with their congressmen, "gasoline rationing" was mentioned more frequently than any other problem except for "high cost of living." By contrast, no other rationing concerns were in the top ten listed. Gallup Poll, "Important Problems," September 3, 1943.

53. Memorandum No. 42, "Public Opinion After Two Weeks of Nationwide Gasoline Rationing."

54. Ibid., 4.

55. Ibid., 9.

56. Steve F. Reznik to W. M. Jeffers, Office of Price Administration, November 7, 1942, File "Detroit, MI, Office of Defense Transportation," Record Group 219, Series 109, box 115, National Archives at College Park, Md.

57. Joseph Eastman, Director of Office of Defense Transportation to Steve F. Reznik, November 20, 1942, File "Detroit, MI, Office of Defense Transportation," Record Group 219, Series 109, box 115, National Archives at College Park, Md.

58. Quoted in Cohen, *A Consumers' Republic*, 70.

59. See Leff, "The Politics of Sacrifice."

60. United Motors Service, "Fill her up . . . MY EYE," *Saturday Evening Post*, October 27, 1945, 44, Reel 10, D'Arcy Collection, University of Illinois Communications Library.

61. United Motors Service, "What's to Stop Us?," *Saturday Evening Post*, December 22, 1945, 42, Reel 10, D'Arcy Collection, University of Illinois Communications Library.

6

Fossil-Fuel Futurity

Oil in Giant

DANIEL WORDEN

But I know where oil is ... it is future, undrilled ...
—Rick Bass, *Oil Notes*

The discourse of sustainability asks us all to imagine possible futures that are more environmentally responsible and less reliant on fossil fuels. This is no small feat, as Imre Szeman has argued, since many visions of a sustainable future rely on technological fantasies, the apocalyptic threat of a barren landscape, and, most important, the wishful desire that a new, environmentally friendly society will not require any radical readjustments to our lives at all.[1] Moreover, the difficulty of imagining life without, or beyond, fossil fuels stems from the fact that fossil fuels themselves connote futurity in late twentieth-century culture. It is difficult to imagine a future without petroleum, in part because petroleum underlies the normative vision of family, work, and social belonging in the late twentieth-century United States. Mowing the lawn, taking a road trip, getting a personal parking space at work, and teaching an adolescent child to drive all connote social belonging and are imbued with a ritualistic, affective charge. It is the ruse of "oil culture" that these activities do not typically strike us as activities having anything to do with drilling and refining, even though they are utterly dependent on the availability and social acceptability of fossil-fuel consumption. One of the tasks of an environmentally aware criticism, then, might be to document the ideology of what I call "fossil-fuel futurity." This ideology functions as a variation on what Lawrence Buell calls the "environmental unconscious,"

"Fossil-Fuel Futurity: Oil in *Giant*" by Daniel Worden. *Journal of American Studies* 46, special issue 2 (May 2012): 441–60. Copyright 2012 Cambridge University Press; reprinted with permission.

a narrative device that disconnects fossil-fuel consumption from the environment and instead places it in the realm of the "merely" cultural.[2] Since fossil fuels are bountiful and available within this imaginary, they do not register as a part of the environment but instead as a medium through which the environment becomes consumable and upon which is constructed a vision of normative family life.

In this essay, I will develop an account of how fossil-fuel futurity operates in the 1950s, in the novel and film versions of *Giant*, and then, briefly, chart the emerging crisis of this ideology in the 1978–91 television series *Dallas* and the 2007 film *There Will Be Blood*. These three texts align with three significant moments in U.S. oil culture. *Giant* was written and filmed during the postwar rise in oil consumption and the imperative to construct the interstate highway system, legislated through the Federal Highway Act of 1956; *Dallas* became a national as well as an international success during and after the 1979 oil crisis; and *There Will Be Blood* was released during the Iraq War.[3] Together, these texts map how oil becomes central to the vision of the good life in the post-1945 United States and the ruptures that changes in oil production and availability, as well as the stability of marriage and gender roles, produce in the vision of the petroleum-consuming family dominant in the late twentieth and early twenty-first centuries. All three texts imagine oil and futurity as inextricably and intimately connected, though the ameliorative connection of oil to the family depicted in *Giant* morphs into less idyllic, though still binary, relations in *Dallas* and *There Will Be Blood*. Accordingly, this essay begins with a reading of *Giant*'s celebratory account of oil culture as leading to pluralistic, familial national belonging, and then moves on to an analysis of *Dallas* and *There Will Be Blood*'s representations of oil as ultimately anathema to family belonging. In so doing, I aim to account for not only the resiliency of fossil-fuel futurity, as, for example, *Dallas* and *There Will Be Blood* continue to bind oil to all family relations, even as oil's effect on the family is explicitly dramatized as pernicious, but also fossil-fuel futurity's crisis, as oil's mid-twentieth-century connotations of middle-class prosperity wane and new modes of representing oil's effect on everyday life might become possible.

Edna Ferber's 1952 novel *Giant* and George Stevens's 1956 film adaption of the same title are key texts in the development of "fossil-fuel futurity," in part because *Giant* explicitly dramatizes the economic and cultural shift in West Texas from a cattle ranching economy to an oil economy. At the center of *Giant* are the Benedicts, who own and operate Reata, one of the largest cattle ranches in Texas. Jordan "Bick" Benedict runs the ranch with the help of other family members, a host of Chicano ranch hands, and Jett Rink, a poor, white ranch hand and car mechanic. While on a trip to Washington, D.C., Bick stops in Virginia to buy a horse from the Lynnton family. There, he meets Leslie; they immediately fall in love, and soon Bick and Leslie marry and travel back to Reata. Leslie is new to Texas and approaches the Benedict's ranch with the spirit of a reformer. She is especially troubled by the racial hierarchy at Reata—the ranch is owned and operated by whites and relies on Chicano workers, who form a kind of peasant underclass in West Texas. *Giant* then focuses on Leslie and Bick's arguments about racial injustice, their concerns about whether or not their children will marry properly and/or continue to work on the ranch, and Jett Rink's rise to incredible wealth when he strikes oil on a small piece of land. Race plays into all three plots. Leslie tries to better the conditions of Reata's Chicano workers; Bick and Leslie's son Jordy marries a Chicana woman, Juana, a union that Bick struggles to accept; and Jett Rink institutionalizes racial discrimination by instructing the workers in his new hotel, built with the profits from his oil empire, to refuse service to non-white customers. The novel and film conclude as the Benedicts lease portions of Reata to Jett Rink's oil company for substantial profits yet distance themselves from Jett's excessive wealth, power, and racism. By the end of *Giant*, the Benedicts have become racially tolerant, generous employers, and they accept both Jordy's marriage and the fact that their children will pursue professions outside of the family ranch, while Jett Rink squanders his fortune by building an extravagant hotel and finds himself lonely and desperately unhappy. In so doing, the narrative draws a distinction between "good" and "bad" oil cultures that are, in fact, mirror images of one another, with the key exception that "good" oil culture represses oil as an industry and represents it as a readily accessible medium for everyday life, while "bad" oil culture broadcasts excessive avarice and hostility to the collective good through villains like Jett Rink and, later, *Dallas*'s J. R. Ewing and *There Will Be Blood*'s Daniel Plainview. In *Giant*, as the oil industry is domesticated in Texas, fossil-fuel futurity governs characters' lives and provides a normative vision of family life that supersedes other modes of belonging based on class and ethnicity.

Originally serialized in *Ladies' Home Journal* and subsequently published as a book, *Giant* was a controversial novel upon its publication in 1952 because of its representation of well-to-do Texans as corrupt, crude, racist despots. In a *New York Times Book Review* article, Lewis Nichols commented, "As has been noted through vibrations felt easily from Houston to New York, a fair share of Texas has failed to find in 'Giant' the substance for amusement. Large chunks of a large state are mad in a large way."[4] The article goes on to quote Ferber, who defends her novel against Texan criticism: "'Giant' is not a novel about Texas . . . It's about the United States. The people just happen to live in Texas. The story is about the effect of men and women on Texas."[5] Later in the article, she explicitly links the novel to social uplift: "To be able to have written anything at all in the last fifteen years is a triumph. Serenity and peace must be there before you write anything to make others feel hope, courage, reason. Writers have become disappointed in love with the human race. After I saw Buchenwald, I

couldn't do anything worth doing for five years."[6] Ferber's reference to the Holocaust—
that after seeing Buchenwald she was crippled by disillusionment—positions her
novel in a tradition of sentimental writing that seeks out affective modes of response
to brutality and social injustice. In *The Female Complaint*, Lauren Berlant documents
the ways in which this genre of sentimentalism, also realized in Ferber's 1926 novel
Show Boat, asks the reader "to see all American sufferers as part of the same survival
subculture."[7] For Ferber, *Giant* makes a humanistic claim of survival. *Giant* depicts
the injuries of World War II through the enlistment and war death of one of Reata's
Chicano residents, Angel Obregon, and the excesses of oil production that fuel the
war effort. After the war, the Benedicts arrive at a new vision of social belonging that
is constructed around the family automobile, which serves not as a sign of the parasitical oil industry but instead as a marker of independent mobility and renewed possibilities in postwar culture.[8] *Giant*'s wealthy, racist Texans and exploited Chicano
workers are foils through which the novel then imagines an idyllic, pluralistic family.

Despite Edna Ferber's own claim that *Giant* is about the United States or, even
more broadly, about the world after the nightmare of the Holocaust, both the novel
and the film trade on regional detail. Texas culture is presented as foreign, feudal, yet
avowedly democratic. The bluebonnet is paradoxically—and tellingly—described as
the "national state flower of Texas" in the novel, one example of many phrasings that
cast Texas as both an independent nation and a microcosm of the United States.[9]
A European visitor remarks that Texas is "like another country, a foreign country in
the midst of the United States."[10] Texans themselves are portrayed in the novel as
intentionally manufacturing this regional distinctiveness, which also connotes egalitarianism and folksiness: "Each was playing a role, deliberately. It was part of the
Texas ritual. We're rich as son-of-a-bitch stew but look how homely we are, just as
plain-folksy as Grandpappy back in 1836. We know about champagne and caviar but
we talk hog and hominy."[11] *Giant*'s Texans view their own history as common knowledge, and unfamiliarity with Texas landmarks such as the Alamo is shameful. For
example, on his private plane, Bick offers to have his guests flown over Spindletop.
A visiting Hollywood starlet does not recognize the oil culture landmark: "'Spindletop?' said Miss Lona Lane, the movie girl. 'Is that a mountain or something? I don't
like flying over mountains very much.' The Texans present looked very serious which
meant that they were bursting inside with laughter . . . It was as though a tourist in
Paris had asked if Notre Dame was a football team."[12] Texas connotes difference, yet
by the end of *Giant*, this Texan uniqueness becomes normative national belonging, a
shared set of codes and practices that unify subjects despite class and racial difference.

Giant's focus on exotic Texan culture also allows for the development of a liberal,
anti-racist allegory.[13] While Bick claims that his treatment of Chicano workers is

benevolent, his wife, Leslie, takes an immediate interest in remedying the living and working conditions of her husband's employees and raises Bick's racial awareness. Leslie's activism is initially thwarted; the only way that Bick is finally swayed to accept Chicanos as equals is when his son Jordy marries the daughter of one of Bick's ranch hands, Juana, and has a dark-skinned son named Polo, after Juana's father. Bick's social awakening, then, occurs not through an awareness of exploitative working conditions and structural racism but, instead, through a new familial connection to his Chicano workers. Embodied in Jordy and Juana's child, family belonging leads directly to pluralist belonging in *Giant*. While the novel leaves Bick's enlightenment merely implied, the film renders it quite explicit as Bick, played by Rock Hudson, gets into a fistfight with a racist cook who refuses to serve Juana and Polo in a roadside diner. Rafael Pérez-Torres has argued that *Giant* "gives voice to a new America, one that struggles with discourses of inclusion and pluralistic liberalism."[14] This discourse of inclusivity and pluralism, of course, is premised on the benevolence, tolerance, and permissiveness of wealthy, white elites such as Bick Benedict. *Giant*'s setting and its representation of Texas as a synecdoche of the United States allows for Bick's racial awareness to become a kind of national imperative, not to treat workers better but to love and accept all members of one' family. This is central to *Giant*'s dramatization of fossil-fuel futurity, as the petroleum industry transforms the unique landscape of Texas and facilitates this pluralist vision of the American family.

While *Giant* is very much a narrative about the emergence and possibilities of racial equality in the postwar United States, there is another narrative strain that is equally important to both film and novel: the emergence of the "oil rich" in Texas, dramatized through the rise of Jett Rink, played by James Dean in the film adaptation. Jett is an overt racist, though he is also portrayed as a class warrior, mumbling in his introductory scene in the film that "ain't nobody king in this country" as Bick orders him around.[15] Jett Rink's newfound oil wealth, though, does not usher in new democratic structures but instead allows him to become even more despotic than the feudal lord and cattle rancher Bick Benedict. Jett Rink represents "bad oil" in the narrative. In the film, Jett Rink strikes oil and immediately drives his rickety Model T Ford to Bick Benedict's ranch house. Covered in oil, Jett laughs maniacally and attempts to flirt with Leslie Benedict, played by Elizabeth Taylor. This monstrous figure, covered in "black gold," represents avarice, racism, and cowardice, as opposed to the nostalgic Bick Benedict, whose major fault seems to be his inability to separate his familial obligations from his ranch. Oil functions here as a kind of blackface, racially coding the oil industry in relation to wholesome, white cattle ranching. As Eric Lott notes, blackface was often a way for white performers and audiences to stage otherness in the nineteenth century: "The black mask offered a way to play with collective fears of a degraded and threatening—

and male—Other while at the same time maintaining some symbolic control over them. Yet the intensified American fears of succumbing to a racialized image of Otherness were everywhere operative in minstrelsy . . . as its 'blackness' was unceasingly fascinating to audiences and performers alike."[16] The oil-covered, lascivious Jett Rink rechannels this nineteenth-century trope toward a representation of oil money's pernicious effects, and *Giant* retains blackface's double function, as identified by Lott, as representative of both fear and desire on the part of white America. While Jett Rink's racialization through oil marks him as a villain, *Giant* goes on to dramatize how racial pluralism successfully excludes Rink from national and familial belonging. Blackface is used, then, to mark Jett Rink as outside of the web of sympathy and paternalism that constitutes *Giant*'s vision of racial pluralism. Not race but oil connotes irreducible difference in *Giant*, and both film and novel ultimately demonize the oil tycoon only to subsume petroleum consumption into a normative vision of familial belonging.

Jett Rink is represented in the novel as an anachronism, doing in the present what families like the Benedicts did in a remote past. The construction of Jett Rink's Conquistador Hotel is described as a modern-day feudal project: "Houses had been razed, families dispossessed, businesses uprooted, streets demolished to make way for this giant edifice. All about it, clustered near—but not too near—like poor relations and servitors around a reigning despot, were the little structures that served the giant one."[17] The novel concludes ambiguously, as Leslie and Bick Benedict discuss the "future" that they have created on Reata Ranch—they are, after all, wealthy because they have leased portions of their ranch to Rink's oil company, subsidizing Bick's now financially precarious ranching—while Jett Rink remains a wealthy tycoon. In the film, however, Jett Rink is last seen alone in his banquet hall, after all of his guests have left; he drunkenly falls over a table and passes out, isolated in his decadent consumption. The film renders a clearer message about the impotence of tycoons in the face of the Benedict family, happy to be at home babysitting their grandchildren. As Bob Dietz, a character in the novel who has graduated from Cornell and practices modern agriculture in West Texas, claims, "Big stuff is old stuff now," registering the obsolescence of tycoons and large-scale industries and the emergence of the middle-class family as the basic unit of national belonging.[18] *Giant* represents a cultural shift from organizing society around "great men" like Bick Benedict and Jett Rink to imagining society as a more democratic series of family units, all of whom contribute to the greater good. The family's mobility, in this imaginary, is premised on the ubiquity of oil and the automobile.

Ferber's novel begins where Stevens's film adaptation concludes, with the oil industry. The novel opens with air travel; all of Texas society has been invited to the opening of oil tycoon Jett Rink's new airport, and they converge on Hermoso, Texas in "an aerial stampede."[19] Ferber's portrait of Texas emphasizes the unsettling developments that have transformed the state from a rural landscape to a jumble of technology, luxury, and poverty:

> High, high [the airplanes] soared above the skyscraper office buildings that rose idiotically out of the endless plain; above the sluggish rivers and the arroyos, above the lush new hotels and the anachronistic white-pillared mansions; the race horses in rich pasture, the swimming pools the drives of transplanted palms the huge motion picture palaces the cattle herds and the sheep and mountains and wild antelope and cotton fields and Martian chemical plants whose aluminum stacks gave back the airplanes glitter for glitter. And above the grey dust-bitten shanties of the Mexican barrios and the roadside barbecue shacks and the windmills and the water holes and the miles of mesquite and cactus.[20]

Texas here, as in the film, is portrayed as a site of radically uneven development, prompted by the modernizing effects of the oil industry. It is a region that stands in a synecdochal relationship to economic history itself, containing feudal ranch owners, aristocratic cotton kings, and emergent oil tycoons in the same historical moment. *Giant* ultimately makes the historical argument that the progressive developments of horticulture, ranching, and fossil-fuel production inevitably lead to democratization. This democratization, however, demands the repression of the petroleum industry and the construction of the nuclear family—as opposed to one's ethnic group, one's extended family, or one's coworkers—as the normative unit in American culture. This normativity is represented in *Giant* by the Benedicts, who organize the family in opposition to the greedy, racist Jett Rink and, by extension, the oil industry. By privatizing life and circumscribing industry as merely a staging ground for affective belonging, fossil fuels become a benevolent medium for the production of familial togetherness.

There is a curious nostalgia in *Giant*, which is in many ways a narrative about the future. For example, Leslie claims that she is not impressed with the private planes that have become the chic mode of transportation in Texas: "'I guess I belong to the generation that still thinks the automobile is a wonderful invention.'"[21] This remark is shortly followed by an explicit connection between the automobile and the airplane: "'That's right,' said Congressman Bale Clinch. 'Here every kid's got a car or anyway a motorbike. And a tractor or a jeep is child's play. Flying comes natural, like walking, to these kids.'"[22] In postwar Texas, the engine has replaced the horse, and the automobile is not enough to satisfy Texans' desire for new modes of recreation and transportation. The consumption of fossil fuels—not only in cars but also in recreational vehicles and airplanes—runs rampant in *Giant*'s representation of contemporary life. The automobile, then, is displaced as a dominant technology and takes on the whole-some, natural associations of horseback riding. Fossil-fuel futurity normalizes the automobile not as evidence of modern industry but instead as an object of nostalgia, saturated with domestic affect.

The demarcation between modern, fossil-fuel culture and the older, seemingly more honest cattle ranch does not persist as cleanly in the novel as it does in the film, mainly because in the novel, Jett Rink does not have a physical confrontation with Bick, does not attempt to woo Bick's daughter, and is not, ultimately, left alone and isolated. In the novel, both beef and oil are portrayed, rightly so, as industries. Neither is removed from greed, exploitation, and environmental degradation:

> So now the stink of oil hung heavy in the Texas air. It penetrated the houses the gardens the motorcars the trains passing through towns and cities. It hung over the plains the desert the range; the Mexican shacks the Negro cabins. It haunted Reata. Giant rigs straddled the Gulf of Mexico waters. Platoons of metal and wood marched like Martians down the coast across the plateaus through the brush country. Only when you were soaring in an airplane fifteen thousand feet above the oil-soaked earth were your nostrils free of it. Azabache oil money poured into Reata. Reata produced two commodities for which the whole world was screaming. Beef. Oil. Beef. Oil. Only steel was lacking. Too bad we haven't got steel, Texas said.[23]

The novel draws a parallel between beef and oil, both of which serve as "stinky" fuels for postwar America. Though the "stink" of oil and beef permeates the landscape, these industries come to facilitate rather than impinge upon everyday life.

Giant casts cattle ranching as a racist enterprise, reliant on exploited Chicano work-ers. Oil is also manipulative, though the petroleum workers, in the novel, are capable of petitioning successfully for better schools and public services. The real harm of oil, according to Uncle Bawley, an old cattle rancher, comes from its effects on society and the consumer: "Oil! What do folks use it for! In the war they were flying around shooting up towns of women and children. Now it's lobsters from Maine. Got to have lobsters. And streaking hell-bent in automobiles a hundred miles an hour, going nowheres, killing people like chickens by the roadside. Pushing ships across the ocean in four five days. There hasn't been a really good boatload of folks since the *Mayflower* crowd."[24] Bawley's nostalgia soon morphs into a critique of class erosion:

> The girls all got three mink coats and no place to wear 'em. And emeralds the size of avocados. The men, they got Cadillacs like locomotives and places the size of ocean

liners, and their offices done up in teakwood and cork and plexiglas. And what happens! The women get bored and go to raising pretty flowers for prize shows like their grammaws did and the men go back to raising cattle just like their grampappies did a long time ago. Next thing you know mustard greens and corn bread'll be fashionable amongst the Twenty-One Club when they go to New York. My opinion, they're tired of everything, and everybody's kind of tired of them. They made the full circle.[25]

Bawley yearns for a more authentic life, before the consumerism so emblematic of postwar America becomes ubiquitous. Bawley's nostalgia for the ambitions of the "Mayflower crowd" are juxtaposed in the novel with the emphasis on small business propagated by Bob Dietz, an experimental farmer convinced that "here in Texas maybe we've got into the habit of confusing bigness with greatness. They're not the same."[26] Bob is an advocate of small-scale farming, and his vision of a nation comprising small, family-run businesses takes a strange turn as Bob continues: "Why at Cornell, in lab, they say there's a bunch of scientists here in the United States working on a thing so little you can't see it—a thing called the atom. It's a kind of secret but they say if they make it work—and I hope they can't—it could destroy the whole world, the whole big world just like that. Bang."[27] In his speech about how to shift the United States from obsolete, large-scale industry to more efficient, small-scale businesses, Bob's desire for small farms curiously turns into fear of nuclear war. Through Bob, *Giant* establishes a connection between small-scale businesses, the middle-class family, and large-scale disaster. The family, bearer of normative values in both private and public spheres, must be protected against a world-ending force, a force that, in Bob's speech, emerges from the same focus on the small scale that leads the family to prominence over and above the monopolistic drives of industrialization, which *Giant* depicts as corrupt and anti-American. Discarding older forms of greatness and fearful of an apocalyptic future, *Giant* positions the family as the nation's only hope.[28] This move—seeing in small-scale endeavors predictions of nuclear war—positions the postwar family as looking toward a brilliant future while also fearing inevitable destruction. Families connote perseverance and heroic effort, and Uncle Bawley and Bob Dietz share a core insistence that the family is central to the nation's future. Bob's vision of the future—straddling a stark dichotomy between a technologically sophisticated, anti-federalist agrarianism and the world-extinguishing effects of the atom bomb—articulates what Fredric Jameson has identified as a central ideology in late capitalism: "the obliteration of difference on a world scale," which entails "a vision of irrevocable triumph of spatial homogeneity over whatever heterogeneities might still have been fantasized in terms of global space."[29] *Giant* portrays Texas as a heterogeneous space only to unite all of Texas's disparate populations and industries under the banner of national family values. That is, by opposing the feudal cattle ranch to the modern oil field, as well as the tyranny of industry to the difficulties of collective belonging, *Giant* dramatizes the erosion of differences as a transcendent experience of consciousness-raising that places the nuclear family at the center of affective bonds. The "good life" demands the destruction of racial and class differences, yet this erosion of difference relies on the egalitarian affordability of fossil fuels. In a period where many in the United States experienced an increase in the accessibility of college education, a proliferation of managerial and middle-class employment opportunities, the growth of suburban housing developments, and, eventually, the erosion of gender and racial discrimination, the consumption of petroleum plays a central, if hidden, role.

At the end of the film, after Jett Rink's banquet celebrating the opening of his Conquistador Hotel, the Benedicts drive back to Reata. When their private plane flies over, a plane that they have earlier refused to travel in, they wave and realize how glad they are to have embraced the "simple life" of driving. Behind the wheel, Bick remarks, "Just as soon as the durn thing lands I'm going to sell it. This is for me, boy, the simple life. No more of this high-flying nonsense." In this moment, the family car is cast as an intermediary between the horse and the airplane, a happy compromise between the individuality of horseback riding and the excessive luxury of the well-furnished private jet that imbues the modern technology of the automobile with nostalgia, just as *Giant* the film also feels nostalgic in its use of the modern techniques of Hollywood cinema. In her study of French culture in the 1950s and 1960s, Kristin Ross argues that the automobile and film "reinforced each other. Their shared qualities—movement, image, mechanization, standardization—made movies and cars the key commodity-vehicles for a complete transformation of European consumption patterns and cultural habits."[30] As Ross goes on to argue, this transformation is by and large an "Americanization" of Europe, and in *Giant*, an intimate connection between consumption, family belonging, and the automobile is forged through the disavowal of oil's centrality to postwar U.S. culture.

In the novel's concluding chapter, Leslie and Bick discuss the future of the Benedict family name. Leslie comments: "I thought, as we were driving along toward home— Luz and Juana and little Jordan and I—I thought to myself, well, maybe Jordan and I and all the others behind us have been failures, in a way. In a way, darling. In a way that has nothing to do with ranches and oil and millions and Rinks and Whitesides and Kashmirs. And then I thought about our Jordan and our Luz and I said to myself, well, after a hundred years it looks as if the Benedict family is going to be a real success at last."[31] This recognition of "failure" values the "Benedict family" over and above "ranches and oil and millions." This revaluation is key to fossil-fuel futurity: industry and economics are displaced by the affective bonds of the family and the promise of futurity that the family represents. Leslie's remark is followed by Bick's realization, as he looks at his wife: "The lines that the years had wrought were wiped away by a magic hand, and there shone the look of purity, of hope and of eager expectancy that the face of the young girl had worn when she had come, twenty-five years ago, a bride to Texas."[32] This youthful gaze is represented in the film's final shots of the Benedict grandchildren. The Anglo and Chicano toddlers fade into one another, accompanied by a nursery rhyme version of "The Eyes of Texas Are Upon You." The film concludes, like the novel, with a gaze into futurity. These eyes confirm not only the triumph of racial tolerance but also the hidden triumph of oil as the mediator of futurity.[33]

Oil's function as a vanishing mediator between industrialism and family life often renders it an absent presence in American culture.[34] As Amitav Ghosh argues in a review of Abdelrahman Munif's *Cities of Salt* novels, there is an absence of art, literature, and culture about oil, for "the history of oil is a matter of embarrassment verging on the unspeakable, the pornographic."[35] In *Giant*, though, it is clear that oil is not so much absent from U.S. culture as it is fractured and subsumed into a vision of a divided social field, constituted of, on the one hand, disavowed, avaricious businessmen and, on the other, normative, happy families. Subsequent representations of oil employ this "fossil-fuel futurity" logic as well, though in the television drama *Dallas* and the film *There Will Be Blood*, oil's subsumption into demonic industry and domestic normativity becomes increasingly unstable, as petroleum consumption and the family both become widely recognized problems in the late 1970s. Following the 1973 and 1979 oil crises, oil connotes scarcity, and the American family loses its idyllic, paternal stability following the successes of the women's liberation movement and the passage of no-fault divorce laws in most states by the early 1980s. Accordingly, the pillars of fossil-fuel futurity—the availability of petroleum and the happily reproductive, middle-class family—become unstable by the late 1970s, throwing the connections between oil and the future into crisis.

First aired as a miniseries in 1978, then picked up as a series for thirteen more seasons, from 1979 to 1991, *Dallas* details the lives of the Ewing family. The Ewings own a ranch and an independent oil business, and the show mixes family and business problems, often blurring the distinction between the private and the public.[36] *Dallas* draws heavily on the iconography of *Giant*, especially in its representation of the moral divide between ranching and oil. The Ewing family home doubles as Southfork Ranch, where the good-natured son, Bobby Ewing, played by Patrick Duffy, often works alongside the ranch-hand beefcake Ray Krebbs, played by Steve Kanaly. Krebbs is later revealed to be the illegitimate son of patriarch oilman Jock Ewing, played by Jim Davis, in one of the show's typical plot twists. The Ewing matriarch, Miss Ellie, played by Barbara Bel Geddes, grew up on the cattle ranch and views ranching as a

wholesome industry, especially in relation to the oil business. Ewing Oil, founded
by Miss Ellie's husband, Jock Ewing, is controlled by the Ewing's eldest son, and the
show's anti-hero, J. R. Ewing, played by Larry Hagman. J.R. shares not only the initials of Jett Rink but also the unchecked ambition and greed of his predecessor.

Ewing Oil serves as mere background for domestic conflicts during the initial
miniseries that launched *Dallas* in 1978.[37] The oil business provides a rationale for the
Ewing family's wealth and serves as a justification for why the wealthy family, including
two adult sons, their wives, and a third Ewing son's daughter Lucy, played by Charlene
Tilton, all live in the same large home. *Dallas* develops multiple plots and subplots
around its central cast of characters. For example, the Ewing family has a long-standing
feud with the lawyer and politician Cliff Barnes, played by Ken Kercheval; Bobby and
Pamela, played by Victoria Principal, wrestle with their inability to have a child; J.R.
and Sue Ellen, played by Linda Gray, go through ups and downs in their dysfunc-
tional marriage; Jock and Miss Ellie deal with health problems; Lucy has many ill-
fated love affairs; and, Ray Krebbs struggles with his unhappy bachelorhood. The
oil business remains largely a subplot to these more central narratives until Ewing
Oil jeopardizes the family itself, in the plot that leads to the show's famous "Who
Shot J.R.?" cliffhanger.[38] This oil narrative reflects contemporary anxieties around the
global oil industry, exacerbated by the 1979 oil crisis as J.R. risks the Ewing fortune by
pursuing oil in an unstable foreign country.[39] Leading up to the cliffhanger, J. R. Ewing
has started building oil wells in an unnamed Southeast Asian country, which end up
posing a number of financial difficulties for J.R. Ultimately, he mortgages his family's
home, Southfork Ranch, to fund the construction of the wells, without the knowledge or approval of the rest of the Ewing family. J.R.'s risky business deal is eventually
brought to light. The Southeast Asian oil wells do not come in on schedule, and
Ewing Oil is soon in danger of defaulting on its loan and therefore losing Southfork
Ranch. This crisis is averted when the oil wells start producing, right when they need
to in order to save Southfork. Following his triumph in Southeast Asia, J.R. is tipped
off that his oil wells are in further danger due to an impending coup d'état in the
Southeast Asian nation where Ewing Oil has drilled. J.R. then quickly sells these oil
wells to close business associates, right before the wells are nationalized during the
coup. J.R. has, then, betrayed both his family, by mortgaging their home without their
consent, and his business community, by selling them oil wells soon to be nationalized. Following all of these developments, in the season finale, an unidentified person shoots J.R. The next season picks up with J.R. being rushed to the hospital, and
the subsequent episodes play up the possible suspects, which include J.R.'s wife, Sue
Ellen, as well as his wronged business associates.[40] In the lead-up to his shooting,
J.R. has jeopardized the Ewing family's home and put the family in crisis, and he has

violated the fraternal bonds between the members of the Texas oil cartel to which the Ewings belong. This disregard for familial and fiscal stability leads to the shooting, as J.R. is symbolically punished for his pursuit of wealth over and above family.

Eventually, Sue Ellen's sister, Kristin Shepard, played by Mary Crosby, confesses to the crime, which she committed after her affair with J.R. ended and she was left without the promise of further Ewing wealth and glamour. Despite J.R.'s nefarious business practices, it is ultimately his entanglement with family—his bad behavior as a husband and a lover, his affair with his wife's sister—that results in the shooting. In *Dallas*, the oil business corrupts even the family, dooming the Ewings to a life of intrigue, betrayal, and duplicity. Fossil-fuel futurity registers not through the disavowal of the petroleum industry, as it does in *Giant*, but instead through the recognition that the family, itself, is just as exploitative and manipulative as the petroleum industry that empowers it. Nonetheless, both the family and the oil business are represented as necessary, even sacrosanct, institutions, to be preserved and protected at all costs, despite their obvious flaws. Along with its clear reference to the 1979 oil crisis, *Dallas*'s representation of the family as always on the verge of collapse reflects another crisis in the 1970s United States, the sharply rising divorce rate.[41] As Nancy Cott notes, beginning in the 1970s, "the divorce rate rose . . . furiously, to equal more than half the marriage rate, portending that at least one in two marriages would end in divorce."[42] In *Dallas*, divorce is discussed, or threatened, by every Ewing couple, making it clear that the family is no longer a stable entity, insulated from crisis by normative belonging. Many of the women in *Dallas* are also frustrated by their husband's understanding of gender roles; Pamela Ewing wants to have a career, which goes against her husband Bobby's desire for Pamela to stay at home and have children, while Sue Ellen is unsatisfied as a stay-at-home mother, turning to alcohol and affairs as solace. The vision of the happy family in *Giant* has clearly faltered in *Dallas*, a show produced in the wake of the feminist critique of normative domesticity and the restrictions it places on women.[43]

The pursuit of oil profits and the manipulation of family loyalties are coextensive in *Dallas*, and both practices are discussed in the show as natural and innate, a matter of inheritance, rather than a set of economic, ethical, and political practices. For example, in the concluding episode to the original 1978 *Dallas* miniseries, Bobby Ewing's wife, Pamela, announces that she is pregnant. In a drunken attempt to apologize for treating Pamela as an outsider, J.R. accidentally causes Pamela to fall from the attic of Southfork's barn, and Pamela suffers a miscarriage. Later, when Pamela is resting in bed, Jock Ewing enters the room and apologizes for the Ewing family: "Us Ewings, we're just not easy to live with, as you've found out. We've had things our way for so long that maybe, well, maybe, it got in the way of our being just people."[44] Family, routed through oil, becomes something impersonal, even inhuman. In *Dallas*, then, the family's ties to oil become deterministic and fatalistic. The family is bound, however unhappily, to oil, and there is nothing to be done about it other than to acquiesce to its hold on everyday life. In *Dallas*, fossil-fuel futurity no longer offers a happy vision of family belonging but instead a claustrophobic union of oil and family, a union that exacerbates both the instability of the oil industry and the fragility of marriage.

Dallas represents the resiliency of fossil-fuel futurity, as an oil crisis is explained away as a family crisis and, historically, the 1979 oil crisis makes no lasting impact on fossil-fuel consumption in the United States. Contrary to *Dallas*'s focus on the family as both the cause and solution to oil crises, Paul Thomas Anderson's *There Will Be Blood* dramatizes a possible rupture between family and oil. Daniel Plainview, the film's central character played by Daniel Day-Lewis, is another independent oilman. The film begins in the early twentieth century, as Plainview works as a lone prospector, then quickly jumps ahead to Plainview's larger scale oil business. In an essay on the limitations of representing oil in American culture, Peter Hitchcock criticizes *There Will Be Blood* for casting oil as a kind of individual pathology rather than a part of political or economic history.[45] The film's focus on Daniel Plainview is indeed the major revision that Anderson undertakes of his source material, Upton Sinclair's 1927 novel *Oil!* While Sinclair's novel has more characters and a clearly political message in comparison to Anderson's *There Will Be Blood*, both the novel and film respond to and critique privatization. Sinclair's *Oil!* is written, in part, as a response to the 1924 Teapot Dome Scandal and the "privatization of public goods . . . typical of the Harding administration."[46] Featuring a character named "H.W.," in 2007 an invocation of the Bush presidencies both past and present, *There Will Be Blood* is also about privatization, and the way in which the intertwining of privatized oil and the "private sphere" of family relations produces the violent yet curiously noble figure of Daniel Plainview.[47] In my reading, it is precisely the focus on the individual, oil, and family relations that gives *There Will Be Blood* resonance as a narrative about oil. Hitchcock describes the connection between blood and oil, kinship and crude, as a "metonymic disjunction," and it is the "disjunction" that the film introduces between oil and blood that marks a departure from oil's role in *Dallas* and *Giant*.[48] *There Will Be Blood* presents a challenge to the viewer: can a notion of familial belonging—"blood"—be recuperated from or even exist without oil? The film seems to present a rather bleak answer to this question, as it concludes with Plainview alone and drunk in his mansion after rejecting H.W. and beating the evangelical minister Eli Sunday, played by Paul Dano, to death with a bowling pin.

While *There Will Be Blood*'s conclusion gestures to oil's persistence—Plainview successfully guards his fortune—it is also a meditation on how oil and blood become equivalent in our social imaginary. *There Will Be Blood* begins without dialogue, as a single figure, Daniel Plainview, mines silver. The first complete sentence spoken in the film is, "There she is," as Plainview finds silver in a well that he has just dynamited, resulting in not only the setting loose of ore but also Plainview's own broken leg.[49] The discovery of ore leads to Plainview's reentrance into society, as he drags himself back to a town to verify his claim. In its early sequence, the film associates Plainview with the earth—he is a primal creature, and culture only enters upon the discovery of natural resources. The film, then, dramatizes fossil-fuel futurity as an ideology— natural resources are explicitly portrayed as the material upon which a set of norms and a vision of the future are constructed. *There Will Be Blood's* difference from *Giant* and *Dallas* lies in the fact that the oil business is not subsumed into the family but remains visible and irreducible to "natural" family relations.

Following this opening primal scene, Plainview gains an oil claim and a family simultaneously. While working in the well, a worker is killed. Plainview lives, covered in oil in a way that mirrors the demonic, oil-soaked Jett Rink in *Giant*. The dead worker leaves behind an infant son, and Plainview adopts the son. Plainview's adopted son, H. W. Plainview, played by Dillon Freasier as a child and Russell Harvard as an adult, becomes central to the way that Plainview positions himself as an oilman. After two scenes of Plainview with the infant H.W., the film again jumps ahead in time, as Plainview tries to acquire oil leases on a newly discovered oil field in California.[50] H.W. stands right next to him, and part of Plainview's pitch to the landowners is that "I'm a family man, and I run a family business. This is my son and my partner, H. W. Plainview. We offer you the bond of family that very few oilmen can understand." Daniel and H. W. Plainview have no blood relation, only an oil relation. The "bond of family" is, in fact, produced not through conventional kinship relations but through the family oil business. In *Dallas*, oil complicates family relations that are already strained by the threat of divorce. In *There Will Be Blood*, oil becomes the only family relation.

Early in the film, H.W. is deafened permanently when he sustains a blow to the head as the Bandy oilfield violently comes in. H.W.'s deafness serves as a figure for the political lesson of the film—the process of understanding oil's claims on our everyday lives might require something more than unmasking the oil industry as corrupt and exploitative, something that has become all too familiar. From Jett Rink to J. R. Ewing and Daniel Plainview, we have come to expect to view oilmen as monstrous anti-heroes, and these devilish oilmen are a means by which the oil industry can be dis-avowed while oil itself is endowed with futurity.[51] *There Will Be Blood* gestures to the persistence of futurity's connection to oil in its final sequence, as an adult H.W. leaves his father's oil business to start his own oil wells in Mexico. When H.W. tells his father of his plans, Plainview reacts angrily and tells H.W. that he is not Plainview's legitimate child. H.W. leaves as his father screams, "You're a bastard from a basket." Following this exchange, the film concludes with Plainview's murder of the evangelical preacher Eli Sunday. The murder signifies the extinguishment of the troubled connection between capital and religion represented by the two characters. In the end, capitalism triumphs in Plainview's single-minded pursuit of profit and monopoly; it is void of the hypocritical moral and spiritual trappings of Sunday's ministry. By leaving Plainview before this bloody confrontation and pursuing oil outside of the family, H.W.'s independent oil wells represent a refusal of fossil-fuel futurity; only by rejecting oil's claim on family can H.W. hope to escape the control of his father. When read in relation to the Iraq War, *There Will Be Blood*'s political message becomes clear: oil promises to secure the future, yet the pursuit of oil replaces family, society, and reason.[52] The film negates George W. Bush's widely reported personal rationale for invading Iraq to depose Saddam Hussein: "After all, this is the guy who tried to kill my dad."[53] Serving U.S. oil interests as well as the president's sense of familial obligation, the Iraq War becomes, through the lens of *There Will Be Blood*, just another inevitable, violent effect of fossil-fuel futurity's binding of the oil industry to family relations, a connection that is severed at the film's conclusion as Daniel Plainview and H.W. part ways.

What *There Will Be Blood* dramatizes, then, is the constructedness of oil's connections to futurity. In *There Will Be Blood*, oil persists as kinship, even when kinship itself—or "blood"—vanishes. In this sense, H.W. illustrates the political challenge of the film and a rupture in the ideological workings of fossil-fuel futurity. If H.W. can refuse family to pursue his own oil wells in Mexico, then it is conceivable that we, the audience, can inversely form alternative practices and modes of belonging that are not underwritten by fossil fuels. The way to imagine life without oil might be to do what H.W. does: stop listening to oil's claims on us; turn a deaf ear to the intimate connection forged between oil, the American family, and futurity.

In *Giant*, the avaricious oilman, Jett Rink, is displaced by a pluralistic family, the Benedicts. This family ideal, though, is premised on the consumption of Jett Rink's oil and consolidated through the use of the family automobile, a sign of middle-class prosperity in the 1950s United States. The consumption of fossil fuels, then, is a hidden but necessary component of *Giant*'s vision of the national family. In *Dallas*, as petroleum consumption becomes a problem during and after the 1979 oil crisis and the patriarchal, reproductive American family becomes unstable after an unprecedented rise in the national divorce rate and the inroads toward gender equity made by the feminist movement, the avaricious oilman, J. R. Ewing, is not displaced from but is instead at the center of the family. Oil is inextricable from and incapable of being subsumed within the Ewing family. While *Giant* displaces the oil industry while normalizing petroleum consumption, *Dallas* acknowledges the presence of the oil industry in everyday life and constructs that presence as ultimately nefarious but necessary. In both *Giant* and *Dallas*, there is no future without oil, though each text imagines that future in very different terms, as idyllic family happiness and liberal pluralism in *Giant*, and as a continual cycle of betrayal and redemption in *Dallas*. *There Will Be Blood* marks a contemporary crisis in fossil-fuel futurity, as the family oil business undoes family altogether. Allegorizing the conjoined oil and familial reasons for President Bush's invasion of Iraq, the film dramatizes the disjunction of oil and family, demonstrating how oil's hold on our imagination of family belonging is distortive and parasitical. Through this reading of fossil-fuel futurity's movements through the 1950s to the present in three exemplary texts, we can glimpse the possible imagining of a future without fossil fuels, an act that will also require rethinking ideals of family belonging that have been determined by oil.

NOTES

1. See Imre Szeman, "Oil Futures," in *Fuel*, ed. John Knechtel (Cambridge: MIT Press, 2009), 18–35; and "System Failure: Oil, Futurity, and the Anticipation of Disaster," *South Atlantic Quarterly* 106, no. 4 (Fall 2007): 805–23.

2. See Lawrence Buell, *Writing for an Endangered World: Literature, Culture, and Environment in the U.S. and Beyond* (Cambridge: Harvard University Press, 2001).

3. For accounts of the postwar boom in U.S. oil consumption and the 1979 oil crisis, see Daniel Yergin, *The Prize: The Epic Quest for Oil, Money, and Power* (1990; New York: Free Press, 2009), 391–412, 523–42, 656–725. For an account of the Iraq War, see Peter Maass, *Crude World: The Violent Twilight of Oil* (New York: Alfred A. Knopf, 2009), 136–64.

4. Lewis Nichols, "Talk with Edna Ferber," *New York Times Book Review,* October 5, 1952, 30.

5. Ibid.

6. Ibid.

7. Lauren Berlant, *The Female Complaint: The Unfinished Business of Sentimentality in American Culture* (Durham: Duke University Press, 2008), 69.

8. For analyses of the automobile and the interstate highway system in the postwar United States, see James J. Flink, *The Automobile Age* (Cambridge: MIT Press, 1988), 358–76; Cotten Seiler, *Republic of Drivers: A Cultural History of Automobility in America* (Chicago: University of Chicago Press, 2008), 69–104; and Rudi Volti, *Cars and Culture: The Life Story of a Technology* (2004; Baltimore: Johns Hopkins University Press, 2006), 104–13.

9. Edna Ferber, *Giant* (1952; New York: Perennial, 2000), 29.

10. Ibid., 31.

11. Ibid., 7.

12. Ibid., 21.

13. For readings of *Giant*'s racial politics, see Lee Bebout, "Troubling White Benevolence: Four Takes on a Scene from *Giant*," *MELUS* 36, no. 3 (Fall 2011): 13–36; June Hendler, *Best-Sellers and Their Film Adaptations in Postwar America: From Here to Eternity, Sayonara, Giant,*

Aunt Mame, Peyton Place (New York: Peter Lang, 2001), 115–52; and Rafael Pérez-Torres, "Chicano Ethnicity, Cultural Hybridity, and the Mestizo Voice," *American Literature* 70, no. 1 (March 1998): 153–76.

14. Pérez-Torres, "Chicano Ethnicity, Cultural Hybridity, and the Mestizo Voice," 158.

15. *Giant* (1956; Warner Brothers DVD, 2003).

16. Eric Lott, *Love and Theft: Blackface Minstrelsy and the American Working Class* (New York: Oxford University Press, 1995), 25.

17. Ferber, *Giant*, 35. In a chapter on the Texas oil industry, the journalist and labor activist Harvey O'Connor notes the function that books like *Giant* had on the oilmen after which Jett Rink was modeled: "The lives of these men range from the uncouth to the gaudy; perhaps none has reveled more in the public eye than Glenn McCarthy whose opening in 1949 of the Shamrock Hotel in Houston set new records in slapstick lavishness. The postwar era of the Texas oil millionaires has been described in many a novel splashed with the adjective 'fabulous,' and none more observing probably than Edna Ferber's *Giant*. But these people enjoy everything—even seeing their lives portrayed in novels which leave nothing to the imagination." See Harvey O'Connor, *The Empire of Oil* (1955; New York: Monthly Review Press, 1962), 204.

18. Ferber, *Giant*, 350. The ideological function of "fossil-fuel futurity" is evident here since oil production and consumption grew rapidly in this era, while *Giant*'s Bob Dietz views large-scale industry as obsolete rather than emergent. As Daniel Yergin notes, between 1948 and 1972 "the numbers—oil production, reserves, consumption—all pointed to one thing: Bigger and bigger scale. In every aspect, the oil industry became elephantine." Daniel Yergin, *The Prize*, 524.

19. Ferber, *Giant*, 1.

20. Ibid., 1–2.

21. Ibid., 17.

22. Ibid., 22.

23. Ibid., 366–67.

24. Ibid., 378.

25. Ibid., 378–79.

26. Ibid., 350.

27. Ibid.

28. For an account of the family's priority in Cold War America, and especially how that vision of the family changed normative masculinity, see James Gilbert, *Men in the Middle: Searching for Masculinity in the 1950s* (Chicago: University of Chicago Press, 2005), 106–63.

29. Fredric Jameson, *The Seeds of Time* (New York: Columbia University Press, 1994), 27.

30. Kristin Ross, *Fast Cars, Clean Bodies: Decolonization and the Reordering of French Culture* (Cambridge: MIT Press, 1995), 38.

31. Ferber, *Giant*, 401–2.

32. Ibid., 402.

33. For a reading of an earlier novel written by Edna Ferber, her 1931 *American Beauty*, that emphasizes how Ferber strives both to retain the power of and reimagine the family and its role in the nation, see Susan Edmunds, *Grotesque Relations: Modernist Domestic Fiction and the U.S. Welfare State* (New York: Oxford University Press, 2008), 95–122.

34. In his account of "The Oil Ecofilm," for example, John Shelton Lawrence mentions only three films: *Tulsa* (1949), *Giant*, and *On Deadly Ground* (1994). See John Shelton Lawrence, "Western Ecological Films: The Subgenre with No Name," in *The Landscape of Hollywood Westerns*,

ed. Deborah A. Carmichael (Salt Lake City: University of Utah Press, 2006), 34–36. For a reading of *Tulsa*, see Peter C. Rollins, "*Tulsa* (1949) as an Oil-Field Film: A Study in Ecological Ambivalence," *The Landscape of Hollywood Westerns*, ed. Deborah A. Carmichael (Salt Lake City: University of Utah Press, 2006), 81–93. For an account of the concept "vanishing mediator," see Fredric Jameson, "The Vanishing Mediator; or, Max Weber as Storyteller," in *The Ideologies of Theory: Essays, 1971–1986*, vol. 2: *The Syntax of History* (Minneapolis: University of Minnesota Press, 1988), 3–34.

35. Amitav Ghosh, "Petrofiction: The Oil Encounter and the Novel," *New Republic,* March 2, 1992, 29.

36. For a comparison of *Dallas* to *The Godfather*, based on how both represent the family as inextricably bound to business, see Mary S. Mander, "*Dallas*: The Mythology of Crime and the Moral Occult," *Journal of Popular Culture* 17, no. 2 (Fall 1983): 44–50. *Dallas* was immensely popular in the United States and Europe. For a reception study of *Dallas* in the Netherlands, see Ien Ang, *Watching Dallas: Soap Opera and the Melodramatic Imagination*, trans. Della Couling (1985; New York: Routledge, 1989). For a reading of *Dallas* in relation to the city of Dallas, Texas, and especially the television show's relationship to Dallas as the site of the Kennedy assassination, see Harvey J. Graff, *The Dallas Myth: The Making and Unmaking of an American City* (Minneapolis: University of Minnesota Press, 2008), 13–17.

37. The miniseries is collected as Season One in *Dallas: Complete First and Second Seasons* (1978–79; Warner Brothers DVD, 2004).

38. The cliffhanger episode that concludes with J.R. being shot is "A House Divided," *Dallas: The Complete Third Season* (1979–80; Warner Brothers DVD, 2005).

39. Tied to the Iranian revolution's effects on oil production and compounded by the Three Mile Island meltdown and the Iran hostage crisis, the 1979 oil crisis led to increasing concerns about dependency on "foreign oil" in the United States and the question of U.S. global dominance. See Yergin, *The Prize*, 657–96.

40. The episodes are "No More Mister Nice Guy Part I," "No More Mister Nice Guy Part II," "Nightmare," and "Who Done It?" These episodes are collected in *Dallas: The Complete Fourth Season* (1980–81; Warner Brothers DVD, 2006).

41. The rise in the divorce rate is often attributed to the unilateral and no-fault divorce laws passed in all but two states in the United States between 1970 and 1983. For readings of the divorce rate and the debates about the cause of its rise in the late twentieth century, see Leora Freidberg, "Did Unilateral Divorce Raise Divorce Rates? Evidence from Panel Data," *American Economic Review* 88, no. 3 (June 1998): 608–27; and Justin Wolfers, "Did Unilateral Divorce Raise Divorce Rates? A Reconciliation and New Results," *American Economic Review* 96, no. 5 (December 2006): 1802–20.

42. Nancy Cott, *Public Vows: A History of Marriage and the Nation* (Cambridge: Harvard University Press, 2000), 203.

43. This critique was most popularly articulated in Betty Friedan, *The Feminine Mystique* (New York: W. W. Norton, 1963).

44. "Barbecue," *Dallas: The Complete First and Second Seasons*.

45. Peter Hitchcock, "Oil in an American Imaginary," *New Formations* 69 (Spring 2010): 81–97.

46. Kevin Mattson, *Upton Sinclair and the Other American Century* (Hoboken, N.J.: Wiley and Sons, 2006), 141.

47. For a reading of *There Will Be Blood*'s connections to free market economics, see Michael W. Clune, *American Literature and the Free Market, 1945–2000* (New York: Cambridge University Press, 2010), 27–52. For a psychoanalytic reading of *There Will Be Blood* and Paul Thomas Anderson's other films, see Julian Murphet, "P.T. Anderson's Dilemma: The Limits of Surrogate Paternity," *Sydney Studies in English* 34 (2008): 63–85.

48. Hitchcock, "Oil in an American Imaginary," 96.

49. *There Will Be Blood* (2007; Paramount DVD, 2008).

50. For a history of oil in California, of which Upton Sinclair had firsthand knowledge, see Paul Sabin, *Crude Politics: The California Oil Market, 1900–1940* (Berkeley: University of California Press, 2005).

51. The trope of the devilish oilman goes back at least to Ida M. Tarbell's *History of the Standard Oil Company*, which focuses on the business practices and personal character of John D. Rockefeller. Tarbell's *History* was originally serialized in *McClure's Magazine* from 1902 to 1904 and published as a two-volume book in 1904. See Ida M. Tarbell, *The History of the Standard Oil Company*, briefer version, ed. David M. Chalmers (1966; Mineola, N.Y.: Dover, 2003). For historical accounts of oilmen that take into account this mode of representation, see Roger M. Olien and Diana Davids Olien, *Oil and Ideology: The Cultural Creation of the American Petroleum Industry* (Chapel Hill: University of North Carolina Press, 2000); Roger M. Olien and Diana Davids Hinton, *Wildcatters: Texas Independent Oilmen* (1984; College Station: Texas A&M University Press, 2007); and Brian Black, *Petrolia: The Landscape of America's First Oil Boom* (Baltimore: Johns Hopkins University Press, 2000), 1–12.

52. This reflects the 2003 Iraq War, and especially the Bush administration's unclear motives for invasion. As Peter Maass notes, "America's desires were so influenced by Iraq's inebriating crude that Washington could not think straight about the reasons for invading [Iraq]." "Neither [Vice President Dick] Cheney's motives nor the motives of the administration he served can be distilled into one word. WMD, democracy, religion, Oedipus, oil—America was like a drunk fumbling with a set of keys at night." See Peter Maass, *Crude World*, 159, 161. For a psychoanalytic reading of the Iraq War, the Bush administration's conflicting reasons for invasion, and the war's ideological stakes, see Slavoj Žižek, *Iraq: The Borrowed Kettle* (New York: Verso, 2004).

53. John King, "Bush calls Saddam 'the guy who tried to kill my dad,'" *CNN.com*, September 27, 2002, http://edition.cnn.com/2002/ALLPOLITICS/09/27/bush.war.talk/.

7

"Liquid Modernity"

Sundown *in Pawhuska, Oklahoma*

HANNA MUSIOL

> We tend to ignore ... history. We are unable to put the pieces together.
>
> —Immanuel Wallerstein, *World-Systems Analysis: An Introduction*

> Literary form [is] the most profoundly social aspect of literature.
>
> —Franco Moretti, *Graphs, Maps, and Trees: Abstract Models for Literary Analysis*

In his work on globalization and contemporary cultural transformations, Zygmunt Bauman argues that in the "'developed' part of the planet," people now live in a "liquid" phase of modernity, marked by the radical instability of cultural forms and social institutions, a reality "never before encountered."[1] To Bauman and other scholars,[2] this era of radical uncertainty is closely tied to transformations within labor structures and to the organization of the state in the Global North, deriving in large part from post-1970s financial deregulation and the emergence of electronic communication.[3] This essay focuses on the Osage reservation in Oklahoma during the first half of the twentieth century, where oil culture produced surprising symptoms of "liquid modernity," well before the invention of the Internet or the collapse of the Bretton Woods system of financial exchange. Rather than seeing such instances of liquidity as accidental and unrelated, I argue that examining petromodernity in Pawhuska can help us rematerialize, historicize, and reterritorialize accounts of cultural, material, and political instability. The very transformation of "information" into a global commodity through electronic distribution has a complicated history, which cannot be fully understood away from the developments of energy industries and the violent conditions under which raw materials are extracted from actual physical, geopolitical places rich in mineral resources.[4] Similarly, the haphazard, transnational flow of capital has a much longer history than the one that starts with the dissolution of the Bretton Woods system in the seventies or with the advent of electronic communication. Thus, I want to use the 1934 bildungs-roman *Sundown* by John Joseph Mathews, a celebrated Book-of-the-Month Club writer and an Osage American tribal historian,[5] to trace the "solid" to "liquid" narrative of global modernity along the oil flow back to early twentieth-century Osage oil culture.

Mathews's only work of fiction, *Sundown,* is a peculiar bildungsroman about the life of Chal Windzer in the aftermath of oil discovery and extraction on the Osage reservation in Oklahoma, and Mathews's use of the novel of formation is an impor-tant literary and intellectual intervention. First, at a time when the significance of material resources to the American economy and American national imaginary was growing, Mathews uses the generic convention of progress to narrate oil's social "for-mation," alongside the tale of Chal's dissolution—his failure, that is, to realize the very development that the novel's genre implies. Second, Mathews's literary represen-tation of the oil boom as the Great Frenzy—a period of widespread violence and cul-tural chaos—complicates traditional accounts of capitalist economic development, which define late-stage capitalism and its financial deregulation and IT boom as an era of radical cultural "fluidity."[6] By drawing attention to the material and social fluid-ity of oil culture and the force with which it affected the Osages, *Sundown* points to a specific Osage American colonial, and thus always already transnational, history of petromodernity. The novel, I suggest, demonstrates that colonialist practices of "prim-itive accumulation,"[7] which Karl Marx envisioned as a prelude to the "solid" phase of modernity, in fact enable and often coincide with both "solid" and "liquid" manifesta-tions of "hydrocarbon civilization."[8] In other words, I contend that Mathews's novel *Sundown* allows us to think about "solidity" and "liquidity" not as chronological stages of modernity but as its expressions, often occurring simultaneously, if in different geopolitical spaces. In my view, contemporary "liquidity" or "uncertainty" does not, as Richard Sennett argues, "exist nowadays without any looming historical disaster,"[9] a luxurious claim to make about the 1980s or the 1990s. What his observation sug-gests, however, is that denying ongoing "historical disasters" their material reality, and de-emphasizing the connection between econo-political developments in the Global North and the "liquefying" of cultures and people in (neo)colonial sites of "extrac-tion,"[10] is instrumental to the contemporary understanding of cultural "liquidity." Mathews's work, I argue, is of particular importance because it shows how in the early twentieth century oil extraction and its violence simultaneously produce social sta-bility in American industrial centers and cultural fragmentation on the Osage reserva-tion. To demonstrate my point more clearly, let me turn to the "oily" matter of Osage petromodernity as narrated in Mathews's *Sundown.*

The novel focuses on Chal Windzer, an affluent Osage American man whose life coincides with the birth of the petroleum culture on the Osage reservation in Oklahoma in the early twentieth century. Despite his wealth and multiple educational and career opportunities off the reservation, Chal passes through life unable to comprehend or to influence the rapid transformations occurring around him, or to see his place in the dangerous world of the Great Frenzy,[11] or Reign of Terror, as the Roaring Twenties were called on the reservation. Chal and other mixblood and fullblood Osages struggle to "adjust" themselves to the new reality of oil culture,[12] to "civilized ways of talking and acting" (112), but find no happiness "amid the clanking, screaming of steel, groaning bull wheels and rasping cables; amid the chuffing and coughing of operation; amid the rumbling and roaring of passing trucks" (239).

The title of the novel indicates that petroleum, which animated both Henry Ford's factory culture and Pawhuska in Osage County, also effected a "sundown" for Osage traditions, for a way of life, and for the prairie, "where oil and salt water had killed every blade of grass" (250). Oil discovery certainly made the Osage suddenly, and briefly, "the richest tribe in the world."[13] However, the sudden overproduction of oil also had a devastating economic, environmental, and social impact on reservation culture, producing, the novel argues, the richest but also the most chaotic and bloodiest chapter in twentieth-century Osage history.[14] Nonetheless, petroleum both on and off the reservation was undeniably a "vibrant matter,"[15] crucial to the newly emerging American auto and energy industries and to national identity, and it left multiple traces, seeping into or, at times, flooding the symbolic sphere of American culture.[16] In popular and high art of the period, oil was an American emblem of the era, a key "power-thing" that made the United States a major technological and political power and a symbol of national vitality.[17] Since its force was often feared, petroleum was a leitmotif of sorts in such novels as Upton Sinclair's novel *Oil!* (1927) and such studies as *The United States Oil Policy* (1926),[18] which offered scathing indictments of the oil industry, capitalism, and greed. Yet oil's creative power also dazzled consumers of the arts in diverse genres, media, and venues. The opulence of industrial exhibits and promotional materials at World's Fairs and Expositions across America were legendary, and petroleum-industry-sponsored visual and literary productions quite common.[19] Thus, despite ideological resistance to it, oil culture was indispensable to intensified aesthetic experimentation across genres and media, experimentation we often associate solely with the domain of high, rather than energy, culture.

Sundown, however, is a conventional bildungsroman set in the context of the Osage oil boom, and the novel allows us to consider the actual and very complex conditions under which petroleum culture emerged on the Osage reservation. I would argue that if we understand what forces destabilized Osage cultural forms and social institutions, and what made Osage modernity in Oklahoma so "fluid" and "uncertain," we may be able to see what links Mathews's oily "liquid" times to our supposedly "immaterially" fluid modernity.[20] Mathews shows that it is not simply the physical and social plasticity of oil alone that produces "liquid modernity" in Pawhuska, Oklahoma. Nor is the capitalist overexploitation of resources the sole culprit. *Sundown* emphasizes the transnational, colonial conditions of Osage oil culture, which, paradoxically, financially enriched the Osage in the early twentieth century and further upset their traditions, "deranging" their political and cultural institutions. The novel uses the conventional bildungsroman's development to indicate oil's flexibility and Chal's failure to *become* in such circumstances.

COLONIAL DESIGNS AND OSAGE PETROMODERNITY

Nothing prepared the Osage for the oil boom that made them rich.[21] The new land that the Osage moved to in the nineteenth century was not, as C. B. Glasscock put it, "worthless acreage,"[22] but it was still "a land no white man would have,"[23] and at the time of the Removal Act of 1870, "no white man" was aware that it hid mineral resources that would generate millions of dollars for the tribe. Decimated by violence, illness, and poverty, the Osage were "in danger of extinction" by the time the Osage Removal Act forced them to cede their land in Kansas and move to a new location in Oklahoma (there were more than 12,000 Osage in 1815, but fewer than 3,500 by 1860).[24] Later, on the new reservation, the Osage experienced federal mismanagement of their tribal fund, as well as the ongoing "civilizing" efforts of the U.S. government.[25] Thus, the oil discovery[26] and subsequent Osage oil boom were referred to as the "most ironic practical joke ever played . . . upon the government and taxpayers of the United States."[27] However, the Osages' ability to take advantage of the oil wealth in the first place had less to do with the accidental discovery of oil on their new land than with their own efforts to negotiate a special agreement as a condition of accepting the Dawes Allotment Act, which forced most Native American tribes across the United States to privatize, or "allot" to individuals, communally owned tribal lands. Although the Osage were unable to exempt the tribe from allotment, they forestalled the act's passage to a time much later than that enforced on other tribes, by which point they were aware of the presence of oil on their reservation land. In 1906, in a radical legal and political maneuver, they agreed to privatize their "surface" land but not its subsurface mineral resources.[28] Thus, while ceding the communal ownership of their land, the Osage gained an "underground reservation," and the tremendous profits derived from oil were divided equally among all tribal members, or descendants of members listed on the 1906 Osage census.[29]

In the post-Allotment era, there were many attempts to restrict the access of Osage to their newfound wealth. Oil companies, federal and local governments, and the widely abused reservation credit system consistently indebted and defrauded Osages. The so-called guardians assigned to Osages who had been deemed legally "incompetent" took personal advantage of estates in their care, and various species of speculators, drug dealers, and other "exploiters of oil" who "had come [to the reservation] like Cortés, to get rich," further profited from Osage wealth.[30] Yet Osage oil royalties until the 1930s were immense, ranging from $10,000 to $15,000 annually per person,[31] and the communalization of mineral resources collectively pulled the entire tribe, not just the lucky few (as was the case with other American tribes), out of poverty. Osage financial independence seemed to frustrate the progressive rhetoric of America's "civilizing mission" and the widely accepted racial and cultural hierarchies that never placed Native Americans on the receiving end of colonial or capitalist largesse. Native Americans were often viewed as unfinished social subjects who needed to be "civilized" first in order to become consumers and political participant subjects, and in the oil era the public longed for the comfort of a prior "Golden Age," a time before "Indians [rode] in limousines [and] squat[ted] in blankets among Grand Rapids furniture."[32]

Form and Progress: The Matter of Mathews's Bildung

One of the most resonant and persistent questions in literary scholarship concerns the relationship between literary forms and the cultural work they perform,[33] and it is particularly important to consider the significance of Mathews's use of the conventional bildungsroman for his only fictional work.[34] Contemporary scholars emphasize that the novel of formation centers on the processes of "individuation" but also offers models for the exploration of processes of "socialization"[35] that make people social subjects.[36] Joseph Slaughter and Marianne Hirsch point out that the bildungsroman's ongoing appeal transcends literary vogues precisely because the genre "has viable social work to perform" for marginalized social subjects in the twentieth and twenty-first centuries, even if it had lost its "social . . . appeal" for "the Anglo-European white male (the ostensibly already-incorporated and capacitated citizen)."[37] That is, despite its emphasis on conventionality, progression, linearity, and the protagonist's individual or social development, the novel of formation continued to enjoy wide popularity and perhaps even gained new audiences in the early twentieth century, telling stories of New Women, New Negroes, or New Americans becoming political and social subjects.[38] Moreover, indigenous American literary history itself had a rich tradition of oral and written "life histories," autobiographies or travelogues that had informed twentieth-century bildungsromans by Native American authors.[39] Meanwhile, I see additional reasons for the bildungsroman's popularity in the early twentieth century. During that period, various kinds of nonhuman objects (new modern commodities, commodified natural resources such as oil, and "exotic" artifacts) began to enjoy a growing political, economic, and cultural significance.[40] Thus, the popularity of the novel of formation in twentieth-century American literary culture, I argue, is also linked to the bildungsroman's ability to articulate a growing socialization and politicization of nonhuman matter as well as the social becoming of marginalized human subjects.[41]

While Mathews's earlier projects were praised as great "studies of the Indian mind,"[42] *Sundown* was criticized for its "decidedly inarticulate quality" in its presentation of "a young Indian's attempt to adjust himself to a fundamentally alien civilization."[43] Readers saw the protagonist's inability to express himself and to "progress" as a formal failure of Mathews's bildungsroman. However, Mathews certainly did not employ the genre to validate the "Western" concept of progress. Nor did he allow his human characters to form, as "progress" for them was inseparably tied to the physical, economic, and cultural violence of colonization. Instead, he deployed this literary convention precisely to narrate the impossibility of social becoming for Osage and Osage American social subjects and to contrast it with the rapid cultural, social, and political incorporation of oil engendered by American colonialism. Chal never "becomes"—he neither fully "Indianizes" nor "civilizes"—and his cultural duality collapses within the narrative apparatus of the genre that expects him to progress and "achieve" things.[44] Oil, in contrast, manages to "become" different things: commodities (cars, fashion, cosmetics), "magical" gushers on the prairie (*Sundown,* 73), environmental pollution—all things that suddenly appear on the reservation that are made, bought, or consumed by oil and oil wealth. Oil also transforms Chal and other characters in the novel, and in less than three decades changes the entire Osage community from "vanishing Indians" or "noble savages" into icons of the new wealth, who drove Cadillacs, used morphine, and spent money decadently.[45] And since the terms of Osage allotment tied Osages' genetic and national identity to an oil entitlement, oil, in the novel, and oil ownership, literally, penetrate and fracture the identity of mixed and fullblood Osages alike.[46] In this context, *Sundown* is a bildung of the "social life" of oil,[47] of its physical but also social and political force and plasticity. As such, it becomes a narrative of the Osage "queer world" (20), as Chal calls it, of frenzied modernity, "liquid" in both the material and Baumanian senses, with petroleum as its emblem, not its cause. That is, oil is the character whose bildung reveals the "liquefying" force of colonial modernity, and the novel of formation serves as a technology to extract and contain that history. However, the novel's form also works as a terrifying "veneer" (13, 14) that constrains and unravels, rather than develops, the life of its human protagonists: Chal, his family, and his Osage and American friends.

Sundown begins with the birth of an Osage American man whose father, John Windzer, names him Challenge in hopes that he will "challenge . . . the disinheritors of his people" (4). Oil is another great hope for Chal's father, and Mathews announces its presence on the reservation in the same scene. John travels to a reservation store to boast of the birth of his son, only to have his impromptu celebration interrupted by local fullbloods and mixbloods who argue that the "guv'mint" and the Reservation Oil and Gas company might try to "git a finger in" the Osage oil and "ruin the [Osage] Nation" (8). This early passage foreshadows oil's future key role in the affairs of the tribe, the state, and the prairie. After the Allotment, John is "thrilled . . . in the atmosphere of growth and progress" (49), expecting "something cataclysmic and revolutionary" (49), some "petro-magic," to improve the lot of the Osage.[48]

Meanwhile, on and off the Osage reservation, John's son, Chal, lives a confused existence, wandering from place to place, and finding no satisfaction in education, work, sex, or drinking. Chal drifts from one life phase to another without joy or purpose, yet paradoxically his journey seems predictable and predetermined. Mathews depicts Chal's childhood on the reservation, his first encounter with Christian religion, his Americanization in a reservation school (and later at university), his tenure in the Air Force, and, finally, his return to the reservation. These conventions, categories of "being Native" and "becoming American," were familiar "bildungs" to Native Americans across the United States at the time,[49] and it is not surprising that Chal consistently views his own development and socialization through the prism of American colonialist rhetoric. Since such rhetoric (and colonial practices of forced assimilation associated with it) literally outlined the strategies of social incorporation for Native Americans,[50] for oil-rich Chal to form as a social being means, for instance, subjecting himself to the pedagogical experiments of his reservation school teacher, Miss Hoover, who loves "teach[ing] little Indian minds" (26) and "see[ing] them open like flowers"—undergoing the reeducation of "blankets" (110), as Americans refer to Chal and his friends at a U—and owning modern commodities. Oil royalties and a hefty inheritance from his father give him instant access to such "civilizing" things and activities—money, clothes, his roadster, airplanes, as well as education, golf, and sex with white women—and yet, toward the end of the novel, Chal is never really "doin' anything" apart from "slumming" (241). Despite following the logic of the civilizing mission—despite his American education, his career as a pilot, his wealth, his sexual exploits, and active social life—Chal is "growing angrier in his futility" (310). Unable to connect to either his Osage or his non-Osage peers or to situate himself in the new reality of an oil-boom culture, he feels forever lost and incapable of the great deeds his parents expect of him. The novel concludes as Chal, after one of his weeklong drinking binges, announces in a half-conscious state that this time he intends to go to Harvard Law School to become a "great orator" (311). Soon after, he passes out, and *Sundown* ends.

Sundown describes a sunset for Chal and other Osages on the reservation, for whom material wealth, reindigenization, and assimilation all fail to fulfill the promise of meaningful cultural participation. In this sense, Mathews writes an antibildungs-roman. Yet in classic bildungsroman fashion, Mathews also shows the beginning of the many new lives of oil, and this other character (described in parallel with the story of Chal's development) seems much more vibrant and more alive. Chal's blood seems to flow much slower through his veins than the oil that is "carried out of the old reservation . . . through large pipelines and long trains" (241).[51] Its flow makes and unmakes people, brings wealth and produces the world's biggest field of derricks in the middle of the prairie. Petroleum seems to have the vitality—the force, flow, and direction—that Chal dreams of with envy yet can never achieve.

Introduced as a key character in the first chapter, oil is narrated as physical and social. It is the material mineral that comes from underground and the one protagonist that can form and become: a force that marks the symbolic horizon for reservation inhabitants. Mathews constantly points to petroleum's intervention in the reservations' natural and social environments. Derricks, which initially "spring up among the blackjacks" (78),[52] soon tower above them "like some unnatural growth" (62), eventually redrawing the prairie's frontier with "sterile forests set against the sky" (239). The gushers "shoot[ing crude] high into the air" (73) dazzle tourists and Osages alike. Oil prices, oil leases, "oil bizness" (183) are town news and soon world news as well (239). And just as oil wells, oil ponds, and derricks transform the physical landscape, oil attracts new people to the reservation—"representatives of oil companies, geologists, lease men, and oil field employees from the ends of the earth" (77), as well as "men and women with criminal tendencies" (77)—and remakes the social landscape there. Soon, it literally "fabricates" with people on the reservation, giving them names—baby Osage Oil DuBois, for instance (86)—and new identities.[53]

Petroleum's material fluidity and social flexibility in Mathews's novel challenge popular assumptions about the role of wealth in "white" and Native American cultures. After all, part of the ideological rationale behind the Dawes Allotment Act was that private property ownership would finally "civilize" savage Indians. Oil indeed gives the Osage access to supposedly "civilizing" commodities, and does so, ironically, by letting them skip the stage in which they were to engage in a white-man-approved type of labor—farming—to generate wealth. Yet oil wealth betrays the Osage, and Mathews narrates the socialization of oil, the new all-powerful commodity, while simultaneously showing that in the context of colonialist and capitalist exploitation—and even under the collectivist redistribution of wealth, post-Allotment—similar possibilities are unavailable to Chal. Mathews seems to argue that the bildungsroman does not have the capacity to narrate the successful "progress" of an Osage American human subject, because prior to the 1934 Indian Reorganization Act, no literary or political convention imagines an Osage American as capable of becoming a "fully capacitated social subject" without sacrificing his or her Osageness.[54]

Thus, *Sundown* traces the development of oil culture but does not validate or legitimize it. Instead, it forces readers to focus on what makes oil "become" social, on the colonial "designs" of allotment that give it political and cultural power, and on what is left behind in the process—the material and human social debris that the oil flow produces, and that form the new Osages and Americans of the 1910s and 1920s. This includes the always-confused Chal, his misguided father, his fullblood mother, and his friends: peyote worshiper Sun-on-His-Wings, morphine-addicted Running Elk, as well as prostitutes, drug dealers, and oilmen.[55] The framework of the novel of formation works, then, as a narrative technology demonstrating oil's vitality rather than Chal's socialization. If anything, *Sundown* should make us reconsider assumptions about progress and modernity. The arbitrariness of economic and cultural development on the Osage reservation, which in the 1910s and 1920s is a space of colonial exploitation with oil at its center rather than people, reveals the terror within projects of "civilization," of destructive but always forward-moving technological progress, and of the cultural and "material weight" of oil extraction, distribution, and transformation into commodities.[56]

FLUIDITY, HISTORY, MODERNITY

By turning his attention to the way Osage oil makes cultural and economic forms fluid—by pointing to oil's own transnational flows and the conditions of oil's own transformation into new things[57]—Mathews complicates the easy chronology of the stages and transformations of capitalist modernity.[58] Through his attention to the messy fluidity of petroleum itself, Mathews urges us to consider a very material history of technological, economic, and political developments, one that is attentive to the oily matter of crude petroleum that enables these developments and also to the messy matter of colonial projects that accompany the physical extraction of mineral resources in Mathews's and our times.

In Mathews's time, the widespread application of new technologies and forms of labor organization and control, such as Taylorism and Fordism, as well as the legal and political interventions that strengthened the power of state or federal governments, indeed produced shifts within the social structure of American society. The emergence of new masses of wage laborer–consumers produced "solidity" and certainty in cultural, social, and political forms. However, Mathews's novel dramatizes the ways in which these developments fail to produce order in all geopolitical spaces and for all social subjects at the same time; in fact, his work reveals the ways in which this solidity within "core industries" and societies not only is synchronous with but always depends on "liquefying" cultures elsewhere.[59] Thus the territorial and fiercely violent "extraction" of humans as well as natural resources—defined as "primitive accumulation" or "accumulation by dispossession"[60]—might be best understood not as a stage that precedes or follows the consolidation of the capitalist system, but as an ongoing colonialist project upon which the economic "world-system" depends.[61]

In the case of Pawhuska, the Osage experience of "liquid modernity" had much to do with the place the Osage hold in the history of American colonization of the continent.[62] Thus, the chaos and frenzy of the 1920s and 1930s in Pawhuska produce solidity elsewhere; they unravel precisely at a time when white male workers in other parts of the United States are disciplined by Frederic Winslow Taylor and Henry Ford. Within Taylorist ideology, workers perform their routines at Ford's conveyor belts, work a predictable number of hours, receive predictable wages, experience a modicum of stability, and thus are empowered as a class of worker-consumers.[63] In contrast, the Osages, simultaneously outside of and incorporated into the national United States, like residents of many colonized, semiautonomous territories then and now, never enjoy such a "solid," predictable phase of modernity. On the Osage reservation, Taylorism, Fordism, and state-sponsored social and political architecture, whether represented by the U.S. Bureau of Indian Affairs, the Department of the Interior, or the Osage Tribal Council, never produce order and stability.

Mathews's attention to oil and the "progress" it brings challenges the logic that places the Osage and other Native Americans in a state of "cultural [and economic] childhood."[64] In the local, reservation world in *Sundown*, the "oily matter" of petroleum desynchronizes social time, showing how unevenly oil culture affects the Osage, their "Negro chauffeurs," the "virginal" white girls (248) who sell dope to get rich off Osage oil, the Osage women, and the criminals. None of the people or the physical spaces engulfed by Osage oil's power can resist its destabilizing force, and Mathews sees these changes as yet another sign of oil-induced, violently territorial, and ultimately colonialist pandemonium. On a macroscopic scale, oil, as depicted in *Sundown*, is more than an "actant" or "intervener."[65] Petroleum diffuses, linking bodies and commodities across national and racial borders in unusual configurations, and its power to radically alter Osage culture in the short span of thirty years stems from its material properties and from the emergence of technologies that discover new uses for it; from the econo-legal provision of the 1906 Allotment Act; and from federal policies designed to exploit, if not actually persecute, Native Americans.[66] Oil gains its force because colonial institutions and legal frameworks imbue it with power and render rampant overexploitation of Osage wealth acceptable. And, as Nancy Lee Peluso and Michael Watts compellingly argue and document, the same dirty oil, extracted not in the virtual space of the Internet but in specific physical locations—in Ecuador, in the Black Sea region, or in the Gulf, and often under much more exploitative conditions than those on the Osage reservation—enables our own "liquid times."[67] Thus, in *Sundown*, Mathews not only questions the possibility of progressive and linear articulations of Osageness or Osage Americanness in a novel of formation but also challenges a certain kind of cultural narrative that relegates entire regions and peoples, along with their experiences of economy, culture, and politics, to the realm of historical irrelevance. Narrating supposedly "solid times" through the prism of colonial exploitation and petromodernity, Mathews accounts for modernity's colonial expressions, manifest differently in areas of petroleum extraction than in the globally connected "core" and "peripheral" spaces of political and economic significance.[68]

NOTES

I thank Elizabeth Maddock Dillon, Guy Rotella, and an anonymous *Journal of American Studies* reader, as well as Daniel Worden and Ross Barrett for their thoughtful comments on earlier drafts of this article. I also thank Kristen Ebert-Wagner for her editorial advice.

1. While I borrow the phrase "liquid modernity" specifically from Zygmunt Bauman, *Liquid Times: Living in the Age of Uncertainty* (Cambridge: Polity Press, 2007), 1, 69, 79, several other studies of late capitalism emphasize fluidity and uncertainty as key markers of post–World War II capitalism. See Richard Sennett, *The Corrosion of Character: The Personal Consequences of Work in the New Capitalism* (New York: W. W. Norton, 1998), 30; and *The Culture of New Capitalism* (New Haven: Yale University Press, 2006), 13, which explicitly invokes Bauman's "liquid modernity." Bauman's earlier project examines the relationship between "order building" and its "waste" and "chaos" more directly; see Zygmunt Bauman, *Wasted Lives: Modernity and Its Outcasts* (Cambridge: Polity Press, 2004), 1–33.

2. Nancy Fraser, "From Discipline to Flexibilization: Reading Foucault in the Shadow of Globalization," in *Scales of Justice: Reimagining Political Space in a Globalizing World* (New York: Columbia University Press, 2010), 116–30; David Harvey, "Neo-Liberalism and the Restoration of Class Power," in *Spaces of Global Capitalism: Towards a Theory of Uneven Geographic Development* (New York: Verso, 2006), 7–68. Harvey, however, is extremely attentive to the ways that capitalism migrates its crises across national borders and how neoliberalism enables exploitation of environments, resources, and people. See David Harvey, *The New Imperialism* (New York: Oxford University Press, 2003).

3. For Sennett, "flexible capitalism" is explicitly a late twentieth-century phenomenon, and he does not examine in detail the "liquid" dimensions of colonial modernity that predate and facilitate the establishment and dissolution of the Western welfare-state model and the increased flow of transnational capital. Sennett, *Corrosion of Character*, 9.

4. Imre Szeman offers a brilliant discussion of globalization as seen not only through the prism of the seemingly weightless transfer of capital and information across national borders but also in the context of the very tangible, heavy "transfer of material weight" of oil, in Ursula Biemann and Imre Szeman, "Forced Transit: A Dialogue on *Black Sea Files* and *Contained Mobility*," in *Political Typographies: Visual Essays on the Margins of Europe* (Barcelona: Antoni Tàpies

Foundation, 2007), 13–45 (35–36). For a comprehensive discussion of the violent "hydrocarbon capitalism" and the role that abundance of nonrenewable resources plays in the projects of neocolonial economic exploitation, see Michael Watts, "Petro-Violence: Community Extraction and Political Ecology of a Mythic Commodity," in *Violent Environments*, ed. Nancy Lee Peluso and Michael Watts (Ithaca: Cornell University Press, 2001), 189–212, at 189.

5. John Joseph Mathews, *Sundown* (New York: Longmans, Green and Co., 1934). Mathews produced several documentary, nonfiction book projects but only one novel. All subsequent parenthetical citations refer to the 1934 edition.

6. See Bauman, *Liquid Times,* and Sennett, *Corrosion of Character.*

7. Karl Marx, *Capital: A Critique of Political Economy*, trans. Samuel Moore and Edward Aveling (New York: Modern Library, 1906), 784–87.

8. Michael Watts, "Petro-Violence," 191.

9. Sennett, *Corrosion of Character*, 31.

10. See the discussion of "colonial and (precolonial) violent 'extractions'" of humans and rubber in Ecuador in Watts, "Petro-Violence," 202.

11. Louis F. Burns, *A History of the Osage People* (Tuscaloosa: University of Alabama Press, 2004), 439.

12. Although these are obviously racist categories, I use them here to refer to historical terms with legal and cultural significance, as they defined degrees of legal and cultural belonging and of Osageness.

13. "Oil Makes Osages of Northern Oklahoma Richest Indian Community in the World," *New York Times*, May 21, 1923, 1; "Auctions of Oil Land Enrich Osage the American Indians," *Los Angeles Times*, March 4, 1934, 15.

14. For accounts of this gruesome period, see Burns, *History of the Osage People*, 439–42; Lawrence J. Logan, *The Osage Indian Murder Plot: The True Story of a Multiple Murder Plot to Acquire the Estates of Wealthy Osage Tribe Members* (Frederick, Md.: Amlex, 1998); Declassified FBI files about the Osage murder cases are available at http://foia.fbi.gov/foiaindex/osageind .htm. See also Michael Watts's discussion of more recent instances of "petro-violence" in Ecuador and Nigeria in Watts, "Petro-Violence," 207.

15. Jane Bennett, *Vibrant Matter: A Political Ecology of Things* (Durham: Duke University Press, 2010).

16. Michael Watts calls it petroleum's "mythos." Watts, "Petro-Violence," 191.

17. See the following on commodities and American national culture: Phil Patton, *Made in the USA: The Secret Histories of the Things That Made America* (New York: Grove Weidenfeld, 1992).

18. John Ise, *The United States Oil Policy* (New Haven: Yale University Press, 1926), criticized the deregulation and predatory nature of oil extraction. In his view, both led to waste and overexploitation of oil resources, caused environmental damage, and lowered oil prices to below production-level costs.

19. John Joseph Mathews, *The Osages: Children of the Middle Waters* (Norman: Oklahoma University Press, 1961), was sponsored by his oil tycoon friends.

20. I borrow this word from Maurizio Lazzarato, "Immaterial Labor," in *Radical Thought in Italy: A Potential Politics*, ed. Paolo Virno and Michael Hardt (Minneapolis: University of Minnesota Press, 1996), 133–47.

21. C. B. Glasscock, *Then Came Oil: The Story of the Last Frontier* (Indianapolis: Bobbs-Merrill, 1938), 147.

22. Ibid., 146.

23. Matthews, *The Osages*, 776.

24. Burns, *History of the Osage People*, 335, 243.

25. The "civilization" of the Osage involved, among other things, forcing Osage men to take up farming, an occupation they viewed as "unmanly"; educating Osage children in reservation schools; and altering gender roles and family structures to conform to the Western model of the family unit. For fascinating reading on this subject, see Mark Rifkin, *When Did Indians Become Straight? Kinship, the History of Sexuality, and Native Sovereignty* (New York: Oxford University Press, 2011).

26. The first commercial well appeared in 1897. Mathews, *The Osages*, 772.

27. Glasscock, *Then Came Oil*, 146.

28. Burns points out that "never before had American law allowed Indian allotment on the basis of separating the surface ownership from mineral ownership." See Burns, *History of the Osage People*, 400.

29. Terry P. Wilson, *The Underground Reservation: Osage Oil* (Lincoln: University of Nebraska Press, 1985), 74–75.

30. Mathews, *The Osages*, 776.

31. Glasscock, *Then Came Oil*, 147. William Shepherd's "Lo, the Rich Indian" cites the instance of an Osage woman who made nearly as much money as President Wilson in the 1920s. Surely his *Harper's* readers found this information shocking. See William Shepherd, "Lo, the Rich Indian," *Harper's Weekly* (November 1920), 723–34.

32. "Osages before Oil," *Time*, November 7, 1932, n.p. http://www.time.com/time/magazine/article/0,9171,744704,00.html.

33. Jane Tompkins, *Sensational Designs: The Cultural Work of American Fiction, 1790–1860* (New York: Oxford University Press, 1985); see also Jacques Rancière, "The Politics of Literature," trans. Julie Rose, in *The Politics of Literature* (Malden, Mass.: Polity Press, 2011), 3–30.

34. The bildungsroman was an unusual choice for Mathews, whose literary career began in 1932 with the nonfiction *Wah'Kon-Tah: The Osage and the White Man's Road* (Norman: University of Oklahoma Press, 1981), in which he transcribed the memoirs of the Indian Agent Laban J. Miles, and ended in the 1960s with the magnum opus *The Osages*, a six-hundred-page collective tribal history. Fascinated by indigenous traditions, he dedicated most of his life to preserving and reconstructing what he considered vanishing Native American lore. He documented indigenous oral and visual culture in his tribal history, helped found the Osage Museum—the first tribally owned museum in the country—and received two Guggenheim fellowships to research First Nations' cultures in Mexico in 1939 and 1940. For a discussion of Mathews's early work, see Susan Kalter, "John Joseph Mathews' Reverse Ethnography: The Literary Dimensions of Wah'kon-Tah," *Studies in American Indian Literatures* 14, no. 1 (2002): 26–50. For Mathews about his return to the Osage reservation, see Guy Logsdon, "John Joseph Mathews: A Conversation," *Nimrod* 16, no. 2 (1972): 70–89.

35. Joseph R. Slaughter, *Human Rights, Inc.: The World Novel, Narrative Form, and International Law* (New York: Fordham University Press, 2007), 19.

36. In Sennett's view, people see "their lives as narratives . . . [of] not so much what will necessarily happen as of how things should happen." Sennett, *Culture of New Capitalism*, 23.

37. Slaughter, *Human Rights, Inc.*, 27; Marianne Hirsch, "The Novel of Formation as Genre: Between *Great Expectations* and Great Illusions," *Genre* 12, no. 3 (1979): 293–311.

38. Note that Mathews is part of a larger twentieth-century literary tradition of "life writing." See, for example, James Weldon Johnson, *The Autobiography of an Ex-Colored Man* (New York: Sheldon, French, and Company, 1912); Anzia Yezierska, *Bread Givers* (1925; New York: Persea Books, 2003); Mourning Dove, *Cogewea, the Half Blood: A Depiction of the Great Montana Cattle Range* (1927; Lincoln: University of Nebraska Press, 1981).

39. A. LaVonne Brown Ruoff, *American Indian Literatures: An Introduction, Bibliographic Review, and Selected Bibliography* (New York: Modern Language Association of America, 1990), 4–75; Robert Allen Warrior, *The People and the Word: Reading Native Nonfiction* (Minneapolis: University of Minnesota Press, 2005).

40. Patton's *Made in the USA*; for a discussion of "things" and their political power, see Bruno Latour, "From Realpolitik to Dingpolitik: Or How to Make Things Public," in *Making Things Public: Atmospheres of Democracy*, ed. Bruno Latour and Peter Weibel (Cambridge, Mass.: MIT Press, 2005), 4–31.

41. *Sundown* is not an exception here. Many early twentieth-century bildungsromans centered on the unsuccessful individuation and socialization of human persons and the socialization of modern things. For example, consider the contrast between the difficult lives of female protagonists and the vibrant lives of artifacts and commodities in Nella Larsen's *Quicksand* (1928; New York: Penguin, 2002); and between Aunt Delilah and her famous waffles in Fannie Hurst, *Imitation of Life* (New York: Collier, 1933).

42. Eda Lou Walton, "The Osage Indians," *Nation* (January 8, 1933), 156. Interestingly, although national presses consistently emphasized Mathews's Indianness, Mathews was only part Osage. During the era of intertribal conflicts between fullbloods supporting collectivization of resources and mixbloods favoring assimilation into the U.S. national culture and privatization of land, an Osage-American mixblood, educated in the United States and Europe, would often be treated with suspicion by tribal elders and feel alienated from traditional Osage culture. In short, white readers would describe Mathews's identity as Osage, following the one-drop rule of racial identity; for Osages, and for Mathews specifically, one's "Indian" identity did not depend simply on one's genetic traits but was defined in complex ways, as a matter of heredity, kinship, or cultural identification, or in relation to one's views concerning tribal property rights. For a discussion of Mathews's and the novel's protagonist's cultural identity, see Carol Hunter, "The Protagonist and a Mixed-Blood in John Joseph Mathews' Novel: *Sundown*," *American Indian Quarterly* 6, no. 3/4 (1982): 319–37; see also her discussion of the schism between "bourgeois" mixbloods and fullbloods (65) living in traditional camps in, "The Historical Context in John Joseph Mathews's *Sundown*," *MELUS* 9, no. 1 (Spring 1982): 61–72.

43. *New York Times Book Review*, November 25, 1934, 19–20.

44. For a different interpretation of Chal's alienation as modernist posturing, see Christopher Schedler, "John Joseph Mathews: Tribal Modernism," in *Border Modernism: Intercultural Readings in American Literary Modernism* (New York: Routledge, 2002), 41–52.

45. Articles in national publications aimed to shock readers with descriptions of Osages' outrageous purchases. What Watts calls "petro-magic," or "wealth without effort," was not expected to materialize on an Indian reservation. See Watts, "Petro-Violence," 205, 193. Whereas it was good style for white Americans to overspend and overconsume, an Osage "Brave pay[ing] $2000 for Clock to Chime His Reservation Tepee" was a national scandal. "Osage Indians Face Depletion of Oil Incomes," *Christian Science Monitor*, November 2, 1929, 1.

46. Mathews often mentions "a strange chemical running through [various characters'] veins," and it is unclear if he means whiskey, morphine, or oil. *Sundown*, 14.

47. Arjun Appadurai, ed., *The Social Life of Things: Commodities in Cultural Perspective* (Cambridge: Cambridge University Press, 1986).

48. Watts, "Petro-Violence," 205.

49. I use the term "bildung" after Sennett as a "convention of life." Sennett, *Culture of New Capitalism*, 89.

50. *Sundown* was written before the 1934 Indian Reorganization Act reversed, or at least ameliorated the impact of, federal policies of forced assimilation and privatization of Native American lands.

51. See also Michael Watts's reflection on how such "liquid mobility" of oil simultaneously signals abundance and vitality as well as "evacuative despolation" and "environmental threat." Watts, "Petro-Violence," 205.

52. Mathews calls the region the "blackjacks" because of its blackjack oaks.

53. Alain Pottage and Martha Mundy, eds., *Law, Anthropology, and the Constitution of the Social: Making Persons and Things* (Cambridge: Cambridge University Press, 2004), 1.

54. Slaughter, *Human Rights, Inc.*, 27.

55. The use of morphine and marijuana in addition to whiskey was widespread on the reservation. Wilson, *Underground Reservation*, 156. Shepherd mentions "mescal"—by which he most likely meant peyote—and alludes to heroin use. Mathews often references ceremonial peyote use in *Sundown* (83, 162, 184, 277).

56. Biemann and Szeman, "Forced Transit," 36.

57. Oil is fluid and transnational matter; it is extracted on the Osage reservation by transnational corporations, and the oil-rich land is at once American and Osage and, before the Removal Act, had belonged to different Native American *nations*. Yet oil also "produces a state" and valorizes its power. Watts, "Petro-Violence," 204.

58. See Fredric Jameson, *Postmodernism; or, The Cultural Logic of Late Capitalism* (1991; Durham: Duke University Press, 2003), 1–66.

59. See the discussion of "unequal exchange" and "core" and "peripheral" products in Immanuel Wallerstein, *World-Systems Analysis: An Introduction* (Durham: Duke University Press, 2004), 28.

60. Marx, *Capital*, 784–87, and Harvey, "Accumulation by Dispossession," in *New Imperialism*, 137–82. Harvey updates Marx's term to "accumulation by dispossession" precisely to describe the post-1970s world. Naomi Klein refers to similar contemporary colonialist exploitation as "disaster capitalism." Naomi Klein, *Shock Doctrine: The Rise of Disaster Capitalism* (new York: Picador, 2008). While deeply indebted to these scholars, I emphasize colonial, human, and material "extractions," not as a stage but as an ongoing technology of accumulation. Watts, "Petro-Violence," 202.

61. Wallerstein, *World-Systems Analysis*. See also David Harvey's discussion of global geographies in "The Geography of It All" and "Creative Destruction on the Land," in *The Enigma of Capital: And the Crises of Capitalism* (Oxford: Oxford University Press, 2010), 140–214.

62. Bauman, *Liquid Times*, 1.

63. Gary Cross, *An All-Consuming Century: Why Commercialism Won in America* (New York: Columbia University Press, 2000), 20–21.

64. Robert Dale Parker, *The Invention of Native American Literature* (Ithaca: Cornell University Press, 2003), 29.

65. Bruno Latour, *Politics of Nature: How to Bring the Sciences into Democracy*, trans. Catherine Porter (Cambridge: Harvard University Press, 2004), 75.

66. The Bureau of Indian Affairs, the Justice Department, and the local police neglected to investigate Osage murders until the 1930s. They also turned a blind eye to the speculative economic practices on the reservation as well as the abuse of the legal system of "restricting" Indians. Watts would call such practices "an incomplete decolonization." Watts, "Petro-Violence," 192.

67. Nancy Lee Peluso and Michael Watts, "Violent Environments," in *Violent Environments*, ed. Nancy Lee Peluso and Michael Watts (Ithaca: Cornell University Press, 2001), 3–38; and Bauman, *Liquid Times*.

68. Wallerstein, *World-Systems Analysis*, 1–41.

8

From Isfahan to Ingolstadt

Bertolucci's La via del petrolio *and the Global Culture of Neorealism*

GEORGIANA BANITA

I intend not to write a historical novel but only to create a form.

—Pier Paolo Pasolini, *Petrolio*

What interests me most in the cinema is the possibility it provides for showing very small, regional concerns by an array of bastardized, international, raw materials.

—Bernardo Bertolucci

Pier Paolo Pasolini's novel *Petrolio*, published in fragmentary form in 1992, almost twenty years after the death of the celebrated poet and director, focuses loosely on the meditations of a character in the employment of ENI (Ente Nazionale Idrocarburi), the Italian state-owned oil-and-gas company. But Carlo is not a run-of-the-mill petroleum engineer. His corporate career provides him with a platform for an "intellectual meditation" on the empirical and aesthetic experience of power, more specifically on the "'space' where real power is found."[1] After traveling to the United States, Arab countries, and as far as Tanzania on behalf of his employer, Carlo acquires a spatially circumscribed view of Italy as "a very particular world, one of many parts of a whole and not among the most important." His former nationalist ambitions dissipate like a childhood fantasy. That Italy could ever aspire to inhabit "the center of the world, the navel of the world, seemed to him at a very young age a fable." Yet Carlo cannot fully detach himself from the ancient and poetic national imaginary, distilling from it a desire for power and freedom—for energy itself and the expanse of its dominion, "a freedom as if born of itself and endowed with such real force that a whole part of the historical universe could be rendered immune from the curiosity of conscience."[2]

The grandeur of the poetic language that characterizes Pasolini's literary and filmic work left a lasting imprint on the imagination of his young assistant, Bernardo Berto-lucci, whose debut film, *La commare secca [The Grim Reaper]* (1962), was based on a Pasolini story and was offered to Pasolini first, who then passed it on to his protégé. When he directed the documentary film *La via del petrolio* for ENI, while trying to accumulate funds for his feature films, Bertolucci was, however, one step ahead of his mentor, who did not start work on *Petrolio* until 1975, after the issue had been brought to wide attention by the global oil crisis. *La via del petrolio* ["The Oil Route"] uncov-ers, through a blend of neorealist strategies and the more poetic style that Pasolini himself favored, how nationalist effusion dovetailed with petro-imperial overreach to weave a culture of exuberant potency, almost a corporate cult, around ENI and the future it promised for a dejected nation in the long aftermath of the war.

Bertolucci expresses petroleum's "historical universe" in a cinematic language that was not only innovative in the annals of petro-cinema, but simultaneously held a blissful immunity from the political critique that has shaped the late twentieth cen-tury's energy conscience. Scholars have dismissed *La via del petrolio* as a stopgap phase in Bertolucci's development. The film has not received any critical attention beyond a brief plot summary and a note on its neorealist techniques. As one among several oil-themed works of postwar neorealist cinema, however, Bertolucci's film documents Italy's unique role in petroculture: as a country with its own hydrocarbon reserves, but also one whose postwar development inscribed its national economy into a global arena, an ambitious energy player that stood toe-to-toe with the "Consortium for Iran" oil cartel while retaining a strong sense of Italian cultural involvement in ENI's global energy exploits.

Bertolucci's style in this phase of his career is neorealist in essence, but it also points toward a poetic vision that would dominate his more mature work. After all, he was only twenty-six when he shot the film. The influence of Jean-Luc Godard in particular, but also Michelangelo Antonioni and other neorealist filmmakers, makes this a deriva-tive work in many ways, a fact that may have inspired the junior director to consciously adopt quotation and artistic phylogeny as the structuring motif of his film. The trajec-tory of petroleum is also Bertolucci's own as he makes sense of his craft. Before *La via del petrolio* aired on Italian television, neorealism had already manifested a fascination with the texture of oil and the infrastructure of the petroleum industry. From the oil refineries in the vicinity of Ravenna, around which Antonioni's protagonists aimlessly wander, to the engorging libidinal fantasies of the oil magnate at the center of *Petrolio*, oil saturates the landscapes and aesthetics of neorealist art.[3] Certainly the social and economic ravages of World War II stimulated these filmmakers' interest in the labor regimes, architectural angularity, and environmental impact of Italy's oil-powered economic miracle. Even more salient, however, is the involvement of ENI and its influential manager Enrico Mattei in how Italy's oil interests were portrayed for the Italian public and indirectly for an international audience increasingly fatigued by Mattei's attempts to arm-twist established corporations into granting ENI a share of their global market. *La via del petrolio*, released just a few years after Mattei's death and shortly before the peak of the energy crisis, was commissioned as part of this public image campaign.

The link between the observational, exposing style of cinéma vérité, the dominant documentary aesthetic of the 1960s, and the earlier neorealist movement can be found in Joris Ivens's 1960 documentary *L'Italia non è un paese povero [Italy Is Not a Poor Country]*, cowritten by the neorealist literary heavyweight Alberto Moravia. ENI commissioned the film, but RAI (Radiotelevisione italiana) refused to broadcast the work in its original form, demanding massive revisions to eliminate the images of poverty and squalor that would have undermined Mattei's vision of Italy's blossoming oil economy.[4] While Ivens's focus does not transcend a national scope, the three works commissioned by ENI from the Swiss documentary filmmaker Gilbert Bovay tread new international ground. His first ENI film, *La valle delle balene* ["The Valley of the Whales"] (1965), tapped into the geological basis of the Po Valley's oil boom, yet the more ambitious *Gli uomini del petrolio* ["The People of Petroleum"] (1965) expands into Iran, Egypt, Libya, Nigeria, and Tunisia with less poetic fervor than Bertolucci and even less sensitivity to the political ramifications of the company's incursions into the deserts and jungles of black gold. An especially tone-deaf moment pairs a pathos-filled affirmation of Italy's economic ascent with footage of high plumes of smoke from desert fires, the same fires we see in the opening minutes of Bertolucci's documentary, yet filmed here with such fetishist abandon that we find ourselves not in awe of its religious resonance but repulsed by the fantasy of nationalist supremacy peering through the choking black clouds against, the narrator proclaims, the "transparent sky" of ENI's world dominion. And if the purpose of this industrial anthem— which received the top prize at an industrial film festival held in Genoa—has not become sufficiently transparent, a similarly naive petro-progressivism certainly drowns out any ideological suspicion or skepticism in Bovay's *Africa, nascita di un continente* ["Africa: The Birth of a Continent"] (1968), in which ENI is praised as a godsend for a number of oil-exporting African countries that crave political emancipation.[5]

Bertolucci's film clearly continues this tradition,[6] even though it breaks with both corporate documentary style and neorealism itself by infusing the story of oil with unusually poetic fervor, paving the way for the historicist lyricism of Pasolini and

Francesco Rosi (director of *Il caso Mattei* [*The Mattei Affair*], a 1972 film that details the oil manager's controversial death).[7] *La via del petrolio* is one among only a handful of documentary features and shorts that Bertolucci directed, so he approached it accordingly, in a spirit of dissidence: "Although it was a commissioned piece I tried as much as possible not to be swamped by documentary conventions and whenever I could I swerved right away from the received notions of that type of filmmaking. For instance, I saw the drillmen in terms of pioneers of an archaic West, and the helicopter pilots were anarchic and individualist heroes, like Godard's lonesome would-be's or those of *Only Angels Have Wings*."[8] Such diverse sources grouped together with little regard for national or historical context anticipate the film's eclectic understanding of oil culture and its visual repertoire, as well as its far-reaching narrative scope. In three parts, *La via del petrolio* documents the voyage of crude from ENI's extraction fields in Iran (part 1), through the Suez Canal to Genoa (part 2), and finally its subterranean land route through two major pipelines leading to Switzerland and Bavaria, Germany, respectively (part 3).

I want to examine this film as a hybrid text in which visual style, philosophical reflection, and poetic metaphor converge to carve out a geopolitically uncritical void that elevates petroleum into a potent agent of material change and cultural transformation at the interface of imperialism, global capitalism, and supranational alliance. This reading of the film scrutinizes its cultural origins in Italian neorealism and its role as an early index for discerning the transnational scope of oil culture.[9] Tracing the shifting thresholds of petroleum between human and nonhuman, between symbolic agent and intractable thing, Bertolucci unveils the unexpected resonance of petroculture with traditions and discourses of the Western world that previous filmmakers interested in the oil route had not been able to reclaim. *La via del petrolio*, I suggest, is a groundbreaking testament to how cinema as an art form and the industry of filmmaking overall can reflect a reality drenched in the promise of energy and petro-industrialization, a document that is as informative in its depiction of the global oil industry as it is provocative in its aesthetic form. Bertolucci offers a model for integrating economic and cultural history that in turn reshapes our understanding of the interplay between oil and culture in the twentieth century and beyond.

THE THOUSAND AND ONE NIGHTS OF OIL

The first part, *Le origini* ["The Origins"], depicts oil production in Iran and is, at least at first glance, pure style. The commentary is replete with cultural allusions and the camera moves with overemphatic self-awareness from buildings to people and to sand dunes in an attempt to encapsulate petroculture's oriental roots. It is worth noting, however, that Bertolucci himself intended this mythical sequence as a personal paean to Enrico Mattei, head of AGIP (Azienda Generali Italiana Petroli) and president of ENI—the oil shah, as it were, of postwar Italy.[10] For Daniel Yergin, Mattei emerged from the ruins like a "new Napoleon" to take control of the nation's operating companies and to embody its vision of resource-fueled resurrection. Yet the partnership that Mattei struck with the shah of Iran did not secure a central position for Italy in the global oil market, given that in the small areas of partnership only insignificant amounts of crude were found. It is no accident, then, that *La via del petrolio* includes none of the gushing spectacles that oil films otherwise thrive on. Instead, Bertolucci erects a poetic monument to the Italian oil magnate that fully reflects Mattei's own relations to exporting countries: vociferously anti-imperialist and sympathetic to their struggles for independence, yet also calculating and self-serving.[11]

That the journey of petroleum begins here has to do with Iran's sixth place among world oil producers at the time of filming.[12] *Le origini* acts as an establishing shot for the entire movie, assembling thematic, structural, and aesthetic features that organize the film as a whole. The epic, mythological span of the story spun along familiar cultural routes, aerial views, and the contrast of darkness and light—masterfully derived from the opposition of black crude and desert sun at the source—are used throughout the film to sketch the path of the black gold as it traverses sand, water, and snow-capped European mountains. In a prelude to the film, even before the first section title appears on the screen, Bertolucci shows gas fires in the desert and the shots fired by local workers to set these gases alight in a symbolic starting shot for the journey ahead. The flames above the sand convey a sense of heat and energy saturation that tilts the soil and the oil reserves of the desert along with it, toward far-off, fuel-hungry nations.

In these primal scenes, Bertolucci employs images of springs, sources, and cradles of civilization. The exalted, overarticulated commentary (authored by Bertolucci himself) propels the film forward with an almost mechanical relentlessness that matches the pressure of the valves pumping the pipelined oil across the sharp inclines of the Alps as it makes its way across Europe. The voice-over mines the history of Iran in a belabored attempt to embed the oil route into archaic cultural circuits. Persian shepherds, for instance, regard the gas fires "with a sense of the familiar and the sacred."[13] Bertolucci even aligns the origins of petroleum with the philosophy and history of Zoroastrianism, whose beginnings he tentatively locates in the same areas from which oil was extracted in the 1960s. Citing the teachings of Zarathustra in the sixth century BC, the film describes fire as creator of the world and of light—already foreshadowing the dialectical link of energy and globalism that fuels the oil route from Isfahan to Ingolstadt. Every shot of this section dramatizes the tension between the shock of oil's unfamiliarity and the need to assimilate it into recognizable physical and aesthetic material—to route it through cultural pipelines as well as those of steel and plastic.

Another familiar constellation into which the film inscribes petroleum is imperial history. The narrative follows the displacement of the Persian Empire by its subterranean equivalent, from which "nature's energy" emerges to encompass and fuel the world. This transfer of power adumbrates the reach of petro-imperialism at a time when its full implications were not fully in sight, yet it also replaces the political agency of empire with a natural force—not man-made but self-propelling. Bertolucci enacts this dehumanization of the oil world by focusing on wells, pipes, tankers, and refinery towers at the expense of the people who man and operate them. The metallic, pulsating music—whose industrial sound emerges naturally from the metal contraptions— adds to this effect, as does the aerial perspective of long tracking shots over oil rigs in the Persian Gulf (Figure 8.1).

The camera surveys the landscape from a disembodied perspective at heights that the human eye cannot reach without the assistance of oil-fueled machines. Yet the inherent paradox of this framing does not evince the critical force of more recent aerial views over the petroleum industry in the photographic work of Emmet Gowin or Edward Burtynsky.[14] A key postphotographic function of the helicopter gaze in this film has to do with the specific, cinematically productive motion that the aerial shot

FIGURE 8.1. Aerial views of oil rigs underline the vertical axis organizing the first part of *La via del petrolio.*

can dramatize. *La via del petrolio*, a film whose very title is premised on movement, routing, and shifting grounds, swings between the axes that different forms of mobility create on screen. Aerial tracking provides an effective counterpoint to the slow-motion journey of petroleum along its subterranean, horizontal path. It also reinforces the verticality of the petro-imaginary especially in neorealist culture, which, as many scholars have pointed out, often traces the upright growth of the industrial landscapes and especially cityscapes of postwar Italy.[15]

On the horizontal axis, the camera sets petromobility against the slower or even soporific speeds of local lifestyles. Oil is extracted in explosive bursts from the ground and piped undersea to processing units, while above ground the desert teems with purposelessly strolling figures and their mules. We learn later on that oil does not move faster through a pipeline than a human does on foot (at an average speed of 4 kilometers an hour), but Bertolucci wants to stress the contrast between a languorous culture and its dynamic contribution to world energy. More generally, he insists on distinguishing between oil time and human time, either by locating the flowing crude under the feet of aboriginal flaneurs or by separating oil workers from the routines of daily life as they unfold in their small Italian hometowns, away from the busy oil fields of Iran.[16]

Like *La commare secca* [*The Grim Reaper*] (1962) and *Prima della rivoluzione* [*Before the Revolution*] (1964), the two feature films that Bertolucci directed prior to his documentary for ENI, *La via del petrolio* is shot completely in black and white, the kind of film stock that reflected the cash-strapped economic conditions of filmmaking in postwar Italy. Its austere documentary aesthetic ties with the neorealist movement's interest in poverty, as well as with its clear-eyed attention to changes in the human psyche amid harsh labor conditions and the disintegration of social bonds. The visual polarities of the oil route are served especially well by this stark framing.[17] In the opening section, a wide palette of grays and ashen hues suffuse the screen to visually reenact the sensory withdrawal to which workers in the desert would have been subjected. The chromatic minimalism also allows for a journey that is almost forensic: tracing white pipes through dark woods, pump stations on snow-covered hills, or refinery towers among the ornate curves of small-town rococo housing in southern Germany.

Meanwhile the oil flows invisibly, weaving its mythical, snakelike weight through the earth. The juxtaposition of images from Isfahan, Genoa, and Ingolstadt are all the more jarring for the lack of any concrete trace of oil. This detail is remarkable in itself. The absence of oil throughout the oil route is part of the film's oblique political statement. "Perhaps I shall make my best films dealing with politics without talking about politics," Bertolucci would comment a few years later and puzzlingly single out *Ultimo tango a Parigi* [*Last Tango in Paris*] (1972) as his most political film.[18] At no time during this long journey does petroleum appear in the frame: in Iran it is yet untapped, and in the port area and inside the tanker we are invited to imagine it filling visible containers, until the pipeline picks it up in Italy and feeds it directly into refineries. Bertolucci translates the film's poetic message on the blindness of petroleum into its screen language. Petroleum never sees the light of day, nor can it be captured on camera.[19] Hence the extensive use of aerial footage, which not only denotes the sheer vastness of oil spaces but also suggests something of the representational blockage posed by crude. Not only petroleum itself but its entire habitat is impossible to explore by conventional documentary methods (road or foot) due to climate, missing infrastructure, and time constraints.

Once the flowing gold has a chance to pool at the first oil depot on its route, the oil tanks clutter the screen in another still captured from above, which may appear to reduce the immensity of the containers but in fact reinforces the stylized mobility of the camera as it tracks its mysterious subject: Oil goes from deep below to great heights along a vertical axis, which dominates the first part of the film. The tall tanks give the director ample opportunity to explore again the contrast of light and darkness along the oil route.[20] We even step inside the tanks to take in the geometry of shadow and sunlight crisscrossing the space where petroleum should be when the camera is not filming.[21] Bertolucci follows petroleum from one container to the next looking away from crude itself as if from blood, as if he were filming a war documentary without gunshots or casualties. Yet the mere fact that the camera has little use for petroleum itself exposes visually the inaccessibility of petroleum for the countries most closely involved in its extraction. Oil, we are told, is a "fortune with deferred usefulness in a society with a medieval economy." The globalism of petroleum has its roots in this misplacement of resources at locations that do not possess the high industrial porosity necessary to absorb it. Indeed the movie seems to guide the viewer away from a flat, impermeable surface that repels the precious liquid once it reaches the surface to the more petroconductive areas of Europe, where countless pipes, tankers, and tanks immediately soak it up. Oil is thus "assimilable in a modernized society, but in a pre-industrial one, it is basically a commodity to be loaded on ships like spices in ancient times." The parallels between the oil route and the Silk Road and spice-trade routes romanticize the image of oil barrels shifted between continents, but they cannot obscure the fact that unlike silk or spices, oil is not a commodity fit for direct consumption. Its mystique, despite the extended footage of heaving bazaars in the opening section of the film, is not tactile or olfactory and the visual imaginary of cinema seems as unable to assimilate oil as the preindustrial milieu from which it is extracted.

At this point Bertolucci shifts his perspective from uninhabited oil exploration deserts to the manned petroleum extraction machinery in Iran's Zagros Mountains, following an AGIP truck to one of the world's highest wells, Mount Rig, at an altitude of 3,300 meters above sea level. Again the vertical axis stabilizes the narrative, although the camera wanders away from the well itself to the workers that operate it. Bertolucci's debt to neorealist aesthetics comes most forcibly to the fore in this section, which follows no other plot structure than the mere observation of these men's working conditions, daily routines, habits, gestures, and the strange distance they maintain to the concrete environment of their employment. They are quartered for a short time in this almost immaterial, transient, and vibrant desert oasis, which they will most surely forget once they have been relocated to another well, another rig, managing a different group of local hands.

Little detail about the workers is worth noting. Their statements are as neatly drilled as an oil well. Everything about them sounds rehearsed, repeated, and mechanical. They are said to be laughing "because innocence will always be innocence," because machines are as docile as sheep, and they have become docile mechanical sheep themselves. They question nothing about their work and seem comically fascinated with the rocks and geological strata through which they are drilling ("You get used to anything," a worker opines later in the film. "Actually you end up with a strong interest in it"). One engineer claims to have dreams about his favorite rocks, and there is something magical—three years before the first manned Moon landing in 1969—about how these workers interact with the arid soil and the promise it contains, the patient wait for the drilling bit to break through subterranean frontiers in much the same way that a spaceship would pass through boundaries of space. Little wonder, then, that the workers assemble the rigs vertically like missile launch pads from parts carried up the mountain on a telpher line.

The extended focus on the Italian workers in Iran, who seem almost fully unaware of and totally uninterested in their human and cultural surroundings, suggests that the lack of cultural transfer separates the oil route from the silk and spice trade of earlier eras. Instead, oil appears to support financial and industrial rather than human migration. The final section of the film stresses all sorts of international, multilingual, and culturally diverse encounters within Europe in an effort to humanize the petroleum world, although as the camera follows a reporter through Italy and Germany and his interaction with the people he meets, the oil scenery pulls back considerably to the point where in Ingolstadt a refinery representative discusses the plant's layout and operations by pointing to a framed map on the wall. From poetry to boardroom-speak, the film never ceases to raise a wall of abstraction between its viewer and the slippery protagonist it purports to describe. Cementing the level of abstraction is the film's central literary allusion, namely, Captain Ahab's hunt for Moby-Dick, which the voice-over situates as overlapping with the search of Colonel Edwin Drake for oil deposits in Titusville, Pennsylvania. Although not mined to the full, the specter of the whaleship *Pequod* and its symbolic search for the white whale as a forerunner to the ambitious transoceanic sprawl of petroculture sets the stage for the film's middle section, which follows the passage of an oil tanker from the Persian Gulf to the port of Genoa.

THE IMPOSSIBLE VOYAGE

The second part, *Il viaggio* ["The Voyage"] opens with what may be called an underground aerial view. A lengthy tracking shot catches what we are made to believe is the surface route under which oil trickles through a maze of pipelines: "The oil arrives from the sea. Silently and out of sight, the pipes extend under the sand of the beaches, reaching the roots of the palm trees, stretching toward the huge tanks in Bargan Shar, where the SIRIP oil hub is located—the Italo-Iranian oil company." Under the skin of the desert the veins of the oil trade pulse invisibly, although they do not nourish the landscape above (despite their tangential contact to tree roots) as much as they collect into a transfusion bottle for foreign export. The notion of oil escaping from the ground is invested for the first time with the idea of danger, anemia, hemorrhage, and every other perilous condition the loss of blood brings about (although oil spills are never a topic). The invisible availability of oil is not only taken for granted; it also adds to its mystification and recalls not so much the spice or silk trade of ancient times as the drug trade of the present. The oil shot is delivered on the quiet. Tankers and pipelines supply the drug like so many syringes and needles.

The journey leads past Abu Rudeis on the Sinai Peninsula, a place where religion has given in to petroleum culture "to meet the new needs of man." Striking images of pious men praying in the shadow of pump jacks paint a scene of a quiet, self-sufficient shore where the waters of the gulf do not mix well with the oil underneath, leaving two cultures in contact and yet entirely unconnected: "Sinai means aridity, desert, and silence. Maybe oil will end an order that has been immobile for centuries and bring, along with industries, pipelines, asphalted roads, the notion of improving the standard of living. Quite a few cities of the modern world, from Caracas to Baku, were depressed areas before the oil era. And the silence, where until now only the Bible echoed, is broken by the songs of the workers." The oil route stretches over precisely this chasm. Its destination in Bavaria is equally silent, yet it is the peacefulness of prosperity rather than the preindustrial, elemental quiet of the desert sand. It is also a deeper discursive silence through which the oil route becomes tacitly naturalized, erasing the traces left on land and water, on lives and livelihoods, near and far. In his study *The Magical State*, Fernando Coronil argues that in the case of Venezuela and other petrostates, "the arduous establishment of state authority was achieved in intimate relation with the exploitation of petroleum."[22] Bertolucci tells the "magical" story of petroleum in reverse, as a transcendent agent of cultural separation and disunion.

The two currents, one pointing to the future, the other leashed to the past, converge in one of the film's most iconic moments: the slow glide of oil tankers through the Suez Canal, along 164 kilometers of waterway connecting the Red Sea and the Mediterranean. As "one of the fixed routes of the world," the Suez guides the path of oil along familiar tracks. Here the reluctant fist of the exporting world opens to let the precious fuel escape from its grasp. The tankers proceed almost stealthily, as if trying to smuggle their cargo without being noticed. Positioned between two sets of houses perpendicularly on a narrow street, the camera lies in wait until the bow of a tanker advances slowly from the right (Figure 8.2). One and a half minutes crawl by as the tanker "glides along silently, with its engines idling and no one on the deck." Again the machines take over and the screen is evacuated of all human presence to drive home

FIGURE 8.2. The oil tanker gliding along the Suez Canal within a few feet of the houses on shore is one of Bertolucci's most memorable images. From *La via del petrolio*.

the invisibility of globalizing forces and the idle habitats they pass by. The unmanned, unpopulated, anonymous natural force fills the screen for a "moment which seems to last an eternity, then the ghost disappears." Ghosting is what Bertolucci does to his subject throughout the film, submerging it under ground and water, pipes and tanks, machinery and constructions. The ghost of oil is all we perceive, dimly and intermittently, which compounds the impression that it cannot be interfered with or stopped: "The column of funnels travels slowly through the desert, hemmed in by the dunes, but never stopping, proceeding without interruption, day and night, in a tunnel of light, air and sand."

Viewers in the 1960s would have been reminded of the Suez Crisis triggered by the nationalization of the canal in 1956. Britain, for whom the canal provided a lifeline to the former colony of India, colluded with France and Israel to mount a military operation to regain the Suez, before the United States intervened to stop the allies, fearing that the oil-producing countries would retaliate with massive embargoes. Imperial power changed hands along the Suez. This vision of the Suez as a symbolic knot that entwines declining imperialist agendas and the emergent forces of global capitalism invokes other representations of the canal in the 1960s, especially David Lean's epic *Lawrence of Arabia* (1962), whose essentializing, orientalist fascination with the pristine Bedouin way of life Bertolucci adopts unchanged, but without the anticolonialist agenda of Lean's film or its perceptive treatment of how oil politics changed the Arab world at the macro-level of national economy, government, and its relationship to the citizenry.[23] *La via del petrolio* never strays from its grassroots view of the oil route. For a documentary, the film is oddly tight-lipped on concrete historical background and political commentary. That the many nations along the oil route are not seen as distinct political entities supports my understanding of this film as a transnational, or rather subnational, geo-genealogy of oil intended less to inform or rally than to instill in its audience a sense of wonder about the subtle presence of things (like oil) we have little time for. For Bertolucci, as we shall see, cinema itself serves the same purpose.

The Suez section features a comical soundtrack that accompanies the slow motion of the ship as it glides seemingly across the sand (we cannot see, due to the narrowness of the canal, the water on which it floats) with almost slapstick hilarity. To underscore the surrealism of the scene, the voice-over recalls Georges Méliès's *Le voyage à travers l'impossible [The Impossible Voyage]* (1904), a satirical "road movie" about a journey commissioned by the Institute of Incoherent Geography, a journey that transports its protagonists and viewers through as many vehicles and settings as *La via del petrolio* (including trains, streetcars, automobiles, and a submarine). Besides accentuating the trick-photography effect of the tankers' sailing through the Suez Canal, the invocation of Méliès and his fantastic journeys says something about Bertolucci's film that might explain its meager resonance with critics and audiences alike. Méliès used precinematic popular arts in cinematic language with the aim of satire and entertainment, without concern for the aesthetic sustainability of his forms.[24] Much the same way that Méliès, as Siegfried Kracauer remarks, "used photography in a pre-photographic spirit—for the reproduction of a papier-mâché universe inspired by stage traditions,"[25] Bertolucci films the oil route within the conventional temporality of successive tableaux and settings, presenting every striking or odd twist on his path as internally motivated rather than magic moments that do not declare themselves as such. This is not petro-magic-realism as much as a form of neorealism that acknowledges its illusions and sees them as inherent to reality rather than as a "film" imposed on the surface of the world, for purposes of allegory or enhancement.[26]

By quoting cultural traditions and tropes that filmmakers of the 1960s were quickly dispensing with and by remaining ignorant of the historical momentum of other oil films released in the 1960s and 1970s, Bertolucci produced a unique artifact that is difficult to compare with other oil films or narratives of the last half of the twentieth century. The film is an example of an experimental work rooted in the past and too unimaginative about the future—of both the oil world and its aesthetic expression. If it fails to transcend the scope of earlier European petro-neorealist productions such as Vittorio de Sica's *Miracolo a Milano [Miracle in Milan]* (1951) or Henri-Georges Clouzot's suspenseful *Le salaire de la peur [The Wages of Fear]* (1953),[27] *La via del petrolio* is something of an impasse, residing in a perpetual twilight at the confluence of documentary and art, datedness and poetic timelessness, realism and enchantment. The film has drawn no scholarly attention whatsoever, has influenced no later oil films—at least no evidence to the contrary exists—but precisely for that reason it stands out as the most faithful dramatization of the unrepeatable moment when oil entered the language of cinema and vice versa. Bertolucci struggles to recover a reality that had not yet set in, a reality devoid of signs that had yet to be introduced in the cultural circulation of meanings and images, just as it was propelled on its global voyage. The effort to trace the global oil route is thus coextensive with the attempt to locate its aesthetic genealogy and envisage its future. "The era of tankers will soon come to an end," the narrator proclaims, suggesting that conventional transporters would soon be replaced by floating nylon tanks known as dragons. "Underwater pipelines have been built using flexible plastic and aluminum. And the maps of the land and seas will be rolled up, like the maps of the world seen in 18th century prints." Global maps rolled into pipelines and crumpled into nylon floaters create the atlas of a world in which life without oil appears unthinkable. The film is buoyed by the unrestrained optimism of this era.

Yet despite their implied interdependency on the ship as on the ground, man and oil share a fragile alliance. In multiple shots of the tanker, AGIP's *Venezia*, the mechanics of oil remain impersonal, as if people were just along for the ride. In images of the apparently self-steered ship, people are crammed in a corner of the shot, diminutive and powerless, moving gingerly across the body of the steel behemoth like parasites on larger fish (Figure 8.3). The transfusion of petroleum from Iran to Italy occurs as if outside human intervention, like an unstoppable black ocean current.[28] No explanations are given for the desirability of petroleum or what the drain of fuel leaves behind, back in Iran and other exporting nations, although the language used invokes colonialism and captivity, as if the tankers were not carrying oil but a primitive prisoner workforce: "Extracted from the ground, imprisoned in the tanker's belly, the oil returns to the ground. Poured into the pipeline, it becomes an underground river. Flowing under plains, lakes, and mountains, until it reaches the ancient heart of Europe." One dead empire lends the energy of its putrefied carcass to the next.

FIGURE 8.3. AGIP's *Venezia* does not appear to require human guidance. We rarely see anyone on deck. Human figures seem tiny and light—remnants of a different, less mechanized era. From *La via del petrolio*.

Toward a European Oil Union

The third part, *Attraverso L'Europa* ["Through Europe"], is essentially a portrait of 1960s Europe and the shared need of its nations for petroleum power. Of course, the European Union traces its origins partly from the European Coal and Steel Community, a supranational six-nation organization formed in 1951 by the Treaty of Paris to bring about a French-German rapprochement and pool European heavy industry in order to diffuse potential military tensions. To prevent a revived Germany from dominating Europe economically, the United States and France sought to impose prolonged controls on the Ruhr energy and heavy industry.[29] Pooling energy resources through a supranational institution that sought to preempt German political ascendancy ironically planted the seed of an integrated Europe that Bertolucci imagines not as a level playing field, but as a sharp incline tilting the flow of imported energy toward a recently reindustrialized southern Germany.

From the fugaciousness of its more expansive sandbox childhood, oil has matured into an "iron-clad river, pumped, measured, controlled, and obedient to valves, filters, and pumps regulated by electronic brains." The source of this discipline lies outside human control. Energy and populations are at the mercy of a mechanical apparatus whose angular protrusions jar the landscape. A journalist takes over the narrative, as the story would otherwise, for lack of a human voice, peter out into inert industrial footage of interest to no one and especially not to the commissioning company, whose main audience is, after all, human consumers. At a pump station in Chivasso in the province of Turin, the journalist Mario Trejo converses with the local operator about the neuroses of Antonioni's characters in *Il deserto rosso* [*The Red Desert*] (Figures 8.4 and 8.5). They stroll around tanks and towers with a mechanical disinterest that introduces the disjunction of energy and humanity driving the less inspired poetics of this entire section.[30] The oil encounter feels global and homogeneous partly because people gradually retreat from the contraptions that guide petroleum toward industrial centers. These contraptions suggest to Trejo the picture of a "termitarium of pipes," which replaces the bustle of human labor we saw at oil-extraction sites with an uncannily vast and heaving network of automatons. Most of the people Trejo meets care little about the pipelines underneath their feet and barely know their whereabouts. The journalist's journey across Europe in cars and even on foot—to fall in step, as it were, with the flowing oil—is equally disembodied, in the sense that he does not take notice of national boundaries but listens to the local accents (a more human petrogeography) to gain his geographical bearings.

Like a madeleine dipped in tea, the pipeline inspires in him a string of involuntary memories and meditations: of people, books, films, and places, of leisurely walks and bucolic dreams of Europe, from Goethe and Heine to Stendhal—all great walkers.

FIGURE 8.4. Bertolucci's central intelligence, a reporter, discusses with a refinery worker the characters of Michelangelo Antonioni's *Il deserto rosso* and their morbid rambles among oil towers and cisterns. From *La via del petrolio*.

FIGURE 8.5. Monica Vitti is dwarfed by a field of identical oil containers outside Ravenna. From *Il deserto rosso*.

"Me on the road, the oil in that meadow. We travel at the same speed, about 4 km an hour." The romantic flair of this moment reveals how the industry is learning to "discreetly become part of the surrounding nature." The remainder of the journey traces the naturalized path of oil "passing unobserved" through the landscape "like a vein of black gold." The stealthy movement remains the same. While the Iranian landscape allowed for complete submersion, in Europe the challenge is camouflage. The pipelines find it easier to disguise themselves since the web of cultural references thickens the closer we move toward regions that boast as much history and symbolism as Sinai: "The gothic sacredness of the German fir trees was enriched by a new, hidden vitality" (Figure 8.6). Precisely what the cathedral of Ulm, which "looks like it should be in an Orson Welles film," has to do with the oil route is not entirely clear, though Welles, of course, incorporated masterful visuals of the hydraulic knifing movements of pump jacks on the U.S.–Mexico border in his late noir *Touch of Evil* (1958). In a range of analogies that stretches far beyond the oil route itself—with detours through the Konstanz harbor and jazz clubs in medieval Meersburg—Bertolucci tells the parallel story of petroleum and Western culture as if building a pipeline through Western

FIGURE 8.6. Bertolucci selects starkly symmetrical, vaguely religious settings suggestive of how petroleum permeates the fabric of modern culture. Here a pipeline winds its course underneath a forest in Germany. From *La via del petrolio*.

civilization, trying to make it fit Europe much the same way that the ENI pipeline seamlessly absorbs into Alpine nature.

By the time we reach Ingolstadt we are back inside, taking history lessons from local journalists ensconced in old buildings that overlook the historical town center around which modern industry and residential districts mushroomed in the late 1950s, when the town was established as a center of the oil-refining industry.[31] Electrifying, capacious, and slightly ridiculous in tone, the final statement ties together the threads of magic and revolution that bind together East and West along the oil route: "This is where my story ends, where rococo meets the oil extracted in Persia, the Sahara or Sinai. The oil transplanted from deserts to snow-filled areas has huge effects on a country's economy, technology and society. It changes the existence of man like the wheel did, the first metals, gunpowder, steam engines and electric engines." In the history of energy innovation, oil marks a radical step. In an era of deposit abundance and corporate upswing, there is no end or "peak" in sight. In a film that completely ignores the fate of petroleum after it has left the processing units, the future dissolves in the smoke of ignition fires of lesser symbolic density, at least for this director, than the gas fires across the Persian desert. There is nothing sacrosanct about filling up a tank. Because it does not pursue the economic shifts facilitated by the oil economy and prefers instead to suffuse the end of the oil route again in visual and rhetorical mysticism, the film ends exactly where it started—in the Persian desert—as if the maps of the world that oil helped reshape had never been fully unrolled.

We are never made aware of it, but Trejo is a foreigner in these parts, even though a journalist for the *Donau Kurier* conveniently addresses him in Italian when he recounts the history of Ingolstadt as a petrochemical global player. The foreign observer as a political participant belongs to the basic neorealist arsenal, but the usual setting in which he makes his appearance is violent, dilapidated, or otherwise productive for an ethical critique. In *Brutal Vision: The Neorealist Body in Postwar Italian Cinema*, Karl Schoonover argues that while classic neorealist films "seek to turn watching from a passive form of consumption into an activity replete with palpable geopolitical consequence," specifically through the "staging of bodily violence," the second generation of neorealist filmmakers (Bertolucci, Antonioni, Pasolini, and Federico Fellini) set out to "question the terms by which canonical neorealism negotiated Italy's international visibility."[32] Schoonover does not interpret explicitly transnational films; he rereads familiar works usually interpreted as national products and relocates them in the postwar geopolitics of international aid. Second-wave neorealists, he contends, question verisimilitude, especially in the representation of suffering, as a heuristic for political critique and engagement.

La via del petrolio records the lives of ordinary people along the oil route with neo-realist precision, using typical modes of neorealist spectatorship. Yet it does not break down the ethical assumptions of the neorealist movement, through which Italy intervened in the rise of a postwar global ethics of human sympathy. The reason for this failure is twofold. Bertolucci dispenses not only with corporeality but also with suffering itself. His oil workers look placidly into the camera as they talk about their families back in Italy. They are far from being crippled with pain and alienation. Oil, were we to regard it as a living being or at least in some way cognizant of its destination, can be said to gurgle away from the violent scene of its hydraulic extraction into certain annihilation: the refinery towers of Bavaria, the gas-station pumps, the ignition engines of automobiles, finally to energy-releasing destruction. So one might develop an empathetic relationship with oil itself, but of course we never actually see it. Instead, the film devises a transnational mode of vision rooted in its own form rather than in a grim reality, taking the oil route as a global Eucharist for a society united in an effort to stitch the wounds of war with pipelines. The body of history is thus the film itself; the reality depicted with realistic means is one in which the filmmaker vigorously partakes.[33] Corporate propaganda and poetry are therefore inextricably conjoined, and it seems impossible to reduce the film to one or the other. It shows, I think, that loosened from a "postwar politics of need,"[34] neorealism experiences a blockage caused by the inability to encode the dehumanized, largely invisible, and deterritorialized mechanics of energy—in other words, by an inability to imagine a world in the absence of empathy.

Italian neorealism showed "how national identity could be shaped and/or redefined by cinema" and it called "to action for national Italian postwar reconstruction" in a way that drew its strength partly from its break with the past of Fascist rule and also from its resistance to a growing supply of American styles and themes.[35] The Italian energy industry at this time played an identical role, promising Italy a place in the sun especially among Middle Eastern concessionaires against the monopoly of Anglo-Saxon companies. *La via del petrolio* stays aloft on a tightrope between these two vectors in postwar Italian social and political life. It is a paradoxical work in two ways. It enlists nation-building neorealist aesthetics in the service of a global homogenizing movement of people, resources, and culture. In doing so it disregards neorealism's key ideological project, that of promoting what the Italian political theorist Antonio Gramsci called a "national popular culture," at a remove from the highbrow sphere of traditional intellectualism. This disregard is evident in Bertolucci's erudition and the way it drowns out the voices of the petro-proletariat described in the film.[36] As a "poiesis," or poetic composition of history,[37] *La via del petrolio* illustrates not only how petroleum fits into established imaginaries (especially around class, labor, and transnationalism) but also how it can be captured aesthetically as a profoundly cinematic (albeit visually hermetic) subject, a way of seeing the world.

The futurity of oil has already received some attention.[38] Bertolucci plays on this notion as well by filming the tankers and pipelines not only with the gritty accuracy typical of postwar neorealism but also with the wondrous awe of a fantasy story harking back to cinema's beginnings as an art form. Other literary and visual representations of petroleum have already focused on its past, stressing oil's geological properties and especially the element of death and putrefaction without which it would not exist. Yet oil is not only a natural substance but also a vehicle of culture. Oil crosses the seas with a distinct cultural cargo in tow, its own aesthetic imprint of which only tiny fragments have so far been decoded. The film's final shots make it abundantly clear that we are to remain in the dark about the rest. The oil route concludes with a mysterious map of the stars—Bavarian refineries gleaming mysteriously in the night. We are back to the trick imaginary of Méliès and at the same time far into the future, where the journey of petroleum appears, and will continue to do so in the decades to come, as one of geopolitical incoherence, absurdity, injustice—ultimately a journey across the impossible.

NOTES

1. Pier Paolo Pasolini, *Petrolio* (New York: Pantheon, 1997), 24.

2. Ibid., 25.

3. The image of black crude served well as the polar opposite of the "white telephone movies" (trivial romantic comedies set in artificial sets) that proliferated during Mussolini's regime, because it instantiated material, everyday life in ways both very physical and politically charged. On the hunger of neorealism for such unglamorous, quotidian objects, see André Bazin, "What Is Neorealism?" in *André Bazin and Italian Neorealism*, ed. Bert Cardullo (New York: Continuum, 2011), 24.

4. On Ivens's unsuccessful efforts to appease the television producers and Mattei himself, see Hans Schoots, *Living Dangerously: A Biography of Joris Ivens* (Amsterdam: Amsterdam University Press, 2000), 262. Though little seen and so far overlooked in the context of oil cultures, the film compresses three key elements of the petro-imaginary, especially as it has manifested itself on film. It begins with a segment about the recovery of methane in the Po Valley (an area Antonioni had already mined with his 1957 film *Il grido*), entitled "Fire in the Po Valley." *La via del petrolio* similarly takes fire as its starting point, although of course fire is the ultimate destination of the journey rather than its incipit. Second, Ivens juxtaposes the olive trees on which rural families depend to the Christmas trees—the assemblies of valves and fittings used to control the flow of oil or gas out of a well. The switch from an agricultural economy to a heavy industrial infrastructure organizes most of the films around the oil industry that were released in the United States from the 1940s to as late as the 1970s. I have discussed a number of these films elsewhere; see Georgiana Banita, "Fossil Frontiers: American Petroleum History on Film," in *A Companion to Historical Film*, ed. Robert Rosenstone and Constantin Parvulescu (Malden, Mass.: Blackwell, 2013), 301–27. Finally, the third part of the film recounts the relationship of a woman with a North Italian oil-rig worker, providing a classic mythic plot that was adapted by

later films in which oil is not only an economic but also a libidinal force capable of catalyzing sophisticated emotional dynamics.

5. For a lengthy analysis of Bovay's work for ENI, see Giulio Latini, *L'energia e lo sguardo: Il cinema dell'Eni e I documentary di Gilbert Bovay* (Rome: Donzelli, 2011). Published with the support of ENI, the book includes an extremely valuable overview of petrocorporate filmmaking in the twentieth century, but the bibliographic detail is not matched by a strong critical perspective on the assembled material.

6. Scholarship on Bertolucci's early films, though not specifically on *La via del petrolio*—which is usually mentioned briefly, without extensive comment—has charted the extent of his indebtedness to neorealism at great length. See, for instance, Robert Phillip Kolker, *Bernardo Bertolucci* (London: BFI, 1985), 11–35.

7. The verbosity of the film must be seen in the context of the Italian neorealist tendency to deprioritize script, dialogue, and literary references. On this issue, see Angela Dalle Vacche, *The Body in the Mirror: Shapes of History in Italian Cinema* (Princeton: Princeton University Press, 1992), 5.

8. Enzo Ungari, *Bertolucci by Bertolucci* (London: Plexus, 1987), 45.

9. My concern doesn't lie with the transnational dimensions of neorealism per se or with its influence on international filmmaking. For an excellent overview of this influence, see *Italian Neorealism and Global Cinema*, ed. Laura E. Ruberto and Kristi M. Wilson (Detroit: Wayne State University Press, 2007).

10. Bertolucci confesses to the eulogistic subtext of the film's opening in an interview included on the DVD edition, *La via del petrolio. DVD. Con libro* (Feltrinelli, 2010).

11. On the rise of Enrico Mattei, his controversial, high-profile personality, and ENI's partnership with Iran, see Dow Votaw, *The Six-Legged Dog: Mattei and ENI—A Study in Power* (Berkeley: University of California Press, 1964); and Daniel Yergin, *The Prize: The Epic Quest for Oil, Money, and Power* (New York: Free Press, 2009), 483–87.

12. By 2013, Iran climbed into fourth place, behind only Russia, Saudi Arabia, and the United States.

13. All citations from the screenplay are based on the English film subtitles included in the Italian-language release.

14. I am referring here to Emmet Gowin, *Changing the Earth: Aerial Photographs* (New Haven: Yale University Art Gallery, 2002), which includes images of coal mines and petrochemical refineries in the Czech Republic, to Edward Burtynsky, *Oil* (Göttingen: Steidl, 2009), and also to Peter Mettler's documentary *Petropolis: Aerial Perspectives on the Alberta Tar Sands* (Dogwoof Pictures, London. 2010).

15. Mark Shiel emphasizes the urban contexts of neorealism in *Italian Neorealism: Rebuilding the Cinematic City* (London: Wallflower Press, 2006). Bertolucci avoids overtly urban settings and allows instead for a petro-city to emerge from the accumulation of petroleum infrastructure at various junctures in the film. The general effect of the film is urban, though neither Isfahan nor Genoa—let alone Ingolstadt—was a large metropolitan center at the time.

16. Bertolucci takes his cue here again from Antonioni's *Il deserto rosso*, where oil workers appear in typically neorealist scenes as the camera settles briefly on their faces, only to swing upward and remain fixed on a bare wall for minutes at a time. Their alienation is unmistakable, as John David Rhodes observes: "These workers are being plucked from an artisanal and ancient mode of work and trading in order to be inserted into the dynamised and more radically globalised

world (oil) economy." John David Rhodes, "Antonioni and the Development of Style," in *Antonioni: Centenary Essays*, ed. Laura Rascaroli and John David Rhodes (London: BFI/Palgrave Macmillan, 2011), 278.

17. In his reading of *Il deserto rosso*, Angelo Restivo sees in the sickly yellows of the gas flares emerging from the refinery towers around Ravenna the traces of capitalist contamination as it gradually envelops the landscape. Black-and-white film makes such chromatic political critique impossible, which supports my view that cinematic form acts as *La via del petrolio*'s chief political suppressant. See Angelo Restivo, *The Cinema of Economic Miracles: Visuality and Modernization in the Italian Art Film* (Durham: Duke University Press, 2002), 143.

18. Cited in John J. Michalczyk, *The Italian Political Filmmakers* (Rutherford, N.J.: Fairleigh Dickinson University Press, 1986), 108, 137.

19. "The oil is condemned to blindness. By the darkness of the geological structures, the huge shadows of the tanks, the tankers, the pipelines, and the refineries. The oil never sees the light of day." The blindness is also that of the people who fail to notice the "huge shadows" of oil infrastructure in their daily lives and refuse to wonder at the mystery of it.

20. The visual contrast points indirectly to the so-called resource curse, to the opposition between the archaic, underindustrialized backgrounds of oil extraction and the blinding light of the riches it suddenly bestows on unsuspecting natives of nations like Iran. The barely felt tensions of the film turn on this gauzy, enchanted notion of the genie in a bottle, the energy it unleashes, and the question of what aesthetic strategies may recapture it on screen.

21. A more international, politico-historical dimension takes over in the film's focus on the simultaneous discovery of oil in Iran and Texas. The face-off that this intriguing link promises never fully materializes and remains in that sense a testament to the film's slightly disorienting location just short of the threshold that would have made its entire cargo blow up (the film was released six years before the oil crisis). Watching *La via del petrolio* at a time when talks on the future of Iran's nuclear program are ongoing and sanctions meant to curb the country's oil exports are still vigorously implemented, we cannot help feeling somewhat deflated.

22. Fernando Coronil, *The Magical State: Nature, Money, and Modernity in Venezuela* (Chicago: University of Chicago Press, 1997), 4. Although it could be argued that only through the mediating power of Mattei's ambitions could postwar Italy gain international independence and economic leverage in the oil market, the film premises the constitution of the Italian state on the global decentralization of oil wealth.

23. On the role of the Suez crisis for the production and reception of *Lawrence of Arabia*, see Steven Charles Caton, *Lawrence of Arabia: A Film's Anthropology* (Berkeley: University of California Press, 1999), 177.

24. An analysis of Méliès's films that focus on the sustainability of his aesthetic strategies can be found in Simon During, *Modern Enchantments: The Cultural Power of Secular Magic* (Cambridge, Mass.: Harvard University Press, 2002), 167–73.

25. Siegfried Kracauer, *Theory of Film: The Redemption of Physical Reality* (Princeton: Princeton University Press, 1997), 33.

26. For a foundational theorization of this concept, see Jennifer Wenzel, "Petro-Magic-Realism: Toward a Political Ecology of Nigerian Literature," *Postcolonial Studies* 9, no. 4 (2006): 449–64.

27. These works can be categorized as neorealist because of their sociological impetus and attention to how the common man interacts with petroleum culture. In later years, Italian films

featuring oil-related plots came to be more closely inscribed within genre conventions. See, for instance, Ruggero Deodato's cannibal exploitation film *Ultimo mondo cannibale [Last Cannibal World/Jungle Holocaust]* (1977), in which a group of oil prospectors crash land on their way to the Philippines and are hunted by a primitive cannibal tribe. In 1971, Brigitte Bardot and Claudia Cardinale starred in the comedy Western *Les pétroleuses [Frenchie King]*, where the two buxom superstars—somewhat past their prime at this stage—fight over an oil ranch in the tradition of the Italian spaghetti western. The photomontage production *Metropia* (2009) tries to extract from the properties of its visual aesthetic a way of looking in a world where natural resources have run out: the drab, somber visuals portray a pan-European subway system in a postpetroleum world (the precise location here is Sweden). When European filmmakers turn to sites of oil extraction overseas, they come up with unrealistic plots of cannibal natives and ravishing beauties in a crude catfight. More qualitative stories have been told about the oil industry at European locations and their effects on the lives of individuals. Although its oil theme appears expository and at best vaguely symbolic, the rig in particular has become a dense metaphorical and visually stylized cinematic trope. Significant examples include Lars von Trier's *Breaking the Waves* (1996) and Isabel Coixet's *The Secret Life of Words* (2005), one a Dogma 95–inspired film about the conjugal life of an oil-rig worker off the coast of Scotland and his religious wife (the rig makes a striking appearance in an especially pious scene), and the other a somber drama that unites an oil worker with a rape victim of the Bosnian War on an offshore rig.

28. The relationship of people and this underground oil river recalls the obscure symbiosis that links the inhabitants of the resource-packed Po Valley with its ancient river. "In what does this feeling concretize itself," Antonioni writes, "we know not; we do know that it is diffused in the air and that it is sensed as a subtle fascination." Michelangelo Antonioni, "Per un film sul fiume Po," *Cinema* (April 25, 1939), quoted in Noa Steimatsky, *Italian Locations: Reinhabiting the Past in Postwar Cinema* (Minneapolis: University of Minnesota Press, 2008), 2.

29. The origins, legacy, and failures of the ECSC are the subject of John Gillingham's excellent study, *Coal, Steel and the Rebirth of Europe, 1945–1955: The German and French from Ruhr Conflict to Economic Community* (Cambridge: Cambridge University Press, 1991).

30. For an analysis of how the imaginary of petroleum inflects the visual and psychosexual dynamics of Antonioni's *Il grido [The Cry]* and *Il deserto rosso*, see Georgiana Banita, "Antonioni's Ölmalerei: Vom unbewussten Rohstoff einer materiellen Filmästhetik," in *Michelangelo Antonioni: Wege in die filmische Moderne*, ed. Jörn Glasenapp (Munich: Fink, 2012), 153–81. While I remain convinced that Antonioni's and Bertolucci's formal experimentation and sheer fascination with petroleum effectively vaccinate them against ecological concerns, recent interpretations of *Il deserto rosso* in particular project on the film a somewhat anachronistic, proto-ecological interest in climate change. See, in particular, Karen Pinkus, "Antonioni's Cinematic Poetics of Climate Change," in *Antonioni: Centenary Essays*, ed. Laura Rascaroli and John David Rhodes (London: BFI/Palgrave Macmillan, 2011), 254–75.

31. For further information on how Ingolstadt became one of Bavaria's greatest petrochemical hubs, as well as on the new pipeline from Trieste that would open after Bertolucci completed his film (the Transalpine Pipeline), see James Marriott and Mika Minio-Paluello, *The Oil Route: Journeys from the Caspian Sea to the City of London* (London: Verso, 2012), 313.

32. Karl Schoonover, *Brutal Vision: The Neorealist Body in Postwar Italian Cinema* (Minneapolis: University of Minnesota Press, 2012), xvii, 186.

33. André Bazin astutely pointed out that neorealism, in spite of its commitment to authenticity, remained profoundly aesthetic: "The real like the imaginary in art is the concern of the artist alone. The flesh and blood of reality are no easier to capture in the net of literature or cinema than are gratuitous flights of the imagination." Bazin, "Cinematic Realism and the Italian School of the Liberation," in Cardullo, *André Bazin*, 37.

34. Ibid., 203.

35. "Introduction," *Italian Neorealism and Global Cinema*, ed. Laura E. Ruberto and Kristi M. Wilson, 3.

36. See Antonio Gramsci, *An Antonio Gramsci Reader: Selected Writings, 1916–1935*, ed. David Forgacs (New York: Schocken Books, 1988), 365–66.

37. Keala Jewell proposes this term to characterize Bertolucci's generic experimentations and the ideological conflicts they inscribe. Keala Jewell, *The Poiesis of History: Experimenting with Genre in Postwar Italy* (Ithaca: Cornell University Press, 1992), 4.

38. See Daniel Worden's contribution to this volume (chapter 6).

PART III

The Local and Global Territories of Oil

9

Aramco's Frontier Story

The Arabian American Oil Company and Creative Mapping in Postwar Saudi Arabia

CHAD H. PARKER

In the late 1940s and 1950s, the Arabian American Oil Company (Aramco) became intricately involved in Saudi Arabian territorial disputes. The story of Aramco and the Buraymi Oasis, located on the eastern borders of Saudi Arabia, was one manifest in everything from border skirmishes to high diplomacy, but also one that included creative storytelling and the selective use of the past for corporate and national ends. Company story makers operated outside the firm's public relations realm in aid of the Saudi monarchy in its border dispute with the British protectorates of Oman and Abu Dhabi. Using history as a tool of border construction, the Arabian American Oil Company attempted to alter the past to secure its present and expand its future. To protect its privileged position as the sole oil company operating in the kingdom, Aramco representatives cautiously supported Saudi claims, pressured the U.S. government, and dissembled the British government. Through the use of historians, anthropologists, and public relations officials in the company Relations Department and through the hiring of lawyers to serve the Saudi government, Aramco boldly charted a diplomatic path of its own and ensured its place as the sole oil company operating in Saudi Arabia.

States are not the only actors engaged in high diplomacy. Nongovernmental actors' involvement in foreign relations, however, remains an understudied historical phenomenon. More recently scholars have begun to assess nonstate actions more critically.[1] Aramco's cultural and diplomatic endeavors in Saudi Arabia provide scholars an opportunity to explore corporate involvement in international affairs in new ways, illustrating more clearly the significant impact oil companies had in driving U.S. interests in the Middle East. No study of mid-twentieth-century U.S.–Saudi relations can be complete without assessing the role of transnational oil corporations.[2] While some historians like Akira Iriye hesitate to include profit-seeking enterprises among nongovernmental entities,[3] Aramco wielded enormous power that allowed it far-reaching influence in Saudi Arabian affairs. The company's placement in what Daniel Yergin refers to as the "postwar petroleum order" that emerged out of World War II established what diplomatic historians call a corporatist framework in U.S. foreign oil policy. After failed wartime attempts to acquire a controlling share of oil interests in Saudi Arabia, construct an Arabian pipeline, and orchestrate Anglo-American management of Middle Eastern oil, the United States, historians have shown, turned to private enterprise to lead the way.[4] Aramco operated in Saudi Arabia like a state, and it included among its employees a host of scholars who worked to define the company's presence as progressive through an exceptionalist narrative, thereby expanding its command of the nation's natural resources.

Narrative serves as a powerful tool to a corporate entity interested in expanding its presence along the frontiers of the Rub al-Khali, or Empty Quarter, as the Americans called the southern Arabian desert, and narrating the new Saudi state into the distant past. By reimagining the historical connections between the land and the people living on it, Aramco's historians were able to recast widely held notions of territorial claims and compel the great powers to grapple with previously "settled" issues.[5] This new and improved narrative served, as James Ferguson phrases it, like a "blueprint [laying] down fundamental categories and meanings for the organization and interpretation of experience."[6] When the company's own publications referred to the "Aramco story," they placed the company, its employees, and the Saudi Kingdom into a carefully constructed narrative that promoted partnership and the company's role in Saudi development.[7] Through support of Saudi Arabia's new territorial claims, the oil company participated in the politics and diplomacy of border creation to secure future oil fields and placate the Saudi monarchy.

As part of its narrative construction, Aramco changed its name in 1944 from the California Arabian Standard Oil Company (CASOC), adopting a moniker that represented to the company a "more appropriate title." It amplified the notion of an Aramco–Saudi "partnership" emerging as a foundation for company relations with the Saudi monarchy.[8] The significance of the name change at the time seemed relatively trivial to American consular officials in the area, but Aramco officials believed it better symbolized their operations.[9] The company's employee handbook, first published in 1950, helped enmesh American employees in this corporate narrative.[10] The "Miracle of American Production"—not far from the Fertile Crescent—relied on American employees who would help deliver civilization to Saudi Arabia. The

"Arabian-American Partnership" required it.[11] Employees represented much more than a slice of the company; they symbolized, the handbook stressed, "the *American* way of doing things." It was important that employees recognized their "responsibility to the American nation."[12] Aramco's connection to economic development factored prominently in the corporate identity. The "modern" American oil business was expertly juxtaposed with the "ancient" oil business of the Middle East dating as far back as 4000 BCE. "The adventurous, restless energies typical of Americans," brought technological advancements in the late nineteenth and early twentieth centuries. The company's work could be measured "by the clock of the centuries" as American values replaced ancient traditions in the Eastern Province, transporting Saudi Arabia into the twentieth century.[13] Aramco symbolized finished development. Saudi Arabia symbolized a nation in transition.

The Arabian American Oil Company carefully crafted its self-image as a partner in Saudi nation-building from the beginning of its concession in the early 1930s through the early 1970s. A positive relationship with the monarchy required building and maintaining goodwill, and while one might argue that this mythmaking served primarily to mask exploitative corporate behavior, the narrative was central to Aramco's interaction with the Saudi monarchy.[14] The company used its massive public relations arm to construct a narrative of Saudi Arabian tradition and to situate the company as a necessary partner in Saudi nation-building. Aramco's representations of Saudi Arabia and itself through various media demonstrated the company's imagined and constructed role in the Arab world. The monarchy, in turn, encouraged this narrative as it sought to consolidate its rule and assert ever-expanding land claims.[15]

The Arabian American Oil Company's public relations arm often boasted in the 1940s that the oil giant was a "unique organization." And in many ways, it was. Aramco, alone, dominated Saudi oil exploration and made a more substantial impact on life in Saudi Arabia's enormous Eastern Province than any other Western institution, including the U.S. government, in the mid-twentieth century. This distinctiveness, expressed in corporate public relations, implied much more, however. It spoke to a self-defined corporate agenda of enlightened capitalism. Much of the energy behind this legend stems from the company's Government Relations Organization, which consisted of divisions for government affairs, local affairs, and research and translation. George Rentz, historian and Arabist with a PhD from Berkeley, led the department from 1946 to 1963 and hired similarly enthusiastic men, significantly the anthropologist Federico Vidal, who penned numerous regional studies for the company and the kingdom, and William Mulligan, who wrote about the company for the corporate magazine and other publications. These men conducted extensive research and produced a bounty of reports on Saudi Arabia and the corporation, carefully crafting a story of corporate enlightenment, one that served multiple objectives. It provided the foundation for the company's engagement with the monarchy, whose continued backing Aramco required to exploit the kingdom for its riches. To fashion and promote its image, the company submitted simplified annual reports to the monarchy and shareholders and published an employee magazine and official histories of both the company and Saudi Arabia. But the project did more than position the company rhetorically alongside the monarchy. It elevated Aramco's status as U.S. representative in the kingdom, entangling it in both local politics and international diplomacy, including Saudi boundary claims on the eastern frontier.[16]

The oil company worked alongside Saudi Arabia's king 'Abd al-Aziz ibn Sa'ud as he turned to science and technology to expand his authority. It was a nation-building dependent on new technologies for communication, transportation, and administration, and Aramco situated itself as an indispensible partner in this social and economic development.[17] To promote the "distinctly American flavor" of Saudi oil, as company publications put it, Aramco constructed its own frontier epic.[18] After initial contact in the deserts of the Eastern Province, geologists began their search for oil in 1933.[19] With the help of local Bedouin, who acted as guides, American oilmen confronted an ancient, backward, society as they probed the desert for riches. Soon after the discovery of oil in marketable quantities, American oilmen became guides themselves, ushering the kingdom into the twentieth century through various development schemes. Petroleum engineers, construction experts, doctors, agricultural specialists, and more began erecting transportation, communication, and industrial infrastructure, transferring not just objects, but the ideas that came with them. This story that Aramco composed culminates with the successful transfer of new Western values. The Saudi Arabian government emerged as a "modern" nation-state, institutionalizing new technologies and ideas and eventually taking control of oil operations. While nationalization of oil operations in other parts of the world meant conflict, Aramco's transfer to the Saudi Arabian government proceeded smoothly, owing in large part to the mutually respectful relationship forged by corporate executives for decades. The story is compelling, for sure, but the image of a corporation with a soul shepherding an ancient civilization into the modern age obscures much of the exploitation occurring on the Saudi frontier, and it ignores the powerful corporate interests at work.[20] Aramco constructed its world through an allegory meant to position the company as a partner in Saudi nation-building, helping promote it as the catalyst for development in Saudi Arabia.[21] This project found partial success, as the company indeed acted as the primary diplomatic representative of the United States for years, but also it demanded constant vigilance, requiring that corporate agents act on behalf of the burgeoning state. With the outset of the Cold War and evolving

U.S. interests in the Middle East after World War II, Aramco's private diplomacy faced challenges on many fronts, and ultimately it surrendered some of its influence to the U.S. government.

Aramco's story has drama, ample explanation, and a certain degree of exceptionalism, all in the service of institution building. This sort of story is what Margaret Somers refers to as an institutional narrative, which is a powerful force in identity construction.[22] Like any exceptionalist account, Aramco's is selective; certain aspects of the past go unreported. While the link between particular stories and identity might not seem readily apparent, the function of this classic tale is to set actors, in this case Aramco's American employees, into an established mythic teleology.[23] But by establishing corporate benevolence as a core of the story, Aramco found itself working toward Saudi ends in order to protect its privileged place as the sole oil company operating in the kingdom.

Saudi Arabia's king 'Abd al-Aziz surprised nearly everyone in 1949 when he declared the Buraymi Oasis, located in the border regions in the kingdom's southeast, part of his domain, but Aramco remained a steadfast supporter nonetheless. On October 14, the Saudi government officially demarcated its eastern borders, about one hundred miles east of British counterclaims. These newly proposed boundaries represented quite an expansion from those drawn in 1935, and they caught Aramco off guard. The company had prepared maps for a 1948 report on oil in the Middle East that placed Saudi Arabia's eastern boundary far west of the new claims.[24] From that point forward, however, the company's public relations team made Saudi Arabia's past present through studies compiled by anthropologist Federico Vidal and Arabist George Rentz. Aramco accepted the new Saudi claims without question, and it worked to prepare the Saudi government for making legal claims. Vidal penned the study *The Oasis of Al-Hasa* in the early 1950s as part of an attempt to understand the people of the nearby al-Hasa oasis. As company operations moved south of Dhahran into the kingdom's interior, through the construction of railroads and new oil fields, it was clear that construction material, labor, agricultural supplies, oil contract work, and other supplies and services would need to come directly from the region. By studying al-Hasa, Vidal hoped to measure the effects the company was having on oasis residents and area Bedouin, and more important, to gauge whether local residents were prepared to enter into employment with Aramco.[25] The study built on an earlier work on the region produced by George Rentz and William Mulligan, which outlined the Saudi influence on the eastern frontier. Rentz's subsequent study of Oman addressed this subject again, this time with a more specific interest in boundaries. Here, he detailed the status of territory that he claimed remained "undefined" or, he said, under control of neighboring governments. While much of the region had been "defined" and had been controlled by others for some time, this study, not incidentally, helped strengthen the Saudi monarchy's claims on territory outside the kingdom.[26]

Using reports from travelers and other evidence, Rentz placed the contested Buraymi Oasis well within Saudi authority, arguing that the British protectorates of Oman and Abu Dhabi had little historic claim to the region. The dispute ignited a conflict between Saudi Arabia and Great Britain, which hoped to preserve its imperial posture on the Arabian Peninsula. British foreign-office officials quarreled with much of Rentz's research, arguing that he misled readers about the nature and extent of control exercised by central Arabian leaders.[27] Meanwhile, Aramco executives played a difficult game in Saudi Arabia, attempting to pressure the U.S. government, mollify the British government, support the Saudi monarchy, and protect its concession. Aramco's selective use of the past indicates the degrees to which corporate diplomacy extended into state and identity formation on the Saudi frontier in the late 1940s and 1950s.

The story of the Anglo-Saudi dispute over the Buraymi Oasis is well known and expertly covered by historians who evaluate the affair according to the political and diplomatic interests of states. Nathan Citino argues that Buraymi was not about expanding oil resources, exactly, but about fundamental differences in the U.S. and British governments' foreign oil policies and "historical experiences in the Middle East." The dispute pitted numerous actors against one another, all with different interests, approaches, and histories in the region. The U.S. and British governments argued from different historical vantage points. Britain maintained a longtime imperial presence in the Middle East, where the United States had no formal historical ties. And while Britain conducted its oil diplomacy through state apparatuses, the Americans relied more on corporate leadership. During the Buraymi dispute, the United States hoped to improve relations with the Middle East and prevent Soviet advancements in the region as the Cold War heated up, whereas the British hoped to retain its presence in the region and standing as a world power. The Saudis had a deep distrust of the British, which supported rival Arab clans for leadership positions in neighboring states. In making these new claims, the monarchy hoped, in part, to dispel any appearance of complicity with Western imperialism in the Middle East, to its close ties to the Americans, who looked suspicious in an era of emerging Arab nationalism. For the states involved in the conflict, Citino maintains, the dispute was less about oil than international and regional politics.[28] Similarly, Tore Tingvold Petersen sees the dispute as a competition between growing American economic interests in the Arabian Peninsula and fading British imperial interests there.[29] From a corporate perspective, however, it was absolutely about oil and power. Aramco's actions during the height of the dispute indicate an interest in securing its place as the sole exploiter of Saudi oil reserves.

The Buraymi Oasis is situated in the east of the Arabian Peninsula near the south-ern border of the present-day United Arab Emirates and the western border of Oman. To the south and the west is the Rub al-Khali, the Empty Quarter as the desert was known in the West. In the 1950s, Buraymi consisted of eight villages, about six miles across, circling its water wells. Settled farmers in the area mainly grew dates, with some fruit and other grains grown as well, and pastoral farmers (mainly Bedouin) raised camels, goats, and sheep. The scarcity of vegetation prevented much more eco-nomic activity than that. For centuries, the oasis served as a strategic outpost and a point of departure for armies staging invasions to the east. And while no oil had been found in the region by 1949, expectations remained high.[30]

The Saudi monarchy's move in Buraymi stirred international alarm, and Aramco quickly positioned the company alongside the Saudi Arabian government as a pro-tector of Saudi interests and a potential guarantor of American protection. Saudi sol-diers and company surveyors descended on the border regions to the east as early as the spring of 1949. Accompanying this team were officials from Aramco's Relations Department. Conducting operations like geological reconnaissance, water testing, and data compilations on the natural and cultural aspects of the region took up most of the relations men's time.[31] This seemingly innocuous activity, however, troubled the British, who on April 21, 1949, dispatched two armed soldiers along with Patrick Stobart of the Foreign Service, a brother of the Shaikh of Abu Dhabi, and an employee of the Iraq Petroleum Company (IPC), jointly owned by British and American firms. Stobart "rather officiously" announced to Paul Combs, a company driller, that the Saudi-American party was in Abu Dhabi territory and must leave. Combs encouraged him to meet with "company men" and Saudi officials at the base camp, at which point the encounter became heated. Saudi soldiers disarmed the two guards, revealing that the guards carried loaded weapons, heightening tensions even further. Unhappy, Sto-bart exclaimed, "This is going to cause a frightful stink!"[32] While this British demon-stration of power on the Saudi frontier appeared weak, before departing Stobart handed Aramco employee Don Holm a note declaring the trespass and advising all parties to leave "before any possible incident occurs between the Saudis and the Subjects of Abu Dhabi."[33] The political problems this could generate would have been quite clear to the Aramco surveyors and relations men, who must have known the oasis's recent history. As late as 1945, the oasis served as a meeting place for the sheiks of Trucial Oman and the British government to discuss ongoing wars between the Sheik of Abu Dhabi and the Sheik of Dubai.[34] This encounter would be the opening salvo in a long dispute that always had the potential to become very violent.

The Saudi Arabian government agreed to enter negotiations regarding its eastern boundaries but insisted that the regions' historical social and economic practices be the basis for talks. The British rejected this stipulation from the start, noting two problems. First, nomadic tribal boundaries made defining territorial boundaries almost impossible. Second, contemporary politics heightened the importance of the area. A British subsidiary of the IPC, it seems, had already begun explorations in the region, angering the Saudi king.[35] Despite these differences, the parties met in Riyadh beginning on August 30, 1949. The Saudis confirmed the extent of their claims, deep within the territory of Abu Dhabi, far too deep for either the sheikh of Abu Dhabi or the British government to accept. Moving forward, arguments centered on the starting point for negotiations, often based on various British treaties signed over the preceding decades that had established the eastern frontier.[36]

The dispute worried American oil executives, who consistently promoted the Saudi position. Manley Hudson, an Aramco lawyer who served the monarchy in its legal disputes, argued to U.S. State Department officials that attempted British intimidation of the Saudis frustrated relations more than anything. The monarchy hoped to negotiate with regional sheikhs, not the British, who the king argued had no legal authority. Many times in 1950, Aramco pushed this line of reasoning, hoping Washington would support Saudi claims.[37] The following February, 'Abd al-Aziz formally requested Aramco's assistance in preparing his legal claim on the region. In particular, the king asked for research specialist and historian George Rentz. Aramco began planning a fact-finding mission, which would include studies of geographical features, tribal considerations, and tax collection.[38] While a year earlier, Hudson stated clearly that "tribal allegiances and the extent of their ranges are too conflicting or indefinite to offer a fully reliable basis for judgment," that was exactly how Rentz and his aids chose to support the Saudi case. Apparently, Hudson's position evolved, signaling corporate thinking, as he later noted that Saudi tax collection in the disputed territory provided sufficient legal claim.[39] This position is important when one considers the meaning of tax collection, historically, on the Saudi frontier.

The first of the Saud family to emerge as a regional leader in the eighteenth century, Muhammad ibn Sa'ud, employed *zakat*, an Islamic tax paid to the Muslim community's leader, as a way to centralize and consolidate his authority in central Arabia. Originally part of the duty of Muslims as one of the pillars of Islam, it eventually became part of permanent administrative governance, and it is a tradition that survived into the mid-twentieth century.[40] A wealthy landowner and local emir, Muhammad ibn Sa'ud's financial backing of commerce and protection of area tribes provided him a position of status and allowed him to collect tribute, or tax, from residents in the region. To enhance his political standing, ibn Sa'ud allied with Muhammad ibn 'Abd al-Wahhab, an Islamic reformer who provided the religious legitimacy to Saud's political authority. As the Wahhab–Sa'ud alliance expanded control in the mid-eighteenth century, tribes continued to show fealty to the emir through the payment of *zakat*. It was, historian Madawi Al-Rasheed notes, "a token of political submission." Taxes and Wahabbi religious austerity helped consolidate local authority in an era when borders meant very little. 'Abd al-Aziz continued this approach in the early twentieth century as he expanded his reach in central Arabia, and *zakat* emerged as one of his most valuable revenue sources before the discovery of oil. This system of taxation served as a key mechanism for how 'Abd al-Aziz centralized and consolidated his rule in what would become Saudi Arabia in 1932, as taxes could then be redistributed to maintain alliances with outlying tribes.[41] Since taxes paid by tribes defined allegiance on the Saudi frontier for centuries, it is no wonder it was such an important part of the Saudi claim on the Buraymi Oasis.

Whatever the historical basis for Saudi claims, the monarchy also hoped the U.S. government might help coax British acceptance of the Saudi position. In a 1950 letter of friendship to 'Abd al-Aziz, President Harry Truman noted his satisfaction that American private enterprise was working to develop the kingdom and added that "the United States is interested in the preservation of the independence and territorial integrity of Saudi Arabia. No threat to your Kingdom could occur which would not be a matter of immediate concern to the United States."[42] Nearly three years later, the king and his son, Crown Prince Saud, evoked this letter to the new secretary of state John Foster Dulles when he arrived in Riyadh as part of a Near East trip. They demanded American support. Dulles maintained an aloof posture, however, stating that it was not his place to dictate to the British and, furthermore, the dispute was not existential but rather just a problem of boundaries. He maintained that Saudi Arabia was not under attack.[43]

In this environment, talks between the Saudis and British proceeded slowly with tensions often running high. The British called for international arbitration and the Saudis requested either U.S. mediation or submission to the United Nations. The Americans shuttered at a mediating role between two allies, one with oil and the other a key to its regional and global security interests, and they were anxious about the dispute going to the Security Council, where State Department officials worried the Soviet Union might score a propaganda victory against Western imperialism. Additionally, the United States had been unfavorable to arbitration, seeing it, too, as an opportunity for Soviet propaganda, but it soon became the most agreeable option. With no clear end to the dispute in sight, in August 1952 'Abd al-Aziz sent Saudi emir Turki bin 'Abdullah ibn 'Utayshan with a forty-man occupying force to the Hamasa settlement within the Buraymi Oasis, escalating tensions further.[44] The British responded with a blockade, hoping to prevent resupply of Turki's party, and they began to discuss military options. 'Abd al-Aziz still held out hope for American mediation, but the United States only offered to request that the British lift the block-ade and discontinue provocative Royal Air Force (RAF) overflights. The British agreed to end the flights, but local forces loyal to the Sultan of Muscat mobilized soon after, as did more Saudi troops. Meanwhile, Turki, it seemed, had been bribing local tribes since his arrival in August. The British, for their part, played a similar game, reportedly sending a Desert Locust Control Mission to the Oasis from nearby Sharja.[45] While the British continued to push arbitration, the Saudis now introduced a new plan: a plebiscite of local residents, something the king believed he would win handily. Britain and Saudi Arabia eventually reached a standstill agreement on October 26, 1952, hoping to ease tensions for a time while a permanent settlement could be reached. But this too caused conflict, as Turki used the calmer atmosphere to continue offering bribes in return for support of Saudi claims.[46]

Buraymi's importance to British imperial policy expanded rapidly in the early 1950s. This was the start of a difficult period for the British in the Middle East, which would end in the surrender of regional hegemony to the United States. The refusal of the British government–controlled Anglo-Iranian Oil Company (AIOC) to negoti-ate a fifty-fifty profit-sharing agreement on oil revenues with the Iranian government, coupled with the rise of Gamal Abdel Nasser and Arab Nationalism in Egypt that led to the Suez Crisis, seriously challenged the British position in the Middle East. With Buraymi came a possibility to secure future oil rights—something the British refused to give up on easily—while maintaining a degree of international prestige.[47]

In defense of its own strategic interests, the U.S. government privately sided with the Saudi monarchy, but publically it came to support arbitration as a time-honored—and probably the best—diplomatic solution. Clearly Washington's interests remained divided. Aramco held concessionary rights in Saudi Arabia, but other U.S. compa-nies, including two of Aramco's parent companies, held rights through British syndi-cates, including the IPC, the Bahrain Petroleum Company (Bapco), Superior Oil Company, and the Kuwait Oil Company.[48] Further, U.S. State Department officials remained unconvinced by Saudi arguments, but they warned embassy officials not to share this information with the Saudi monarchy.[49]

Saudi Arabia and Great Britain finally agreed on arbitration on July 31, 1954, but tension remained high. The agreement called for an Arbitration Tribunal to define a permanent boundary between Saudi Arabia and Abu Dhabi. The committee was to have five members, one nominated by Britain, one by Saudi Arabia, and three with no ties to either. Both agreed to halt expansion of oil operations beyond the Aramco and IPC teams already in place.[50] Additionally, each side was to remove its forces, except fifteen soldiers.[51] A change in government in London, however, altered the British approach. New Prime Minister Anthony Eden looked to take a tougher line than his predecessor Winston Churchill, as British power waned in Iran and Egypt.[52] So when arbitration began the next summer, it ended almost immediately after the British delegate resigned the tribunal in disgust over alleged Saudi tampering. The British charged the Saudis with bribing local leaders, improperly influencing the Pakistani delegate to the tribunal, and plotting a coup in Abu Dhabi. While the first charge is true and the second likely, there is scant evidence of the third.[53] But as a result, arbitration fell apart, and on October 16, 1955, the new British government seized the oasis with force, declaring a new border unilaterally.[54]

The military occupation did little to ease tensions, and it frustrated U.S. officials, who hoped the British would see that American oil interests benefited the entire West. The British had no intention of relinquishing their position, one Prime Minister Eden maintained was enormously important if Britain were to remain a regional power. It did not stop American Pure to return to arbitration, and events elsewhere in the Middle East soon took precedence for the British, as the Suez Crisis heated up. Regardless of their differences on Buraymi, Anglo-American relations before Suez seemed positive, except for one issue. A July 1956 British policy paper called Aramco "the greatest obstacle to Anglo-American harmony in the Middle East."[55] Anglo-Saudi friction remained for years to come. Saudi support for rebels in Oman in 1957 led to further military incursions by the British; meanwhile American oil executives assured U.S. and British officials that they were uninvolved.[56] Diplomatic relations were restored in 1963, with plans to settle the Buraymi dispute once and for all.[57] Over a decade later, Saudi Arabia finally signed a treaty with the United Arab Emirates that defined the eastern borders of Saudi Arabia.[58] Buraymi would not be part of Saudi Arabia.

Aramco's role in the dispute, while seemingly small, speaks volumes to its corporate diplomacy. The company, to be sure, maintained a distance from the dispute early, claiming it simply served at the pleasure of the Saudi monarchy. George Ray, a lawyer for the company, declared that oil explorations done on behalf of the king were about pleasing him, not corporate political interests.[59] If the Saudi government wanted an area explored, the company argued, it was wise to do so. In 1952, Aramco published Rentz's study of the southern and eastern borders of Saudi Arabia. Much like the corporate image making before it, Rentz used history to situate the company as a diplomatic partner, this time through boundary making. Aside from providing readers with a description of the region, Rentz himself stated in 1951, "This study is an attempt to bring to light the truth regarding areas where boundaries on the mainland happen to be in dispute." Further, he hoped it would "be of value to those who are actively engaged in the endeavor to settle the boundary problems that now exist."[60]

Rentz placed the Buraymi Oasis well within Saudi authority, claiming that the Trucial sheikhs, and therefore the British, had no historic claim to the region, and his history provided the basis for the Saudi arguments a few years later at the aborted arbitration. A look at the arbitration documents illustrates how the company and monarchy defined the disputed region's future by looking to a constructed past. The Saudi position rested on its claim that Buraymi was a Saudi possession for centuries, that the Saudi family protected the oasis over that time period, that when family control "lapsed," the region fell into disorder, and that effective rule had not been established by another power. Saudi Arabia argued that the British had not provided anything positive to the area and that the people wanted to affirm their connection to Saudi Arabia. Saudi Arabia, furthermore, has experience governing nomadic peoples, and the Saudi State respected the Bedouin way of life. The British saw it as "savage."[61] The fact that 'Abd al-Aziz himself looked to settle Bedouin to help consolidate his rule as far back at the 1920s and that the Saudi government continued to devise settlement programs during the dispute failed to make it into the "official" history presented by the government.[62] A true understanding of area tribes and "traditional loyalties," the Saudis argued, is almost unknowable to most Westerners who must visit the region to appreciate Bedouin culture. There was, the argument went, a centuries-old cultural, religious, political, and economic unity between Saudi Arabia and the Buraymi Oasis.[63]

Rentz and the Saudi government argued that Buraymi's link to the Saud dynasty dated to the mid-eighteenth century. The history offered for arbitration argues that the Arabian Peninsula's past is one of nearly consistent Saud family reign, since Muhammad ibn Saud in 1765.[64] This modern Saudi government in the making, as it is described, well before such states would have been in existence, established residence in the disputed region and collected taxes from area tribes. The Saudis argued, "The continuity of the Saudi State has been preserved intact during the past two centuries," remaining "in the hearts of the people," even when formal control eroded.[65] The fact that no Saudi state of any kind existed until the twentieth century seems an inconvenience, at best. This historical reality was ignored as the oil company constructed a more usable past for the monarchy's diplomatic maneuverings. By the twentieth century, Rentz argued, Buraymi was all but independent, if somewhat tied to various rulers along the Trucial Coast. Rentz marshaled evidence from travelers who described the oasis as independent and others who suggested the region relied on the Saudi monarchy for protection and paid tribute to Saudi tax collectors into the 1930s. With Rentz's work, Aramco bolstered the Saudi monarchy's claims by providing historical foundations of continuing influence in the region and the basis of the Saudi legal claim at arbitration.[66] Rentz's history, however, embellished the nature and extent of Saudi control and permanence on the Arabian Peninsula. In fact, most histories of the region placed the oasis outside Saudi control.[67]

Aramco executives attempted to distance the company from the dispute publicly, but they maintained their interest in the possibility of more oil resources. Britain and Saudi Arabia began working on the organization of an arbitration agreement, one that allowed for continued oil exploration by the IPC. This stipulation worried Aramco executives.[68] Company vice president James Terry Duce stated Aramco's position that these activities actually exacerbated the dispute, and that the company had not entered the region since 1949.[69] He continued to assert that the company had no interest in the area and had no intent to search for oil there. The company's holdings, he asserted, were "already sufficiently large."[70] But as both U.S. and British diplomats knew, Aramco consistently protected its concession and its role as the sole oil company in Saudi Arabia, neither of which it planned to relinquish easily.[71] Duce told British and American diplomats what they wanted to hear, while at the same time he assured the Saudi monarchy of the company's real intention to support its claims and produce Saudi oil. When asked about Duce's statements to the British, Vice President Floyd Ohliger claimed ignorance. He stated emphatically that Aramco would not surrender any concession rights and, moreover, would be delighted to explore Buraymi for oil.[72]

Aramco's attempts to define the eastern frontiers of Arabia were a result of two of the company's primary interests: extending its concessionary rights and providing diplomatic support to the Saudi monarchy. While Duce and Ohliger acted as double agents, of a sort, they consistently voiced the company's support for Saudi interests to the monarchy. The identity of the company as an independent actor—with American ties—could not be questioned following the Buraymi dispute. The foundation for this identity emerged through a complex set of actions and representations, all of which served to establish the company as a partner in Saudi nation-building. The dispute, however, shed light on an emerging problem in the Middle East. Colonialism, British Prime Minister Winston Churchill said to President Eisenhower in 1954, no longer posed the same threat. What had replaced it was "oilism," and Aramco represented one of its most powerful and influential midcentury agents.[73]

Aramco's diplomatic support for Saudi boundary claims, combined with its rhetorical strategies, forged the company's position as an independent agent in Saudi Arabia.[74] Creating a positive image was a corporate strategy that allowed the company to legitimate its social and moral role.[75] Aramco sought to be much more than a foreign oil enterprise, but its primary objective remained protecting its concession and future profits. To do so, it projected an image to the world and the Saudi monarchy that it was a partner in Saudi nation-building. In the end, Aramco's vision did more than promote partnership. It required it. And Aramco marshaled its Relations Department to the king's side in hopes of expanding its resource base.

NOTES

1. James Scott, *Seeing Like a State: How Certain Schemes to Improve the Human Condition Have Failed* (New Haven: Yale University Press, 1998); Matthew Connelly, "Seeing Beyond the State: The Population Control Movement and the Problem of Sovereignty," *Past & Present* 193 (November 2006): 197–233; Erez Manela, "Pox on Your Narrative: Writing Disease Control into Cold War History," *Diplomatic History* 34 (April 2010): 299–323. Many others have been working to incorporate nongovernmental actors in the story of U.S. foreign relations. See also John Gripentrog, "The Transnational Pastime: Baseball and American Perceptions of Japan in the 1930s," *Diplomatic History* 34 (April 2010): 247–73; Daniel Maul, "'Help Them Move the ILO Way': The International Labor Organization and the Modernization Discourse in the Era of Decolonization and the Cold War," *Diplomatic History* 33 (June 2009): 375–86; Jason Pribilsky, "Development and the 'Indian Problem' in the Cold War Andes: *Indigenismo*, Science, and Modernization in the Making of the Cornell–Peru Project at Vicos," *Diplomatic History* 33 (June 2009): 387–404.

2. I use the term "transnational" to describe the Arabian American Oil Company as it denotes the way relationships operate above, below, beyond, and through traditional notions of the nation-state. See Ian Tyrrell, "Reflections on the Transnational Turn in United States History: Theory and Practice," *Journal of Global History* 4 (2009): 453–74; Melani McAlister, "What Is Your Heart For? Affect and Internationalism in the Evangelical Public Sphere," *American Literary History* 20 (Winter 2008): 870–95; David Thelen, "The Nation and Beyond: Transnational Perspectives on United States History—An Introduction," *Journal of American History* 86 (December 1999): 965–75.

3. Akira Iriye, *Global Community: The Role of International Organizations in the Making of the Contemporary World* (Berkeley: University of California Press, 2002).

4. Daniel Yergin, *The Prize: The Epic Quest for Oil, Money, and Power* (New York: Simon and Schuster, 1991); David Painter, *Oil and the American Century: The Political Economy of U.S. Foreign Oil Policy, 1941–1954* (Baltimore: Johns Hopkins University Press, 1986); Aaron David Miller, *Search for Security: Saudi Arabian Oil and American Foreign Policy, 1939–1949* (Chapel Hill: University of North Carolina Press, 1980); see also Michael B. Stoff, *Oil, War, and American Security: The Search for a Nation Policy on Foreign Oil, 1941–1947* (New Haven: Yale University Press, 1980); Irvine H. Anderson, *Aramco, the United States, and Saudi Arabia: A Study of the Dynamics of Foreign Oil Policy, 1933–1950* (Princeton: Princeton University Press, 1981); Stephen J. Randall, *United States Foreign Oil Policy, 1919–1948: For Profits and Security* (Kingston: McGill-Queen's University Press, 1985).

5. For a discussion of how narratives call into question widely held beliefs, see Andrew Menard, "Striking a Line through the Great American Desert," *Journal of American Studies* 45 (Summer 2010): 1–14.

6. James Ferguson, *Expectations of Modernity: Myths and Meanings of Urban Life on the Zambian Copperbelt* (Berkeley: University of California Press, 1999), 15; see also James Ferguson, *Global Shadows: Africa in the Neoliberal World Order* (Durham: Duke University Press, 2006);

Margaret R. Somers, "Narrativity, Narrative Identity, and Social Action: Rethinking English Working-Class Formation," *Social Science History* 16 (Winter 1992): 604.

7. Roy Lebkicher, *Handbooks for American Employees*, vol. 1, part 1, *Aramco and World Oil*, rev. ed. (New York: Russell F. Moore Company, 1952), publisher's preface.

8. Lebkicher, *Handbooks for American Employees*, 26.

9. Clarence J. McIntosh to McIntosh family, March 1–2, 1944, Clarence J. McIntosh Papers, box 2, folder 8, Georgetown University Special Collections, Washington, D.C.

10. The handbook was again published in 1952 and then updated versions emerged in 1960, 1968, and 1980.

11. Lebkicher, *Handbooks for American Employees*, ii.

12. Ibid., iv.

13. Ibid., 2–7.

14. Robert Vitalis, *America's Kingdom: Mythmaking on the Saudi Oil Frontier* (Stanford: Stanford University Press, 2007).

15. Toby Craig Jones, *Desert Kingdom: How Oil and Water Forged Modern Saudi Arabia* (Cambridge: Harvard University Press, 2010).

16. Vitalis, *America's Kingdom*; William E. Mulligan, Biographical Information on George Rentz, box 1, folder 53, Mulligan Papers.

17. Jones, *Desert Kingdom*.

18. Lebkicher, *Handbooks for American Employees*, preface; the material in the original 1950 handbook was also published, in a reorganized fashion, as *The Arabia of Ibn Saud* and distributed to a broader audience upon request.

19. Wallace Stegner, *Discovery! The Search for Arabian Oil*, as abridged for *Aramco World Magazine* (Beirut: Middle East Export Press, 1971), 1.

20. Roland Marchand, *Creating a Corporate Soul: The Rise of Public Relations and Corporate Imagery in American Big Business* (Berkeley: University of California Press, 1998); Vitalis, *America's Kingdom*; Margaret R. Somers, "Narrativity, Narrative Identity, and Social Action: Rethinking English Working-Class Formation," *Social Science History* 16 (Winter 1992): 604; David E. Nye refers to "representational strategies" that emerge to situate corporate identity. See David E. Nye, *Image Worlds: Corporate Identities at General Electric, 1890–1930* (Cambridge: MIT Press, 1982), 152; Robert Vitalis, "Wallace Stegner's Arabian Discovery: The Imperial Entailments of Continental Vision" (working paper, "The Cold War as Global Conflict," International Center for Advanced Studies, New York University, September 2003), 2, 30; Robert Vitalis, "Black Gold, White Crude: An Essay on American Exceptionalism, Hierarchy, and Hegemony in the Gulf," *Diplomatic History* 26 (Spring 2002): 202; Vitalis, *America's Kingdom*.

21. Vitalis, "Wallace Stegner's Arabian Discovery," 2, 30; Vitalis, *America's Kingdom*.

22. Somers, "Narrativity, Narrative Identity, and Social Action," 604. David E. Nye refers to "representational strategies" that emerge to situate corporate identity. See Nye, *Image Worlds*, 152.

23. Ibid., 600.

24. J. B. Kelly, *Eastern Arabian Frontiers* (New York: Frederick A. Praeger, 1964), 143–45; Nathan J. Citino, *From Arab Nationalism to OPEC: Eisenhower, King Sa'ud, and the Making of U.S.-Saudi Relations* (Bloomington: Indiana University Press, 2002), 25–26; John C. Wilkinson, *Arabia's Frontiers: The Story of Britain's Boundary Drawing in the Desert* (London: I. B. Tauris, 1991), x, 237–38; Memorandum of Conversation, Aramco officials and Judge Manley O. Hudson,

April 25, 1950, *Papers Relating to the Foreign Relations of the United States, 1950* (Washington, D.C.: U.S. Government Printing Office, 1950), 5: 45 (hereafter cited as *FRUS* followed by appropriate year).

25. F. S. Vidal, *The Oasis of Al-Hasa* (Dhahran, Saudi Arabia: Arabian American Oil Company, 1955).

26. George Rentz and William Mulligan, *The Eastern Reaches of Al-Hasa Province* (Dhahran: Arabian American Oil Company, 1950); George Rentz, *Oman and the Southern Shore of the Persian Gulf* in *Oman and the South-Eastern Shore of Arabia*, ed. Raghid El-Solh (1952; Berkshire, UK: Ithaca Press, 1997).

27. Research Department of the British Foreign Office, "Comment on Rentz's Book on S. E. Arabia," February 25, 1954, in *Oman and the South-Eastern Shore of Arabia*; see also Kelly, *Eastern Arabian Frontiers*.

28. Nathan J. Citino, *From Arab Nationalism to OPEC: Eisenhower, King Sa'ud, and the Making of U.S.-Saudi Relations* (Bloomington: Indiana University Press, 2002); Tore Tingvold Petersen, "Anglo-American Rivalry in the Middle East: The Struggle for the Buraimi Oasis, 1952–1957," *International History Review* 14 (February 1992): 71–91.

29. Petersen, "Anglo-American Rivalry in the Middle East, 71–91."

30. Alexander Melamid, "The Buraimi Oasis Dispute," *Middle Eastern Affairs* 7 (1956): 56–62; Petersen, "Anglo-American Rivalry in the Middle East," 71–91; Rentz, *Oman and the Southern Shore of the Persian Gulf*, 115.

31. "Log of Activities—William E. Mulligan, on Exploration Field Party Liaison," April 16, 1949, folder 20, "Mulligan, William E., Correspondence, 1946–49," box 11, William E. Mulligan Papers, Archives and Special Collections, Georgetown University, Washington, D.C. (hereafter cited as Mulligan Papers with filing information).

32. William E. Mulligan, Memorandum, April 22, 1949, folder 20, "Mulligan, William E., Correspondence, 1946–49," box 11, Mulligan Papers.

33. Stobart to Holm, April 22, 1949, folder 20, "Mulligan, William E., Correspondence, 1946–49," box 11, Mulligan Papers.

34. Melamid, "The Buraimi Oasis Dispute," 58.

35. Citino, *From Arab Nationalism to OPEC*, 21; W. Taylor Fain, *American Ascendance and British Retreat in the Persian Gulf Region* (New York: Palgrave Macmillan, 2008), 63; The Iraq Petroleum Company (IPC) was a jointly owned subsidiary of multiple firms, among them British and American companies. Arranged in the 1920s, Standard Oil of New Jersey and Standard Oil of New York were just two of a collection of American firms that shared a 23.75 percent share. Royal Dutch/Shell (British), the Anglo Persian/Iranian Oil Company (British, with government shares), and the French company CFP each held 23.75 percent. The remaining 5 percent went to Calouste Gulbenkian, the Armenian dealmaker who negotiated the consortium. See Daniel Yergin, *The Prize: The Epic Quest for Oil, Money, and Power* (New York: Simon and Schuster, 1991); Memorandum of Conversation, April 20, 1950, *FRUS, 1950*, 5: 38.

36. Kelly, *Eastern Arabian Frontiers*, 145; Citino, *From Arab Nationalism to OPEC*, 21–26.

37. Memorandum of Conversation, January 31, 1950, *FRUS, 1950*, 5: 18; Memorandum of Conversation, April 20, 1950, *FRUS, 1950*, 5: v, 38; Memorandum of Conversation, April 25, 1950, *FRUS, 1950*, 5: 45.

38. Memorandum of Conversation, March 19, 1951, *FRUS, 1951*, 6: 286.

39. Memorandum of Conversation, January 31, 1950, *FRUS, 1950,* 5: 18; Memorandum of conversation, April 25, 1950, *FRUS, 1950,* 5: 45.

40. Kelly, *Eastern Arabian Frontiers,* 294.

41. Madawi Al-Rasheed, *A History of Saudi Arabia,* 2nd ed. (Cambridge: Cambridge University Press, 2010), 14–19, 85, 121.

42. Truman to Abdul Aziz Ibn Saud, October 31, 1950, *FRUS, 1950,* 5: 1190–91.

43. Memorandum of Conversation, May 19, 1953, *FRUS, 1952–1954,* 9: 1, 96, 99.

44. Citino, *From Arab Nationalism to OPEC,* 19, 28.

45. Dispatch, Chargé in Saudi Arabia to Department of State, May 14, 1952, *FRUS, 1952–1954,* 9: 2469.

46. Citino, *From Arab Nationalism to OPEC,* 29–31, Melamid, "The Buraimi Oasis Dispute," 59.

47. Citino, *From Arab Nationalism to OPEC,* 63–64; Melamid, "The Buraimi Oasis Dispute," 58; Tore T. Petersen, *The Middle East Between the Great Powers: Anglo-American Conflict and Cooperation, 1952–1957* (London: Macmillan, 2000), 72; Fain, *American Ascendance and British Retreat in the Persian Gulf Region.*

48. "Jurisdictional Dispute of Saudi Arabia," paper prepared by the Department of State for the London Foreign Ministers Meeting in May, April 20, 1950, *FRUS, 1950,* 5: 40.

49. Acheson to U.S. Embassy in the UK, November 20, 1950, *FRUS, 1950,* 5: 117.

50. *Arbitration Agreement between the Government of the United Kingdom (Acting on Behalf of the Ruler of Abu Dhabi and His Highness the Sultan Said bin Taimur) and the Government of Said Arabia,* Jeddah, July 30, 1954, in Kelly, *Eastern Arabian Frontiers,* appendix A, 281–92; Telegram, Wadsworth to Department of State, July 31, 1954, *FRUS, 1952–1954,* 9: 2614–15 and n. 2.

51. Petersen, *The Middle East between the Great Powers,* 83.

52. Citino, *From Arab Nationalism to OPEC,* 64.

53. Ibid., 81–82.

54. Petersen, *The Middle East between the Great Powers,* 84; Citino, *From Arab Nationalism to OPEC,* 84.

55. Fain, *American Ascendance and British Retreat in the Persian Gulf Region,* 67–68; quoted in Petersen, *The Middle East between the Great Powers,* 58–59.

56. Fain, *American Ascendance and British Retreat in the Persian Gulf Region,* 112–18.

57. Kelly, *Eastern Arabian Frontiers,* 268.

58. Saudi Arabia and United Arab Emirates, "Agreement on the Delimitation of Boundaries," August 21, 1974, United Nations Treaty Collection, http://treaties.un.org/doc/publication/unts/volume%201733/i-30250.pdf.

59. Memorandum of conversation, March 19, 1951, *FRUS 1951,* 5: 286–97.

60. Rentz, *Oman and the Southern Shore of the Persian Gulf,* xxv.

61. *Memorial of the Government of Saudi Arabia: Arbitration for the Settlement of the Territorial Dispute between Muscat and Abu Dhabi on One Side and Saudi Arabia on the Other,* vol. 1 (Cairo: Saudi Arabian Government, 1959), 467–518.

62. Joseph Kostiner, *The Making of Saudi Arabia, 1916–1936: From Chieftaincy to Monarchical State* (New York: Oxford University Press, 1993), 71–72; Joseph Kostiner, "On Instruments and Their Designers: The Ikhwan of Najd and the Emergence of the Saudi State," *Middle Eastern Studies* 21 (1985): 303; A. S. Helaissi, "The Bedouins and Tribal Life in Saudi Arabia," *International Social Science Journal* 11 (1959): 533.

63. *Memorial of the Government of Saudi Arabia: Arbitration for the Settlement of the Territorial Dispute between Muscat and Abu Dhabi on One Side and Saudi Arabia on the Other*, vol. 1, 49–53.

64. *Memorial of the Government of Saudi Arabia*, 97–382.

65. Ibid., 377.

66. Rentz, *Oman and the Southern Shore of the Persian Gulf*, 115–20. *Memorial of the Government of Saudi Arabia*.

67. Research Department of the British Foreign Office, "Comment on Rentz's Book on S. E. Arabia," February 25, 1954, in *Oman and the South-Eastern Shore of Arabia*; see also Kelly, *Eastern Arabian Frontiers*.

68. United States Record of the First Session of the United States–United Kingdom Talks on Middle East Oil, April 5, 1954, *FRUS, 1952–1954*, 9: 799.

69. Memorandum of Conversation, Buraymi Dispute, May 27, 1954, *FRUS, 1952–1954*, 9: 822.

70. Memorandum of Conversation, April 3, 1953, *FRUS, 1952–1954*, 9: 2533; Memorandum of Conversation, April 16, 1953, *FRUS, 1952–1954*, 9: 2535.

71. Moline to Eakens, February 16, 1954, *FRUS, 1952–1954*, 9: 2578.

72. Memorandum of Conversation, "Buraimi Dispute: Aramco's Interests in Latest British Proposals," March 10, 1954, *FRUS, 1952–1954*, 9: 2583–87.

73. Memorandum of Conversation, June 25, 1954, *FRUS, 1952–1954*, 6: 1054.

74. Manuel Castells, *The Power of Identity* (Malden, Mass.: Blackwell, 2004); Charles Tilly, "Citizenship, Identity, and Social History," in *Citizenship, Identity, and Social History*, ed. Charles Tilly, 1–18 (Cambridge: Cambridge University Press, 1995); and David Campbell, *Writing Security: United States Foreign Policy and the Politics of Identity* (Minneapolis: University of Minnesota Press, 1992). Campbell argues that states create identity through the promotion of various dangers that fall outside the proscribed normative self-definition, the boundaries of which are "secured by the representation of danger integral to foreign policy." See Campbell, *Writing Security*, 3.

75. Marchand, *Creating a Corporate Soul*, 5. Richard Drayton, *Nature's Government: Science, Imperial Britain, and the "Improvement" of the World* (New Haven: Yale University Press, 2000).

10

Oil Frontiers

The Niger Delta and the Gulf of Mexico

MICHAEL WATTS

During the late evening of April 20, 2010, a geyser of seawater erupted from the mariner rise onto the derrick of BP's drilling rig Deepwater Horizon, located in deep water almost fifty miles off the coast of southern Louisiana in the Gulf of Mexico. Shortly afterward a slushy combination of mud, methane gas, and water was propelled 250 feet into the air followed by a massive explosion, instantly converting the rig into a raging inferno. In the desperate attempts to control the flow of the gas and fluids, the blowout preventer, developed to cope with extreme erratic pressures and uncontrolled flow (formation kick) emanating from a well reservoir during drilling, failed to function. Two days later the Deepwater Horizon sank onto the ocean floor, resting one mile below the sea's surface. As the rig sank, it ruptured the risers, and a mixture of oil and gas, under extreme pressure, was released into the warm and rich waters of the Gulf. The sinking of the Deepwater Horizon produced a calamity fifteen times larger than the Exxon Valdez tanker spill in 1989, which released almost 11 million gallons of heavy crude oil, laid waste to 1,300 miles of Alaska coastline, and covered 11,000 square miles of ocean. By mid-May 2010, the discharge from the Macondo Well was hemorrhaging at a rate of more than 200,000 gallons per day; surface oil covered 3,850 square miles. When it was all over, almost 170 million gallons had been released and 35 percent of the Gulf Coast affected. Rarely noted in the wake of the blowout was the long history of spills in the Gulf, and the systematic destruction of the coastline, especially in the Mississippi Delta, over a century of development along the Gulf oil frontier.

During the Gulf of Mexico crisis, Royal Dutch Shell released a report on its activities in Nigeria, a strategically and economically important petrostate in the Gulf of

Guinea oil zone, and a major supplier of high-quality "sweet and light" crude to U.S. markets. Shell confirmed that in 2009 it had spilled roughly fourteen thousand tons of crude oil into the creeks of the Niger Delta, the heart of Nigeria's onshore oil industry. Put differently, in the course of *one* year, a *single* oil company was responsible for 4.2 million gallons of spilled oil; in 2008, the figure was close to 3 million gallons. In related figures released in April 2010, the Federal Ministry of the Environment included a tally sheet of 2,045 recorded spill sites between 2006 and 2009. Since the late 1950s, when the oil frontier opened and oil and gas resources became commercially viable, more than seven thousand oil spills have occurred across the Niger Delta oilfields. Cumulatively over a fifty-year period, more than half a billion gallons of crude oil have been discharged in an area roughly one-tenth the size of the Federal Waters of the Gulf of Mexico. As an Amnesty International report put it, this spillage is "on par with [an] Exxon Valdez [spill] every year."[1]

The calamitous events in these two oil gulfs, each centers of exploration and production at different locations in the global oil and gas value chain but with common points of reference in the history of the Black Atlantic, point toward a set of family resemblances I want to explore.[2] An examination of the history of these two oil-supply zones offers an opportunity to explore not just the instabilities, risks, and contradictions of the global oil and gas assemblage, but also the cultural and political economic dynamics of "oil frontiers."[3] Both the Niger Delta and the Louisiana coastal and offshore Gulf of Mexico oil fields are exemplary instances of oil frontiers, understood not so much as zones of pioneer settlement and crucibles of democratic values forged by the battle between wilderness and rugged individualism (the canonical view of frontiers derived from Frederick Turner), but rather as particularly dense sorts of spaces—at once political, economic, cultural, and social—in which the conditions for a new phase of (extractive) accumulation are being put in place.[4] They are both exemplary of frontier capitalism with speculative, spectacularized, and violent forms of enclosure, dispossession, and profit-taking. Oil frontiers, as we shall see, are involved in the manufacture of the conditions of possibility for the creation and capture of oil rents.

Frontiers possess quite specific sorts of properties and qualities.[5] Historically frontiers have been associated with imperial and commercial advance typically into geographical border zones in which populations are presumed (or constructed) to be scant or "primitive," property rights are weak or entirely absent, and resources (land, minerals, forests) are largely, if not entirely, unexploited for commercial purposes—in short, frontiers are, in this Whig account, zones of contact between "barbarism" and "civilization." Viewed more expansively, frontiers can be seen, as Markoff puts it, as "boundaries beyond the sphere of routine actions of centrally located violence pro-ducing enterprises,"[6] places where no one has an enduring monopoly on violence.[7] Whatever the resource specificity of frontier dynamics—cattle or soy frontiers in Amazonia, oil frontiers in Angola—and whatever their particular racialized mixing of populations, economies, and cultures (the maritime and piratical frontiers of the Black Atlantic or the *estancia* culture of Brazil), questions of law, order, rule, authority, profit, and property are all subject to intense forms of contestation and opposition. The much-vaunted "wildness" or "disorder" of the frontier is, in fact, an expression of forms of economic and social organization that created "classes specialized in expedi-ency whose only commitment was to preserve the order that made possible the profit-able utilization of such expediency."[8] Pillage and routinized plunder was one expression of this expediency—always site-specific with their own cultures of violence—but it could also be expressed in symbolic rather than physical forms of violence, and through the coercive operations of power associated with what Markoff calls "political nego-tiation" among (typically) a rough-and-tumble mix of actors and agents: vagabonds, outcasts, rogue capitalists and shady state representatives, smugglers, and captains of economic violence.

In this sense, if frontiers are deprived of particular sorts of historical and cultural association (the American frontier, the Chinese or British Imperial frontier), then they become pervasive, indeed recursive forms of what one might call "economies of violence." They are, as James Ron suggests, specific types of "institutional setting" where centralized political authority is thin, where formal rules are elastic, and where state power is despotic and coercive rather than infrastructural.[9] Frontiers are dialec-tical spaces, not simply zones of demarcation. They are spaces of interaction between civil and political institutions, in which "the exercise of governance and the powers of civil authorities are closely guarded and geographically delimited."[10] As social spaces they are material but also constituted through practice, representation, and acts of imagination.[11] Frontiers are, nevertheless, distinctive and complex social spaces: ter-ritorial but at the margins or outer reaches of particular social and political conven-tions; they are dynamic rather than immutable; their life histories do not proceed unambiguously from opening to closure; they can emit, as Redclift says, clues about identity, difference, citizenship, and nation-building.[12] As Markoff notes, we find them in Sri Lankan tea plantations and in U.S. prisons, and, we might have added, on American or African oil fields. Here, he says, "we find oppositional definitions of groups, we find claims of order and claims of freedom, and we find definitions of vio-lence as they create it."[13] Broadly understood, then, frontiers are particular sorts of economies of violence in which the politics of dispossession figure centrally.[14]

Defining the "Oil Frontier"

Africa beckons as an oil frontier, says Daniel Layton, head of a Houston-based energy investment firm. In a resource-constrained world, Layton observes, finding large fields become more challenging. In spite of the "turmoil with insurrection, piracy and terrorism," Africa remains attractive as one of the "last oil frontiers."[15] Control Risks, a global risk consultancy, cites East African oil and gas as *the* "new frontier," a vast swath of territory composed of relatively unexplored basins extending from Somalia to Mozambique. Offshore reservoirs, in particular, are ripe for the plucking, harboring, according to the U.S. Geological Survey, more than 14 billion barrels of oil. Africa has no monopoly over oil frontiers, of course: the Falklands, Siberia, and especially the challenging waters of the Arctic attract the interests of the oil majors, smaller independents, and wildcatters alike. Much is at stake in a purportedly Peak Oil world (the "end of easy oil," as the industry puts it). On August 2, 2007, a Russian submarine with two parliamentarians on board planted a titanium flag two miles down under the North Pole. At stake were the lucrative new oil and gas fields—by some estimations 10 billion tons of oil equivalent—on the Arctic sea floor.

These sorts of discoveries are dubbed by the oil and gas industry as "unconventional" sources of energy, one index of which are the new deepwater frontiers. In late 2006, a consortium of oil companies discovered oil at a staggering depth, 150 miles, into the Gulf of Mexico. The test well, Jack-2, delved through seven thousand feet of water and thirty thousand feet of sea floor to tap oil in tertiary rock laid down 60 million years ago. The drill ships—and the production platforms—required to undertake such efforts are massive floating structures, much larger than the largest aircraft carriers and more expensive, costing well over half a billion dollars to construct. In 2007, a vast new Tupi field in Brazilian coastal waters was discovered in two hundred meters of water and below a massive layer of salt in hugely inhospitable geological conditions. One test well cost over $250 million. What is on offer is a great deepwater land grab—or primitive accumulation at seven thousand meters.

But there are other frontiers of discovery, too: the Athabascan oil sands and Albertan bitumen in Canada, and the more dramatic yet extraordinary oil shale and tight oil boom in the continental United States over the last five years. The Marcellus Shale in Pennsylvania and New York, the Haynesville formation in Louisiana, and the Barnett and Eagle Ford plays in Texas—all triggered by two decades of federal support for horizontal, slick-water hydraulic fracturing of shale—have opened up new domestic frontiers from the mid-Atlantic states to the western United States, and farther afield in Poland and France. In North Dakota's Bakken Shale, almost two hundred wells are drilled every month; small communities like Williston, North Dakota, have become boomtowns with some of the highest real estate rents in the nation, Manhattan included. Armies of drillers, oil and gas service companies, land men, to say nothing of lawyers and shady sales representatives, have invaded rural communities promoting the putative benefits of "clean gas." The oil frontier does not presume a new discovery; fields may be reopened and frontiers reinvigorated as a product of technological innovation (enhanced rates of oil and gas recovery), elevated petroleum prices, and by corporate risk portfolios that see large downsides to exploration and production in authoritarian and unstable states in the Caspian, Africa, and the Middle East. Oil frontiers have no simple life cycle; they are dynamic and unstable, they are discovered and rediscovered—opened and reopened—rather than simply open, closed, and depleted.[16] In short, the oil frontier appears as a permanent prospect.

These constantly emerging new frontiers are, as it were, the shock troops of the advancing oil production network, a system held together materially by a vast global oil infrastructure. Close to 5 million producing oil wells puncture the surface of the earth; 3,300 are subsea, puncturing the earth's crust on the continental shelf, in some cases thousands of meters below the sea's surface. There are by some estimations more than 40,000 oil fields in operation. More than 2 million kilometers of pipelines blanket the globe in a massive trunk-network; another 75,000 kilometers of lines transport oil and gas along the sea floor. There are 6,000 fixed platforms and 635 offshore drillings rigs (in December 2013 there are almost 3,500 rigs in operation globally). More than 4,000 oil tankers move 2.42 billion tons of oil and oil products every year, one-third of global seaborne trade; more than 80 massive floating production and storage vessels have been installed in the last five years. This petro-infrastructure also accounts for almost 40 percent of global CO_2 emissions. All in all, there is nothing quite like it.

In the oil sector, a frontier has a specific set of connotations that speak directly to the space-time dynamics that are distinctive to the political economy of the oil assemblage.[17] A geological province, a large area often several thousand square kilometers with a common geological history, becomes a petroleum province when a "working petroleum system" (or "play") has been discovered. At this point, the oil frontier takes on a life within the play. The discovery of a petroleum field begins a process of appraisal and development—namely, drilling many new wells to confirm the extent and properties of the reservoirs and fluids and also whether configuration warrants larger investment to develop the field.[18] The development of the initial fields in a new province is replete with technical uncertainties and risks that collectively shape the ultimate volume of oil that can be recovered. The properties of reservoir rock, the fluids it contains, and the fluid dynamics within the rock are key, but so too are the fluids, which vary in their composition, specific gravity, and viscosity. Uncertainties around each of these field variables translate into uncertainty in ultimate recovery volumes. But peak production from the field, the life of the field, well flow rates, the density of wells required, required capacities of production, storage, and export systems, and when and whether secondary and perhaps tertiary recovery might ultimately be appropriate all vary field by field.

The frontier, in short, refers to the spatial, temporal, and, of course, vertical dynamics in which fields within a province are discovered, developed, and recovered—from primary reserve creation to tertiary recovery from existing "matured" reservoirs. But it also encompasses the extended processes, sometimes consuming several decades, through which the road to "first oil" extends from exploration and development of onshore sedimentary basins through shallow offshore basins and into deep and ultra-deepwater basins today. Recent frontiers include the demanding, risky, and dangerous challenges of deep water; these emerging frontiers take on the enormously expensive commercialization of potentially vast resources of unconventional oils and gas.

The temporal and risk dynamics of the frontier in the life of the commercial play are clear. Each exploration well carries not only the uncertainty associated with the specific prospect being drilled but also the uncertainty of the entire spatial petroleum system. With the development of one or more commercial fields, a frontier becomes "proven" and uncertainty about the presence of hydrocarbons and their character is much reduced based on probability estimates among and between fields. Reduction in risk often induces an influx of new entrants (which includes state companies and smaller independents) that were deterred initially when entry barriers were high. Another frontier emerges—a function of new technologies and aging reservoirs— as aging oil fields attract investments through tertiary recovery. The frontier within and between provinces is extended and dynamic, both geographically expansionary and, as it were, involutionary. But frontiers also have to be secured politically and may become what Edward Barbier calls an "isolated and self-perpetuating enclave with no linkages to the rest of the economy."[19] The enclavization of the economy—in this sense the frontier begins to resemble what Ron calls the ghetto—is, as the Niger Delta reveals, one historical trajectory of the oil frontier.

"Frontiers," as I deploy the term for oil, should not imply only the technical relations of resource exploitation (as the industry understands frontiers); neither do I take the Latin American cattle frontier as its paradigmatic form. Rather, the oil frontier— indeed any resource frontier—marks the construction of a new space of accumulation and the creation of the conditions of existence for the local operation of what I term the "oil complex" or "oil assemblage."[20] Oil frontiers have their own temporalities and spatialities, but like frontiers elsewhere, questions of access to and control of land, property, and rents as a prerequisite for accumulation—shaped by technological considerations unique to oil—are a driving force. Economists who have explored the so-called frontier-expansion hypothesis emphasize a crucial part of the accumulation frontier, what is termed "disequilibrium abnormal rents."[21] Since frontier expansion takes time and resources (some of which may be scarce, like labor), there are disequilibrium periods in which abnormal rents stimulate further investment. The conditions of possibility for such rents typically arise from institutional forms of coerced labor that "repress the returns to free labor," discoveries of resources and windfall gains from speculative activity, and institutions and technologies developed to "accommodate heterogeneous frontier conditions."[22] Oil frontiers customarily exemplify some configuration of all of these preconditions—that is, particular expressions of abstract space, as Lefebvre would say. If barbed wire, the labor-repressive ranch, and violent land speculations created the rents at the cattle frontier, on the oil patch it is the state-directed (and often militarized) oil concession or oil bloc, the corporate joint-ventures, and the infrastructure of wells, pipelines, and flow stations that constitute the armory of the rentier economy. The propulsive character of the oil frontier is, as Massimo De Angelis notes, a quintessential spatial form in the functioning of global capitalism: "Capital's identification of a frontier implies the identification of a space of social life that is still relatively uncolonised by capitalist relations of production and modes of activity."[23] Oil frontier capitalism is a spatialized expression of Marx's notion of primitive accumulation and enclosure, an emergent property, as De Angelis observes of capital's drive to continuous expansion.[24]

Overlaid on the oil and gas assemblage is an astonishing patchwork quilt of territorial concessions. Spatial technologies and spatial representations are foundational to the oil industry: seismic devices to map the contours of reservoirs, geographic information systems to monitor pipeline flows, and of course the map to determine subterranean property rights. Hard rock geology is a science of the vertical, but when harnessed to the marketplace and profitability it is the map that becomes the instrument of surveillance, control, and rule. The oil and gas industry is a cartographer's dream-space: a landscape of lines, axes, hubs, spokes, nodes, points, blocks, and flows. As a space of flows and connectivity, these spatial oil networks are unevenly visible (subsurface, virtual) in their operations.

As a dense economic, cultural, ideological, and sociopolitical space, the oil frontier also embodies particular sorts of temporality—that is to say, frontiers are time-spaces (this is central to Lefebvre's notion of the production of space). In one regard, since oil is finite, the oil frontier will face inevitable closure, and the petrolic landscape will—as photographer Edward Burtynsky's marvelous images show[25]—become fossilized and residual. As we have seen, however, this closure is often extended, drawn out, and uneven because of changing technologies for recovery of oil and gas deposits (frontiers are regularly reanimated and reopened). The fluctuation in the oil markets, which has become a defining quality of the contemporary hydrocarbon assemblage, also superimposes its own temporalities through boom-and-bust cycles and periodic rounds of speculative activity (in land but also labor and construction markets). And not least, as Guido di Tella points out, over time the rentier logic of the frontier may become self-reproducing (there is a temporal logic of oil exploration and production); the institutional politics of the frontier are sedimented around a culture of permanent rent seeking rather than in mechanisms and incentives to ensure that petrowealth is invested productively in the wider capitalist economy.[26]

As a territorial and historical resource, the oil industry is constantly in the business of creating new and refiguring old frontiers—each marked by complex processes of dispossession, compromise, violence, and engagement—and as a dynamic industry the frontiers so created are, as Eyal Weizman puts it, deep, shifting, fragmented, and elastic territories: "The dynamic morphology of the frontier resembles an incessant sea dotted with multiplying archipelagoes of externally alienated and internally homogenous . . . enclaves. . . . [It is] a unique territorial ecosystem (in which) various other zones— . . . political piracy, . . . barbaric violence, . . . of weak citizenship . . .— exist adjacent to, within or over each other."[27] These oil frontiers are where the oil and gas capital touch down, textbook cases of what Lefebvre calls the "hyper-complexity" of global space in which social space fissions and fragments producing multiple, over-lapping, and intertwined subnational spaces with their own complex boundaries, spatial practices, and forms of life.[28]

The oil frontiers I have selected stand at different ends of the global value chain: one in the heart of advanced American financialized capitalism, the other in the neo-liberalized periphery of the Global South. These oil frontiers, one largely onshore, one largely offshore, exhibit nevertheless similar driving forces and dynamics: coercion and conflict, the elasticity of the law, forms of dispossession under circumstances of radical class inequality, powerful alliances between state and capital. But the trajectories along the frontier are a study in contrasts exhibiting their own particularities. If the Niger Delta frontier reveals one explosive trajectory, characterized by massive conflicts, an insurgency, and a pattern of violent accumulation in a disorderly world of corrupt chiefs, powerful politicians, violent state security forces, and robust and often shady alliances between state and capital, the Deepwater Horizon suggests another. The Macondo Well blowout exposes the deadly intersection of the aggressive enclosure of a new technologically risky resource frontier (the deepwater continental shelf in the Gulf), with the production of what one can call neoliberalized risk, a lethal product of cutthroat corporate cost-cutting, the collapse of government oversight and regulatory authority, and the deepening financialization and securitization of the oil market. These two local pockets of disorder, rogue capitalism, and environmental catastrophe within the oil complex point to, and are expressions of, the deep pathologies and vulnerabilities within the operations of imperial oil.

THE ONSHORE FRONTIER:
VIOLENCE AND SPECTACULAR DISPOSSESSION IN NIGERIA

Nigeria is an archetypical petrostate, the eleventh largest producer and the eighth largest exporter of crude oil in the world.[29] The oil-producing Niger Delta in the southeast of the country has provided "sweet" (low sulfur) oil to the world market for over half a century, during which time the Nigerian state has captured close to $1 trillion. Since the return to civilian rule in 1999, Nigeria has been shaken by extraordinarily contentious and violent politics. After a decade of what resembles nothing more than a "violent" democracy, Nigeria confronted not one but two homegrown insurgencies: one emanating from the oil fields, the other from the Muslim heartland in the north.

The vertiginous descent of the Niger Delta oil fields into a strange and terrifying shadow-world of armed insurgency, organized crime, state violence, mercenaries and shady politicians, and massive oil theft casts a long shadow over Nigeria's purportedly rosy oil future. A powerful insurgent group called the Movement for the Emancipation of the Niger Delta (MEND) exploded from the creeks in 2006. Within two years of taking office in 2007, the new administration of President Yar Adua saw oil revenues fall by 40 percent because of the audacious and well-organized attacks on the oil sector. Shell, the largest operator, had alone lost $10.6 billion since late 2005. In the Port Harcourt and Warri regions, the two hubs of the oil industry, there were more than five thousand pipeline breaks and ruptures in 2007 and 2008 perpetrated by insurgents and self-proclaimed militants. According to a report released in late 2008, during the first nine months of 2008 the Nigerian government lost a staggering $23.7 billion in oil revenues as a result of militant attacks and sabotage. By the summer of 2009, Shell's western operations were in effect closed down, and more than 1 million barrels of daily oil production were shut in.

Nigeria's oil complex was born in the mid-1950s, when the first helicopters landed in Oloibiri in Bayelsa State near St. Michael's Church to the astonishment of local residents. A camp was quickly built for workers; prefabricated houses, electricity, water, and a new road followed. In the decades that followed, the Nigerian oil industry grew quickly in scale and complexity: 600 on- and offshore fields, thousands of oil wells, seven thousand kilometers of pipelines, almost 300 flow stations, massive liquified natural gas plants and related infrastructure. By the 1970s oil tankers lined the Cawthorne Channel near the oil city of Port Harcourt like participants in a local regatta, plying the same waterways that, in the distant past, housed slave ships in the sixteenth century and the palm-oil hulks in the nineteenth. The petroleum frontier followed the slave and palm-oil frontiers of the seventeenth, eighteenth, and nineteenth centuries.

Oil converted Nigeria into a petrostate but one constituted by vast shadow political and economic apparatuses in which the lines between the public and the private, state and market, government and organized crime, are blurred and porous. The coastal waters of the delta are comparable to the lawless seas surrounding Somalia and the Moluccas. A study, *Transnational Trafficking and the Rule of Law in West Africa,* by the UN Office for Drugs and Crime, estimates that 55 million barrels of oil are stolen each year from the Niger Delta, a shadow economy in which high-ranking military and politicians are deeply involved. Petrowealth has been squandered, stolen, and channeled to largely political, as opposed to productive, ends. Over 80 percent of oil revenues accrue to 1 percent of the population. According to former World Bank President Paul Wolfowitz, around $300 billion of in oil revenues accrued since 1960 have simply "gone missing." Between 1970 and 2000, the number of income poor grew from 19 million to a staggering 90 million. Over the last decade, GDP per capita and life expectancy have, according to World Bank estimates, both fallen. Oil, in sum, has lubricated—it is a medium for—a catastrophic failure of secular national development.

How was the Nigerian oil complex assembled historically? The onshore oil frontier—the offshore frontier began much later—operated in a distinctive fashion. Oil-bearing lands were nationalized, and leases and licenses awarded (typically with little or no transparency) to oil companies who were compelled to participate in joint ventures with a Nigerian state. A Memorandum of Understanding determined, among other things, the very substantial government take on every barrel of oil produced. Local communities across the delta lost access to their lands—their subsurface rights and often property taken as eminent domain. They were typically compensated (in an ad hoc and often arbitrary fashion) for loss of land rights and for the costs associated with the operations of the industry—for example, oil spillage. Communities deemed to be "host communities" by virtue of having oil within their customary territories or being directly affected by oil infrastructure were to receive "community benefits" from the oil companies that, in the absence of an effective local state, came to be seen as local government. The companies built alliances with local political forces, which in effect meant dealing directly with powerful chiefs and chieftaincy systems marked by the exercise of lineage-based gerontocratic powers (the authority vested in elders). For the better part of three decades, the companies could operate with impunity, cutting deals with chiefs and elders and the political classes, who through direct cash payments, contracts, and community funds acquired considerable wealth.

Two logics came to structure the operations of the Nigerian oil complex. The first was the capture of oil rents by the state through a series of laws and statutory monopolies. In effect, the conversion of oil into a national resource ignited two powerful forces: first it became the basis of differential claims making. That is, citizens could, in virtue of its national character, plausibly claim their share of this national cake as a citizenship right. Second, state appropriation of oil flew in the face of robust traditions of customary rule and land rights. The continuity of so-called customary forms of rule and the authorization of systems of community rule in effect institutionalized a parallel system of governance associated with chieftaincy. Oil nationalization trampled on local property systems and land rights and complicated the already tense relations between so-called indigenes (or first settlers) and newcomers (or settlers). For the Nigeria Delta and its sixty ethno-linguistic groups, the raft of oil laws inevitably was construed locally as expropriation and dispossession—the loss of "our oil." These claims were inevitably expressed in ethnic terms ("our land," "our oil") and marked the emergence of so-called "oil minorities" (a postcolonial invention) not only as a political category but also as an entity with strong territorial claims. The fact that oil companies, as cosignatories to joint ventures with the state, were in turn compelled to pay rent to oil-bearing communities (which typically meant undisclosed cash payments to chiefs, councils of elders, and ruling royal or "big houses"), converted an already contested arena of land rights into a charnel house of violent struggles over "who owns the oil" and on what basis (lineage, clan, ethnicity, first settlers, and so on).

The second logic reflects the institutional mechanisms by which revenues were to be allocated within a complex, multiethnic thirty-six-state federal system. Oil revenues are the main source of public revenue in Nigeria, accounting for about 80 percent to 85 percent of the total receipts. The current vertical allocation is 52.68 percent, 26.72 percent, and 20.60 percent for federal, state, and local government, respectively. These figures confirm, of course, the centralizing effect of capturing oil rents, but the details, hammered out in a raft of revenue commissions over the last half century, are the subject of intense contestation and continuing controversy. The federal center captured a disproportionately large share of the revenues; the states and local governments depend heavily on statutory allocations. Since the 1960s the principles of allocation radically reduced the principle of derivation, by which states producing the resource retained a share of the revenues. Fiscal centralization redirected revenues away from the centers of oil production to powerful nonoil ethnic majority states, especially in the north of the country. The federal center became a hunting ground for contracts and rents of various kinds. Derivation politics (and the loss of revenues cascading within the federal allocation system) inevitably became an axis of contention between the delta and the federal center and laid the basis for what became the delta's clamor for "resource control" in the 1990s.

These twin driving forces within the oil complex produced not one but multiple and overlapping crises of authority.[30] First, the federal state had failed to deliver development and was synonymous with graft and limited capabilities. There is the special case of the security forces—especially the police and army—that became objects of utter contempt and illegitimacy, not only for their corruption but also because of the extent to which so much of the urban and rural violence in both regions was triggered by police brutality. Second, the institutions of customary rule—the chieftaincy systems in the oil fields—were no longer legitimate and effective systems of authority either, and most youth felt excluded from this gerontocratic order. By the 1990s and 2000s, Niger Delta chiefs were not often violently ejected from office by rebellious youth angry at their pocketing of monies paid to them by oil companies for purportedly community development purposes (these often substantial cash payments were actually privately appropriated). And third, even religious authority has been subject to debate and questions as the boundaries between religious practice and state power have been eroded.

The illegitimacy, indeed the ethical and moral bankruptcy, of these multiple and overlapping networks of customary, religious, and modern governance created a vast space of alienation and exclusion, a world in which the armies of impoverished youth were neither citizens nor subjects, a social landscape in which the politics of resentment could fester. Contempt is the ruling ideology. These floating populations—land-poor peasants and unemployed educated youth, equally unable to fulfill norms of personal advancement through marriage, patronage, and work—occupied a social space of massively constricted possibility. Millions were reduced to the level of unfulfilled citizenship, what Murray Last refers to as material, social, and political insecurity.[31]

It was from these dynamics that a welter of violent struggles emerged. According to a UNDP report, there are currently 120 to 150 "high risk and active violent conflicts" in the three core oil-producing states.[32] The field of violence operates at a number of levels. First, there are a number of insurgent groups like MEND engaged in armed struggle against the state and the oil companies. Second, intercommunity conflicts often driven by land and jurisdictional disputes over oil-bearing lands (and correspondingly over access to cash payments and rents from the oil companies) are commonplace. And third, city-based interethnic violence associated with the ethnic delineation of electoral wards and local government councils, the means by which urban ethnic communities can access oil wealth in the form of rents paid by oil companies for land used for oil infrastructure or from the revenue allocation process that fills local government coffers with "excess oil profits." Other communities are torn apart by intracommunity youth violence—the famed city-state of Nembe is a case in point. There, armed youth groups do battle with one another and their chiefs in order to provide protection services to the oil companies and get access to various sorts of standby (a salary for doing nothing) and cash payments doled out in the name of "community development." Some of these struggles were financed by oil theft (bunkering) in which insurgent groups were able to insert themselves (typically as underlings beneath high-ranking military officials and politicians). It produced a class of violent entrepreneurs—not unlike the mafias of mid-nineteenth-century Sicily—and a ferocious battle over oil-bunkering territories.[33]

In many instances, the struggles over land, property, and territory in and around the oil concessions and the footprint of the vast oil and gas industry had the effect of fragmenting space and producing what Lefebvre calls multiple and overlapping spaces (chiefly village territories, clan lands, local government areas, bunkering territories, ethnic claims, and especially tensions over the history of land occupation).[34] These resentments were often directed toward a violent assertion of autochthony, exclusionary attachments and belongings of individuals and groups to particular places and identities, sometimes directed at strangers or nonindigenes, sometimes against state predators.[35] The multiplication of oil spaces, often unruly and deeply conflicted, represents a ferocious struggle over how nation-building is to proceed and in whose name and interest. At the same time, with prices at $100 a barrel, the entire rickety structure can stagger on: the Nigerian government, the rebels, the super majors, the oil bunkerers, and the political godfathers can all get their cut in spite and because of the ungovernability of the entire system. Violent racialized accumulation within the oil assemblage can, paradoxically, be self-producing as an economy of violence.

THE OFFSHORE FRONTIER: NEOLIBERALISM AND THE ACCUMULATION OF SECURITY IN THE GULF OF MEXICO

The outer continental shelf in the Gulf of Mexico is the largest U.S. oil-producing region. Not unexpectedly, the Gulf's oil complex—the assemblage of firms, the state, and communities that shape the character of oil and gas extraction—is massive by any accounting.[36] With more than four thousand currently operating wells, the Gulf accounts for one-third of U.S. crude oil production and over 40 percent of U.S. refining capacity. Over the past century, companies have drilled more than fifty thousand wells in the Gulf, almost four thousand of them in deepwater (more than one thousand feet). In the last fifteen years more than sixty wells have been drilled in the ultradeepwater zone—in more than five thousand feet of water—deploying dynamic positioning systems that use computers and satellites to keep rigs and supply vessels steady in rough seas and high winds.[37] By 2001 deepwater oil production surpassed the shallow-water shelf extraction. There are 3,020 platforms currently operating, but they represent only a small part of the Gulf's oil and gas infrastructure: thirty-three thousand miles of pipeline on- and offshore connected with a network of terminals, as well as a huge capital investment of refineries, storage facilities, shipyards, and construction facilities stringing the coast from Mississippi to Texas. It is a massive industrial cluster directly employing more than four hundred thousand people in Louisiana, Texas, Alabama, and Mississippi, generating $70 billion annually in economic value and $20 billion annually in tax revenue and royalty payments to local, state, and federal governments. The total fixed capital in the Gulf oil complex is now valued at an estimated $2 trillion.

Louisiana's section of the Gulf, which contains many of the nation's largest oil fields, holds more than nine-tenths of the crude-oil reserves in that region. As of 2011, Louisiana was the fifth-largest producer of crude oil and the fourth-largest producer of natural gas in the United States. More than 228,000 oil and gas wells have been drilled in the state over the past century. Louisiana's outer continental-shelf oil and gas production is greater than any other federally regulated offshore area in the United States. The industry's history in the state is synonymous with the history of U.S. offshore frontier development. While the first onshore well was drilled in the state in Jennings in 1901, many of the subsequent developments were offshore in shallow water along the coast. Since 1938 the state's inner and outer continental shelf has become the forcing house for the development of deepwater and ultradeepwater oil and gas production.

Louisiana is America's very own petrostate, a living testimony to the petropopulism and oil-based human and ecological development failures that have typically afflicted oil-producing states in the Global South. Petrocorruption and the shady politics of oil development were there from the beginning, as the oil industry emerged on the backs of an extractive economy (timber, sulfur, rice, salt, furs). Local businessmen snapped up land and threw themselves into a chaotic landgrab backed by Texas drillers and operators with little regard for the law. Wildcatting sprung up with no regulation; leases, especially along the coast wetlands, were allocated behind closed doors. Huey P. Long famously launched his career with an attack on Standard Oil and then proceeded to build his own subterranean oil empire. While senator, Long and his political cronies established the Win or Lose Corporation, which acquired cut-rate mineral leases through the government and resold them at a healthy profit. At the same time, he used oil severance taxes to begin a populist program of public service provision, which integrated a white working class (and subsequently African American petrochemical workers) into a program of economic modernization.

In the 1920s and 1930s, the wetlands leases opened up a new frontier as companies built a sprawling network of roads, canals, platforms, and wells, all of which left an indelible mark on the wetlands.[38] By the 1970s, when oil was providing 40 percent of state revenues, Louisiana ranked at the very bottom of the heap in terms of basic development indicators. According to the *Measure of America Report* on Louisiana, the state currently ranks forty-ninth in terms of human development indexes.[39] Massive inequalities between white and black populations mark all measures of human well-being; infant mortality and homicide rates are comparable to parts of Central America and sub-Saharan Africa. In some oil parishes, well-being is roughly at the average level of the United States in 1950. The report points out that 80 percent of wetland losses in the United States over the last century have been in Louisiana. Tulane law professor Oliver Houck put it this way: "What oil and gas did is replace the agricultural plantation culture with an oil and gas plantation culture."[40]

The history of the Louisiana oil industry is a textbook case of frontier dispossession and reckless accumulation running far in advance of state oversight and effective regulation. Offshore drilling technology was in effect born and nurtured in Louisiana with an assist from Venezuela—the former along the shallow coastal waters, the latter on Lake Maracaibo. Technology quickly developed from oil derricks on piers to stationary, mobile, and, by the 1930s, submersible drilling barges. All of this was propelled by new seismic technologies, which uncovered numerous salt domes across the coast and offshore region. Between 1937 and 1977, almost twenty-seven thousand wells were drilled in the coastal parishes, including shallow offshore. It was the first wave of leases and backroom petropopulism that unleashed a torrent of canal construction, dredging, and pipeline corridor construction (to say nothing of the emerging petrochemical complex in what became "Cancer Alley") and permitted large-scale salt intrusion and rapid coastal degradation.

Offshore development is customarily dated to 1905 in Louisiana but began in earnest in 1938 with a Brown and Root–constructed freestanding structure 1.5 miles from shore in fourteen feet of water, the so-called Creole Field.[41] In 1945 Louisiana offered the first lease sale, and one year later a platform was built five miles out in shallow water and, in a move repeated many times over, drew on the local fishing industry to assist in construction and ferrying workers to the "floating hotels." By 1947, Kerr-McGee had drilled the first well "out of sight of land" using war-surplus barges and other equipment to house drilling and workers, thereby reducing the size and cost of the self-contained drilling and production platform. These first, tentative developments precipitated, however, a titanic seven-year struggle over jurisdiction of the outer continental shelf in which the oil companies supported states' rights (to continue the lax, or rather nonexistent, regulation) over federal claims. This struggle was finally resolved in 1953 through two key pieces of legislation—the Submerged Lands Act and the Outer Continental Shelf Lands Act—which authorized the Secretary of the Interior to offer leases for competitive bidding beyond the three-mile limit. By 1957 there were 446 platforms in federal and state waters, and the rush was on.

In practice, the moving offshore frontier was transformed through four giant waves of frontier development, a quartet of landgrabs and dispossession. The first was almost wholly unregulated during the late 1940s and 1950s prior to and immediately after the resolution of the state jurisdiction question. A second occurred in the wake of the oil import quotas of 1959, which unleashed another round of major leasing; 2 million acres were leased in 1962, in water depths up to 125 feet, more than all previous sales combined. Oil production almost tripled between 1962 and 1968, and deepwater operations had by this time reached 300 feet. The first subsea well was drilled in 1966. As the National Commission on the BP Deepwater Horizon Oil Spill noted, this period was associated with massive hurricane damage and serial accidents, including blowouts, injuries, and helicopter crashes. A 1973 National Science Foundation report noted what was clear to everyone—namely, widespread collusion between industry and government and very little government oversight. The U.S. Geological Survey freely granted waivers from complying with the limited regulations and inspection demands while the regulatory agencies were hopelessly underfunded and understaffed.

The election of Ronald Reagan in 1980 marked not just the third round of leasing but a full-fledged neoliberalization of the Gulf deploying the now-expanded powers of the Department of the Interior. Under the leadership of Secretary of the Interior James Watt came a promise to open up the outer continental shelf to area-wide leasing; he placed one *billion* acres on the block. He began by establishing a new agency in 1982—the Minerals Management Service (MMS)—which created eighteen large planning areas rather than the traditional three-mile square blocks. While Watt subsequently resigned amid controversy and congressional opposition to outer continental shelf development on the East and West Coasts, the Gulf was in fact exempt, and the result was a land rush and massive exploration and production that constituted the third deepwater frontier wave. Seven sales between 1983 and 1985 leased more acreages than all previous leases combined since 1962; 25 percent were located in deepwater, and the lion's share was captured by Shell, the leading innovator and player in offshore technology and production. At the same time, the reforms provided for radically reduced royalties and federal bonus bids, with the consequence that companies paid 30 percent less despite a sixfold acreage expansion. In 1987 the MMS reduced the minimum bid for deepwater tracts, enabling a few companies to lock up entire basins for ten years for almost nothing. The fruits of this frontier expansion were visible a decade later: in that period deepwater wells grew from 4 percent to over 45 percent of all Gulf production. The 1980s provided an ideological resolution to the issue: the environmental enforcement capacities were eviscerated, and what emerged was a "culture of revenue maximization," as the National Commission on the BP Deepwater Horizon Oil Spill and Offshore Drilling put it.[42]

The 1990s proved to be nothing short of a "stampede." Seismic innovations, a new generation of drilling vessels capable of drilling in ten thousand feet of water and through thirty thousand feet of sediments, and new drilling techniques pushed the deepwater frontier to the so-called subsalt plays. Ten years later the Gingrich revolution ushered in another reform to lay the basis for another round of accumulation by dispossession: the Outer Continental Shelf Deep Water Royalty Relief Act of 1995 suspended all royalties to be paid by the companies for five years. In turn, this produced another round of landgrabs in which 2,840 leases were sold in three years. By 2000, deepwater production topped shallow-water output for the first time. At the same time, the ascendant BP was increasingly displacing Shell's hegemony in the Gulf. By using new 3-D seismic technologies, BP had made a series of remarkable discoveries and by 2002 was the largest acreage holder in deepwater. The MMS budget reached its budgetary nadir precisely during this boom. The *Houston Chronicle* reported that over the 1990s there was an 81 percent increase in offshore fires, explosions, and blowouts. In the following decade it increased fourfold.

The final wave of frontier accumulation was triggered by the election of George W. Bush in 2001 and the events of September 11. On May 18, two days after Vice President Dick Cheney's Energy Task Force report was delivered, Bush issued Executive Order 13212 (titled Actions to Expedite Energy Related Projects), the purpose of which was to "expedite [the] review of permits or other actions necessary to complete the completion of such projects." The language was, as a number of commentators pointed out, almost identical to that of a memorandum on the "streamlining" of development in the OCS submitted by the American Gas Association to the Cheney Task Force. The MMS was already laboring under a congressionally mandated rule to limit permit review to an impossibly confining thirty days, but the new order pushed things much further: in its wake, four hundred waivers were granted every year for offshore development. As offshore exploration and production stepped into historically unprecedented ultradeepwater, the permitting process and enforcement were laughable. MMS was not simply toothless and staffed by the sorts of oilmen it was designed to regulate but, according to a 2008 inspector-general report, was a hothouse of, among other things, a culture of substance abuse and promiscuity. To round out the abandonment of anything like supervision, in June 2008 Bush removed the ban on offshore drilling. Oversight deteriorated to the point where the National Oceanic and Atmospheric Administration was publicly accusing MMS of purposefully understating the likelihood and consequences of major offshore spills and blowouts. Watt's new system was nothing more than "tossing a few darts at a huge map of the Gulf."[43]

Shell announced the birth of the new neoliberal frontier in 2009. The Perdido platform, located two hundred miles offshore in water two miles deep, is nearly as tall as the Empire State Building, drawing in oil from thirty-five wells in three fields over twenty-seven square miles of ocean. Sitting atop an "elephant" field rumored to contain as much as six hundred billion gallons of oil, the scramble was on. In similar fashion BP pushed forward on a hugely ambitious program to develop multiple fields in the most demanding and unforgiving of environments, pushing deeper into old Paleogene and Lower Tertiary strata. The likes of Thunder Horse—BP's massive semisubmersible production facility—located in the Mississippi Canyon 252 Lease and the Macondo Well (forty miles distant) represented, as the National Commission on the BP Deepwater Horizon Oil Spill and Offshore Drilling put it, "formidable tests."[44]

Viewed on the larger canvas of the *longue durée* of offshore development, Perdido and Macondo were the expressions of what one might call the accumulation of insecurity, and the neoliberal production of systemic risks in the Gulf of Mexico—each rooted in the politics of substituting technological and financial for political risk. BP and Shell were drilling in five to ten thousand feet of water fifty miles offshore to produce deepwater oil close to the U.S. market offering a regulatory framework that can best be characterized as producer friendly. Especially in the wake of September 11, this produced a much better risk audit than dealing with the Russians in Siberia or the Angolans in the Gulf of Guinea. The Deepwater Horizon catastrophe was overdetermined by the vast accumulation of risks fabricated along the shifting frontier of offshore accumulation.

Nowhere are the links between deregulation and neoliberal capitalism clearer than in the 2011 report by the Deepwater Horizon Study Group.[45] In its devastating assessment, the catastrophic failure resulted from multiple violations of the laws of public resource development and its proper regulatory oversight by a BP safety culture compromised by management's desire to "close the competitive gap" and to save time and money—and make money—by making trade-offs for the certainty of production. Because there were perceived to be no downsides, BP's corporate culture was embedded in risk taking and cost cutting with little guidance provided by public regulatory agencies.

FRONTIERS OF PRECARITY AND SPACES OF INSECURITY

The dynamics along these two oil frontiers—one onshore, one offshore—have family resemblances, but each also highlights quite different, and often contradictory, dynamics in the local operations of the global oil assemblage. Each is a socially produced, globalized, yet fragmented space exhibiting what Judith Butler calls a radical precarity, albeit constituted in quite different ways and existentially along different axes of insecurity.[46] Each frontier also produces a sort of ecological and human slow death, but coupled to a radical turbulence (one political and military, the other financial and economic) that threatens the very operations of the industry itself.[47] Accumulation by dispossession is a defining feature along the oil frontier in these geographically dispersed settings. If the Niger Delta frontier reveals one explosive dynamic in the oil assemblage that resembles a combination of violent accumulation with fragmented sovereignties, the Deepwater Horizon suggests another. The Macondo Well disaster reveals the deadly intersection of the aggressive enclosure of a new technologically risky resource frontier with the operations of neoliberalized risk, a lethal product of cutthroat corporate cost cutting, the collapse of government oversight and regulatory authority, and the deepening financialization and securitization of the oil market. These two local pockets of disorder, turbulence, and catastrophe in the global oil assemblage point to, and are expressions of, the deep pathologies and vulnerabilities in the operations of what passes as Big Oil. If the onshore frontier in Nigeria ends in insurgency, in Louisiana and the Gulf the political story ends with class-action suits, a reorganization of regulatory agencies, but ultimately with the abandonment of President Obama's moratorium on and the gradual resumption of deepwater drilling. In both gulfs, the oil assemblage lurches forward, simultaneously advancing and refiguring the frontiers yet at the same time multiplying—and amplifying—the production of both profit and risk. The capacity of the oil assemblage to absorb disorder— the oil frontier is its leading edge—is one of the enduring hallmarks of contemporary hydrocarbon capitalism.

Notes

1. Amnesty International, *Petroleum, Pollution, and Poverty in the Niger Delta* (London: Amnesty International, 2009), 16. These oil spill estimates are contained in a letter written by thirteen Nigerian and five international nongovernmental organizations (NGOs) in relation to the National Oil Spill Detection and Response Agency (NOSDRA) Bill, presented before the Nigerian Senate in February 2014.

2. See Gavin Bridge, "Global Production Networks and the Extractive Sector," *Journal of Economic Geography* 8 (2008): 389–419.

3. See Martí Orta-Martinez and Matt Finer, "Oil Frontiers and Indigenous Resistance in the Peruvian Amazon," *Ecological Economics* 70 (2010): 207–18; Robert Vitalis, *America's Kingdom: Mythmaking on the Saudi Oil Frontier* (Stanford: Stanford University Press, 2006); Michael Peel, *A Swamp Full of Dollars: Pipelines and Paramilitaries at Nigeria's Oil Frontier* (Chicago: Chicago Review Press, 2010).

4. A theoretical point of reference for the idea of a frontier as a form of socially produced space is the work of Henri Lefebvre (1991)—in particular, in seeing the frontier as an expression

of what he calls, on the one hand, abstract space (a matrix of social action characterized by a formal, quantitative, and geometric set of qualities associated with modern capitalism) and, on the other, as a particular form of globalized state territory—that is, a space is managed, produced, and fragmented through one particular form of capitalism (what he calls the "state mode of production" associated with "neo-capitalism"): see Brenner and Elden (2009). As a social space and an expression of particular sorts of spatial practice, the frontier is also in Lefebvrian terms a space of representation (it is cognized in particular ways by expert knowledge—for example, the oil industry) and a representational space (a set of lived realities and form of everyday life).

5. The work on frontiers is, of course, substantial: see Philip McMichael, *Settlers and the Agrarian Question: Capitalism in Colonial Australia* (Cambridge: Cambridge University Press, 2004); Jonathan Goodhand, "Frontiers and Wars: The Opium Economy in Afghanistan," *Journal of Agrarian Change* 5, no. 2 (2005): 191–216; Keith Barney, "Laos and the Making of a Relational Resource Frontier," *Geographical Journal* 175, no. 2 (2009): 146–59; Jeanne Ferguson and Claude Raffestin, "Elements for a Theory of the Frontier," *Diogenes* 34, no. 1 (1986): 1–18; Richard Hogan, "The Frontier as Social Control," *Theory and Society* 14, no. 1 (1985): 35–81.

6. John Markoff, "Afterword," in *States of Violence*, ed. Fernando Coronil and Julie Skurski (Ann Arbor: University of Michigan Press, 2006), 36.

7. Fredric Lane, *Venice and History* (Baltimore: Johns Hopkins University Press, 1966).

8. S. Beretta and John Markoff, "Civilization and Barbarism," in *States of Violence*, ed. Fernando Coronil and Julie Skurski (Ann Arbor: University of Michigan Press, 2006), 51.

9. Ron distinguishes frontiers from ghettos; the latter are "ethnic or national enclaves securely trapped within the dominant state" but that enjoy some protections from the violence of the frontier. See James Ron, *Frontiers and Ghettos: State Violence in Serbia and Israel* (Berkeley: University of California Press, 2003), 16–17, 192. This gradation of spaces in relation to power is precisely what I am trying to identify in the two cases under study here: the U.S. Gulf of Mexico and Nigeria.

10. Michael Redclift, *Frontiers: Histories of Civil Society and Nature* (Cambridge: MIT Press, 2006), 31.

11. Henri Lefebvre, *The Production of Space* (Oxford: Blackwell, 1978).

12. Redclift, *Frontiers*, 208.

13. John Markoff, "Afterword," in *States of Violence*, 81.

14. I have explored the notion of economies of violence and violent accumulation elsewhere; see Michael Watts, "Blood Oil," in *Crude Domination: An Anthropology of Oil*, ed. Andrea Behrends, Stephen Reyna, and Gunther Schlee (Oxford: Berghahn, 2011), 49–80; Watts, "Economies of Violence," online at http://oldweb.geog.berkeley.edu/ProjectsResources/ ND%20Website/NigerDelta/pubs.html. See also Anton Blok, *The Mafia of a Sicilian Village: A Study of Violent Peasant Entrepreneurs* (Oxford: Basil Blackwell, 1974); and Vadim Volkov, *Violent Entrepreneurs: The Use of Force in the Making of Russian Capitalism* (Ithaca: Cornell University Press, 2002).

15. Jeannie Klever, "Despite Troubles, Africa Beckons as Oil Frontier" (January 7, 2013), at http//:fuelfix.com/blog/2013/01/07/.

16. A recent story in the *New York Times*, for example, describes the reopening of an oil frontier in Nixon, Texas: see Clifford Krauss, "Domestic Oil Boom Drives a Cautious Refining Revival," *New York Times*, March 4, 2014, http://www.nytimes.com/2014/03/04/business/ energy-environment/oil-boom-is-driving-a-revival-in-refining.html?_r=0.

17. Peter Nolan and Mark Thurber, "On the State's Choice of Oil Company," *Working Paper 99* (Palo Alto: Stanford Program on Energy and Sustainable Development, 2010).

18. Ibid.

19. Edward Barbier, *Scarcity and Frontiers: How Economies Have Developed through Natural Resource Exploitation* (London: Cambridge University Press, 2010), 39.

20. On resource frontiers, see Barney, "Laos and the Making of a Relational Resource Frontier." By "oil complex" or "oil assemblage" (I use the terms interchangeably), I refer to the vast institutional fields of oil and gas operations: typically this refers to transnational and national oil companies (both of which are granted precious little agency) and government (the corrupt petrostate). However, key actors in the complex include construction and banking corporations, private and other security forces, local chiefs and forms of customary rule, NGOs and transparency organizations, cultural and social organizations (youth groups), multilateral development agencies, and increasingly the organized local social groups and "enterprises" (insurgents, armed militias, organized crime) that seize upon opportunities to acquire oil rents. See Watts, "Blood Oil."

21. See Barbier, *Scarcity and Frontiers*.

22. Ibid., 21, 23.

23. Massimo De Angelis, "Separating the Doing and the Deed: Capital and the Continuous Nature of Enclosures," *Historical Materialism* 12 (2004): 57–87.

24. Ibid., 72. For another account of extractive frontier capitalism, see Anna Lowenhaupt Tsing, *Friction: An Ethnography of Global Connection* (Princeton: Princeton University Press, 2005), 28.

25. Edward Burtynsky, *Oil* (Göttingen: Steidl; Washington D.C.: Corcoran Gallery of Art, 2011).

26. Guido di Tella, "The Economics of the Frontier," in *Economics in the Long View,* ed. Charles Kindleberger (London: Macmillan, 1982), 210–27.

27. Eyal Weizman, *Hollow Land: Israel's Architecture of Occupation* (New York: Verso, 2012), 7.

28. See Lefebvre, *The Production of Space*; Manu Goswami, *Producing India: From Colonial Economy to National Space* (Chicago: University of Chicago Press, 2004).

29. This section draws on my own work and Kathryn Nwajiaku-Dahou, "Oil Politics and Identity Transformation in Nigeria" (PhD diss., Oxford University, 2005); Nwajiaku, "The Political Economy of Oil and Rebellion in Nigeria's Niger Delta," *Review of African Political Economy* 132 (2012): 295–314; *Oil and Insurgency in the Niger Delta: Managing the Complex Politics of Petroviolence,* ed. Cyril Obi and Siri Aas Rustad (London: Zed Books, 2011); Augustine Ikelegbe, "The Economy of Conflict in Oil-Rich Niger Delta Region of Nigeria," *African and Asian Studies* 5, no. 1 (2006): 23–55; Elias Courson, "Movement for the Emancipation of the Niger Delta," *Discussion Paper 57* (Uppsala: Nordic Africa Institute, 2009); Aderoju Oyefusi, "Oil and the Propensity for Armed Struggle in Niger Delta Region of Nigeria," *Post Conflict Transitions Papers,* No. 8 WPS4194 (Washington, D.C.: World Bank, 2007).

30. Nicolas Argenti, *The Intestines of the State: Youth, Violence, and Belated Histories in the Cameroon Grassfields* (Chicago: University of Chicago Press, 2007); Mike McGovern, *Making War in Côte d'Ivoire* (Chicago: University of Chicago Press, 2011).

31. Murray Last, "Muslims and Christians in Nigeria: An Economy of Panic," *The Round Table: The Commonwealth Journal of International Affairs* 8 (2007): 605–16.

32. UNDP, Niger Delta, Situation Assessment and Opportunities for Engagement (Port Harcourt/Abuja: UNDP, 2007).

33. On Sicilian mafias, see Blok, *The Mafia of a Sicilian Village.*

34. Henri Lefebvre, *The Production of Space,* trans. Donald Nicholson-Smith (Cambridge: Blackwell, 1991).

35. Peter Geschiere, *The Perils of Belonging: Autochthony, Citizenship, and Exclusion in Africa and Europe* (Chicago: University of Chicago Press, 2009)

36. This sections draws on the following work: Dianne Austin et al., *History of the Offshore Oil and Gas Industry in Southern Louisiana,* Volume 1: Papers on the Evolving Offshore Industry (Baton Rouge: Louisiana State University, 2001); Brady Banta, "Money, Resources, and Gentlemen: Petroleum Severance Taxation, 1910–1925," in *Louisiana Politics and the Paradoxes of Reaction and Reform, 1877–1928,* ed. Matthew Schott, vol. 7 (Lafayette: Center for Louisiana Studies, 2000), 624–45; Bob Cavnar, *Disaster on the Horizon: High Stakes, High Risks, and the Story behind the Deepwater Well Blowout* (New York: Chelsea Green, 2010); Deepwater Horizon Study Group, *Final Report on the Macondo Well Blowout* (Berkeley: University of California, 2011); Robert Gramling, *Offshore Development, Conflict, Gridlock* (Albany: SUNY Press, 1996); Tom Shroder and John Konrad, *Fire on the Horizon: The Untold Story of the Gulf Oil Disaster* (New York: HarperCollins, 2011); Peter Lehner with Bob Deans, *In Deep Water: The Anatomy of a Disaster, The Fate of the Gulf, and How to End Our Oil Addiction* (New York, O/R Books, 2011); National Commission on the BP Deepwater Horizon Oil Spill and Offshore Drilling, *Deep Water: The Gulf Disaster and the Future of Offshore Drilling,* The Oil Spill Commission, U.S. Congress, Washington, D.C. (2011), http://www.oilspillcommission.gov/final-report; Loren Steffy, *Drowning in Oil: BP and the Reckless Pursuit of Profit* (New York: McGraw-Hill, 2010); William R. Freudenberg and Robert Gramling, *Blowout in the Gulf: The BP Oil Spill Disaster and the Future of Energy in America* (Cambridge: MIT Press, 2010); Tyler Priest, "Extraction not Creation," *Enterprise and Society* 8, no. 2 (2007): 227–67.

37. Craig Colten, "An Incomplete Solution: Oil and Water in Louisiana," *Journal of American History* 99 (2012): 91–99; Tyler Priest and Michael Botson, "Bucking the Odds: Organized Labor in Gulf Coast Oil Refining," *Journal of American History* 99 (2012): 100–110.

38. Priest, "Extraction not Creation."

39. Sarah Burd-Sharp, Kristen Lewis, and Eduardo Martins, *A Portrait of Louisiana* (New York: Social Science Research Council, 2009).

40. David Hilzenrath, "MMS Investigations of Oil-Rig Accidents Have History of Inconsistency," *Washington Post,* July 18, 2010.

41. Tyler Priest, "A Perpetual Extractive Frontier? The History of Offshore Petroleum in the Gulf of Mexico," in *Nature, Raw Materials, and Political Economy,* ed. Paul S. Ciccantell, David A. Smith, and Gay Seidman (Oxford: Oxford University Press, 2005), 209–29.

42. National Commission, *Deep Water,* 76.

43. Freudenberg and Gramling, *Blowout in the Gulf,* 148.

44. National Commission, *Deep Water,* 145.

45. Deepwater Horizon Study Group, *Final Report on the Macondo Well Blowout* (2011), http://ccrm.berkeley.edu/deepwaterhorizonstudygroup/dhsg_reportsandtestimony.shtml.

46. Judith Butler, *Precarious Life: The Powers of Mourning and Violence* (New York: Verso, 2004).

47. Rob Nixon, *Slow Violence and the Environmentalism of the Poor* (Cambridge: Harvard University Press, 2011); Lauren Berlant, *Cruel Optimism* (Durham: Duke University Press, 2011).

11

Petro-Magic-Realism Revisited

Unimagining and Reimagining the Niger Delta

JENNIFER WENZEL

In memory of Fernando Coronil

Given the overlap in the trajectories of petroleum and publishing in its postcolonial history, Nigeria confirms and challenges some long-standing assumptions in literary studies about how cultural production fosters national imagining. In 1958, two years before Independence, Nigeria exported its first barrel of oil from Port Harcourt. That year was also a seminal early moment in Nigeria's literary exports, with the publication in London of Chinua Achebe's *Things Fall Apart*. The timing of Nigeria's simultaneous entry into global print and petrocapitalisms on the eve of its independence may have been a historical coincidence, but the imbrication of oil and literature in national imagining and international circulation has continued for decades. The Nigerian novel boom followed the contours of the oil boom and bust: the number of novels published each year increased steadily through the 1960s and 1970s, with explosive growth, wild fluctuations, and a crash in the 1980s.[1] From the vantage of the Niger Delta—a space that has been politically marginal yet indispensable to the Nigerian national project—one can see how oil has fueled national imagining from the beginning, and how the contradictions within such a project are perhaps better read as an *unimagining* of national community.

The discovery of commercial oil deposits on the eve of Nigeria's independence fostered a narrative of modernization and national development fueled by petro-promise. Political economist Ukoha Ukiwo recalls that at the moment of decolonization, petroleum, like palm oil (and slaves) in centuries past, appeared as "an angel of history ... a mass commodity [that] presented itself as the Niger Delta's savior."[2] Ukiwo refers to the mid-twentieth-century hope among ethnic minorities of the Niger Delta that oil would secure their economic development and political clout within an emergent federation that favored the Hausa, Igbo, and Yoruba majorities. This sense of petro-promise was not to be realized: oil revenues were never channeled back to the oil-producing states in ratios that reflected the indispensability (or commercial profitability) of oil. Likewise, at the scale of the nation-state, postcolonial Nigeria can be read as a story of de-development, what geographer Michael Watts calls a "catastrophic failure of secular nationalist development" that finds "most Nigerians poorer today than they were in the late colonial period" despite the billions of petro-naira that have flowed through the federal state's coffers.[3]

When Ukiwo describes the discovery of oil as an "angel of history," he seems to have in mind an unexpected savior that brings prosperity, rather than a melancholic Benjaminian angel who is forced to look back on the accumulated wreckage of the past while the storm of history blows him helplessly into the future. But perhaps it is better, when considering how oil shapes narratives of development in the Niger Delta and Nigeria more broadly, to see Ukiwo's viscous angel of history as a *deus ex machina* (or *oleum ex machina*): a miraculous agent, external to a historical narrative, whose arrival makes possible what is otherwise impossible within the narrative's own terms. That is to say, there is something almost antinarrative about the ontology of oil, if narrative is understood as the working out of cause and effect and oil is understood to produce something out of nothing. Not so unlike the new breed of militants in the Niger Delta, whose tactics include sabotage and occupation of oil installations and kidnapping of oil company personnel, oil hijacks the imagination. Oil, as Watts has written, "harbors fetishistic qualities: it is the bearer of meanings, hopes, expectations of unimaginable powers."[4] In a passage that has itself captured the imagination of myriad critics, Ryszard Kapuściński writes about the false promises of oil: "Oil creates the illusion of a completely changed life, life without work, life for free. . . . The concept of oil expresses perfectly the eternal human dream of wealth achieved through lucky accident, through a kiss of fortune and not by sweat, anguish, hard work. In this sense, oil is a fairy tale and like every fairy tale a bit of a lie. Oil fills us with such arrogance that we begin believing we can easily overcome such unyielding obstacles as time."[5] All surplus! all the time! is the fantasy of oil; as Kapuściński writes, oil is "a filthy, foul-smelling liquid that squirts up obligingly into the air and falls back to earth as a rustling shower of money" (34). Oil promises wealth without work, progress without the passage of time: the narrative mode appropriate to petro-promise is not the incremental, developmentalist progress narrative of modernization, but rather the fairy tale of instant transformation at the wave of a magic wand, in which every dream of infrastructure comes true.

In a previous essay, I have argued for reading Nigerian literary history in terms of political ecology, a discipline concerned with the political, economic, and cultural aspects of conflicts over natural resources.[6] Here I revisit that argument in order to further examine how oil has shaped narratives of national development and modernization in Nigeria. Watts has identified an "unraveling—or unimagining" of national community in contemporary Nigeria.[7] The literary texts that I read in this chapter, Ogaga Ifowodo's long poem *The Oil Lamp* and Ben Okri's short story "What the Tapster Saw," suggest that the sense of simultaneity and conviviality among strangers that Benedict Anderson sees the novel and the newspaper fostering in *Imagined Communities* are less salient in Nigeria than a sense of disjointed temporalities, contempt for all-too-familiar ethnic others, and disillusion with a shared national project: the unimagining of national community.[8] In addition to this Andersonian sense at the scale of the nation-state, *unimagining* has other meanings in sites of oil extraction like the Niger Delta, where underdevelopment produces wealth for some and poverty for others. Indeed, the oil enclave fosters a peculiar dynamic perhaps better described as de-development, where development begins and turns backward. At both the national scale of boom and bust, and the local scale of oil enclaves and the infrastructure that supports them, development is undone when prices fall or wells run dry. These material processes of underdevelopment or de-development have a cognitive counterpart in acts of unimagining. In underimagining (or de-imagining) these multiscalar relations that join the local oil enclave to the petrostate and the global oil market, unimagining names transitive processes of unmaking that are not merely analogous to underdevelopment but mutually produced with it, so that histories of immiseration are elided in the process of imagining people as eternally poor and backward, or as the victims of an infernal "resource curse."

But as the fantastic illusions associated with oil might suggest, *excess* is as much a hallmark of the petroleum-fueled imagination as are the failures involved in unimagining national community or de-imagining as a corollary of underdevelopment. What cultural forms abet these distinct but related processes of unimagining, and what cultural forms can work against them? In other words, I'm interested in imaginative modes that grasp the elusive workings of unimagining: acts of un-unimagining, which, for the sake of simplicity and elegance, I'll call *reimagining*. Texts like Ogaga Ifowodo's *The Oil Lamp* and Okri's "What the Tapster Saw" are able to grasp both excess and failure; they reimagine the significance of oil in Nigeria through a particular literary mode that I have called *petro-magic-realism,* which conjoins "petro-magic" (a concept from political ecology) with the literary mode of magical realism. In this essay, I want to revisit this link between petro-magic and magical realism, and to consider further who or what puts the "magic" in petro-magic-realism, in order to better understand its implications for the politics of circulation and reception in a global literary marketplace at a moment when expansionist notions of "world literature" are ascendant. Not unrelatedly, in a contemporary moment when it is difficult to decide whether the more pressing problem with oil is that there is too little or too much of it left to burn,[9] I argue that the literary mode of petro-magic-realism can help to puncture more than a few of the illusions associated with oil, not least the geographic foreclosures of the imagination associated with the "resource curse."

Echoing Ryszard Kapuściński's account of the fantastic illusions associated with oil, anthropologist Fernando Coronil argued that in terms of imagining the nation and producing the state, oil is magical. Complicating the conventional distinction between state and nation, Coronil argued that the nation had two "bodies": "a political body made up of its citizens and a natural body made up of its rich subsoil."[10] Arguing, in effect, for a shift from political economy to political ecology, Coronil posited that *petro-magic*[11] was the particular form that "the metabolism between society and nature" took in the Venezuelan petrostate: "By condensing within itself the multiple powers dispersed throughout the nation's two bodies, the state appeared as a single agent endowed with the magical power to remake the nation" (8, 4). Similarly, a "'seeing-is-believing' ontology" emerged at the height of Nigeria's oil boom in the late 1970s, Andrew Apter argues, when "oil replaced labor as the basis of national development, producing a deficit of value and an excess of wealth, or a paradoxical profit as loss."[12] When oil is figured as the blood of the nation, the state becomes a vampire that "consum[es] this lifeblood of the people—sucking back the money that it pumped into circulation" and destroying "the real productive base of Nigeria, those agricultural resources that not even a state-sponsored green revolution could revive" (269). Petro-magic is one of the forms that petro-violence takes;[13] its illusions of sweet surplus can, for a time, mask the harm that petroleum extraction does to humans and nonhuman nature, turning each into instruments of violence against the other.

Both the sweet lies and the bitter truths of petro-magic feature in Ifowodo's long poem *The Oil Lamp*. The fecundity of the petroleum-fueled imagination is evident in Ifowodo's character Major Kitemo, who boasts of his "excess of zeal": he knows "two hundred and twenty-one ways to kill a man."[14] The actuarial specificity of the number 221 recalls Ifowodo's dedication of *The Oil Lamp* to

The thousand-and-one
gone in the struggle
for a livable
Niger Delta,
a just Nigeria.

This invocation of the "thousand-and-one / gone in the struggle" has resonances both within and beyond *The Oil Lamp*. Most immediately, it references the official death toll of a two-week-long conflagration in October 1998 at the site of a pipe-line breach in the rural village of Jese (or Jesse) in Delta State, the subject of the first section of Ifowodo's long poem. But *The Oil Lamp* also invokes more fabled reso-nances in its epigraph from Salman Rushdie's *Midnight's Children*: "Numbers, too, have significance: . . . 1001, the number of night, of magic, of alternative realities—a number beloved of poets and detested by politicians." The excess of Kitemo's 221 methods of murder is countered by the 1,001 gone—the final digit standing in for those unaccounted for by the official tally at Jese, which, Ifowodo writes, included only "charred remains, whole enough to count" (14). Like Rushdie's *Midnight's Children* and *The Thousand and One Nights* before it, *The Oil Lamp* marshals the power of poets against politicians, so that storytelling becomes a strategy for warding off death.

Ifowodo's invocation of Rushdie links *The Oil Lamp* to the literary mode of "alter-native realities" known as magical realism, which Rushdie has aptly described as "so dense a commingling of the improbable and the mundane!"[15] And yet, beyond the ethical force of pitting the literary imagination against forces of death and damage, *The Oil Lamp* does not at first glance evince the fantastic excesses associated with magical realism. In pondering how to read *The Oil Lamp* as a magical realist text, I have come to focus on Ifowodo's depiction of nature within the unnatural landscape of oil extraction. In the "Jese" section, fields, creeks, and ponds are granted the power of speech to protest as fire races through "a land marked by oil for double torment" (12); trees vainly wave "green scarves for peace" at the onset of repression (20). Else-where, elements of the landscape are personified not as speaking for themselves but as instruments or even agents of petro-violence: the "tea-black water of the lake" ex-plodes, seemingly of its own volition. The "air shrieked" and a tree in the "shuddering forest" is transformed from refuge to weapon as it "cleaved in two," crushing a mother and child as it falls (22). These personifications effect something like magical realism's natural(ized) supernaturalism, and they suggest the multivalence of the subjects, objects, and instruments of the harm of petro-violence.

The most striking "comminglings of the improbable and the mundane" in *The Oil Lamp* involve other kinds of transformations and juxtapositions. "Waterscape," the poem that serves as preamble to *The Oil Lamp,* depicts an "alternative reality" not of fantasy and imagination, but of environmental history: a mangrove swamp seemingly prior to petromodernity, with "ancestral lakes" home to eels, crab, mudskippers, and fish in such abundance that fishermen enjoy more than one meal a day. This water-scape too is personified through metaphors of human culture and industry, its man-grove roots described as hands, tongues, and hair; the black water is "deeper than soot, / massive ink-well." But the presence of humans in this preambular waterscape is markedly different from that in the long poem proper. The "Jese" section contrasts spoiled rivers, erstwhile sources of life and livelihood, with "broken pipes, like the mouth of a river," which form "two brooks of kerosene and petrol" (5). In a quintessential image of de-development, the abandoned, rusting oil "drilling tree" at a "drilled-dry well" mocks the actual trees of the forest (5). Electricity—or *eletiriki*— is a "dream" that "burned bright / for forty years, powered by a plant, till the tree drilled its last barrel" (3). In this blighted landscape of techno-nature, trees of metal and trees of wood are artifacts in a techno-ecosystem produced by and, in turn, ever more vulnerable to capitalism.[16] The fuel crunch effects a particular kind of disenchantment that forces humans to disregard their "dread" of disturbing the "spirits that live in trees" (2). Even the "green twigs" are threatened by the year-long "fuel crunch" that "compelled choice between tree and human, / today and tomorrow" (2). Trees and humans are both caught within the incommensurability between the temporality of nature and that of capitalist exigency.[17]

As its organic elements find their counterparts in the machinery of oil extraction, the Niger Delta's singular landscape of petro-nature is juxtaposed with the global economy: the "visible oil market where dealers / sold in kegs and bottles for naira" (6) is contrasted with the faraway electronic abstractions of "faceless traders / in markets without stalls or hand-made goods" (4). Here magical realism's "commingling of the improbable and the mundane" reflects and produces incommensurability and disproportionality: the aggregate power of drilling trees, pipelines like rivers, brooks of hydrocarbons, and faceless, objectless markets over their "natural" counterparts makes them not of the same order, even if they coexist in the same place.[18]

The force of techno-nature becomes legible in *The Oil Lamp*'s recurrent image of rust, an electrochemical process that tends to be associated less with elemental nature (iron meets oxygen) than with industrial modernization having run its corrosive course: most obviously, the U.S. Rust Belt indexes the economic decline of deindustrialization. In *The Oil Lamp*, the de-development associated with resource extraction is legible in the "rusted sinews of the [drilling] tree, / . . . a promise in rust-flakes" (3). In the fuel crunch, trees of wood are too precious to be allowed to rot; for the "drilling tree" of metal, the experience of rot, the cycle or end of its term of "life," materializes not as falling leaves but as flakes of rust.[19] Kapuściński's magical "rustl[e]" of oil-turned-to-money gives way to the rust of metal-gone-to-ruin. Rust exposes the lies of petro-magic: all that is solid metal melts into flakes floating in sulfurous air. Corrosion of oil pipelines was the most immediate cause of the "damage" at the Jese pipeline fire for which Ifowodo also seeks more systemic causes: "A sickened earth rusted the pipes / and threw up the lie encased in hollow metal" (4). Such corrosion is significantly accelerated in this techno-ecosystem by the acid rain that is a literal precipitate of gas flaring.

Ifowodo uses rust as a defamiliarizing image within the chronotope of petro-magic. Its visual dullness and its association with postindustrial decline provide a necessary counterpoint to a more familiar image, the sickly sheen of oil as the emblem of modernity's Faustian bargain: these creeks must shimmer and die so that others may live, encircled within what Ifowodo calls petroleum's "chain of ease" (4). Indeed, Major Kitemo's allegiances to the Nigerian state mean that, while he can acknowledge "the devastation / that pours oil on rivers to float fish" in the Niger Delta (47), he cannot accept Delta residents'

> ... claim for redress, that it should empty
> the coffers and deny the nation's engine
> its lubricant. Rust would follow; there'd be an end
> to motion and a nation to call our own. (48)

For Kitemo, oil-as-money is the lubricant and fuel that keep the machine of national progress running; he must remain blind to the damage of underdevelopment, signified by the presence of the very rust that he fears.

In rendering visible the lineaments and lies of petro-magic—the commingling of rot and rust, and the disproportionalities of a techno-nature produced by hydrocarbon capitalism—Ifowodo's poem gestures toward the literary mode that I have called *petro-magic-realism*, which combines magical transmogrifications and fantastic landscapes with the monstrous-but-mundane violence of oil extraction, the state violence that supports it, and the environmental harm it causes. Petro-magic-realism is a fantastic literary mode that makes visible the all-too-real effects of petro-magic— read here as a mode of violence that mystifies through the seductions of petro-promise. Petro-magic-realism, in other words, can reveal the secrets behind petro-magic's tricks.

A quintessential example of petro-magic-realism is Ben Okri's short story "What the Tapster Saw," which approximates *The Oil Lamp*'s disjunctures between trees and drilling trees, or rivers and broken pipelines, in its superimposition of a landscape of petroleum extraction over the landscape of an earlier commodity with its own mode of combustion—palm oil.[20] Written in London in 1987, in the midst of the Nigerian oil bust, Okri's story offers a phantasmagoric glimpse into a degraded, privatized landscape where the "signboards of the world were getting bigger"; one signboard reads, "TRESPASSERS WILL BE PERSECUTED."[21] The story's protagonist is one such trespasser, a palm-wine tapster whose work of extracting wine from palm trees in Delta Oil Company territory is threatened by the work of petroleum extraction. Okri's tale of a tapster evokes the fantastic-modern world imagined in Amos Tutuola's novel *The Palm-Wine Drinkard* (1952), which itself draws upon a Yoruba narrative tradition, where the forest is a liminal space peopled with transmogrifying creatures. Such creatures are encountered in Tutuola's novel when the eponymous drinkard, who "had no other work more than to drink palm-wine in [his] life," embarks upon a quest to Deads Town in order to bring back his deceased "expert palm-wine tapster who had no other work more than to tap palm-wine every day."[22] *The Palm-Wine Drinkard* offers a prescient, fantastical depiction of petro-magic's disjuncture of work and wealth. Not unlike the actuarial specificity in *The Oil Lamp,* the drinkard's compulsion for outrageous numerical precision brings the "rationality" of capitalist accounting to the impossibilities and externalities of juju.[23]

In Okri's story "What the Tapster Saw," the sun seems never to set or rise as the earth is bathed in the glow of gas flares, "roseate flames [that] burned everywhere without consuming anything" (189). Amid this mysterious combustion-without-consumption, a talking snake glistens with the beautiful and deadly iridescence of oil spilled on water. In this landscape where boreholes crowd out palm trees, a palm-wine tapster carries on plying his trade despite the ominous signboards; when he falls from one of a "strange cluster of palm trees" (perhaps anticipating Ifowodo's drilling tree?), he spends seven days in a hallucinatory liminal state, persecuted by unseen assailants vaguely associated with the oil company employees trying unsuccessfully to "level the forest" with the help of "witch-doctors" and explosives "left over from the last war." The Delta Oil Company brings in the witch-doctors to "drive away the spirits from the forest," while farmers who had been living amid unexploded bombs "as if the original war was over were blown up as they struggled with their poverty" (186, 189, 188).

Juxtapositions of bombs and bullets, coups and executions, with herbalists and witch-doctors, talking animals and masquerades, in this fictional narrative about the collision of palm and petroleum, yield a petro-magic-realism that situates the magical and violent aspects of petromodernity within an older fantastic tradition and extractive economy. Here "tapping" of palm trees evokes both capital-intensive oil drilling and low-tech illicit "tapping" of pipelines that snake through fields and villages. Okri's tapster is not so much a direct descendant of Amos Tutuola's character as a distant cousin within a broader genealogy that the narrator acknowledges by referring to "mythical figures" that include "the famous blacksmith" and the "notorious tortoise" (193). Direct allusion invokes not Tutuola but rather D. O. Fagunwa, whose story at the beginning of *Forest of a Thousand Daemons* about the hero's father shooting an antelope who turns out to be his wife is echoed in a fragmentary tale told with "curious irrelevance" at the opening of Okri's story (183–84).[24]

Ben Okri is among the Anglophone African authors most frequently mentioned in critical discussions of magical realism as a global literary phenomenon. Although

Tutuola and Fagunwa are seen as precursors of West African magical realism, they are thought to lack sufficient cosmopolitan, ironic distance from the "traditional" or "indigenous" materials that tend to be identified as a primary source of the magic in magical realism.[25] But it's important, I think, to pause before such etiologies and to separate out several possible answers to the question, Who (or what) puts the magic in magical realism? That is to say, at stake in "magical realism" are distinct realms of reality, representation, and reception that are too often conflated with one another—a confusion that becomes urgent in the context of magical realist texts' circulation as commodity exports of the Global South in high demand in the northern hemisphere.

The distinction between the first two realms—reality and representation, or ontology and aesthetics—is evident in the tension between two seminal statements in the theorization of magical realism as a literary mode. For Alejo Carpentier, *"lo real maravilloso Americano"*—the American marvelous real—is a state of being shaped by the complex history and distinctive landscape of the Americas. Carpentier undertakes an accounting of this marvelous reality (compatible with petro-magic's promise of miraculously easy wealth) when he declares that there must be something marvelously real about the new world for Europeans to lose themselves in search of El Dorado: "A certain myth of El Dorado reflects a reality, a myth which is still fed by deposits of gold and precious stones."[26] This insight construes the marvelous real as a peculiar kind of resource curse that strikes European minds but also, in turn, afflicts colonized lands and bodies. By contrast, Franz Roh's 1925 discussion of Post-Expressionist European painting, where he coined the term *"der magischer Realismus"*—magical realism—describes an aesthetic strategy, a mode of representation.[27]

One of my aims in conjoining political ecology's analysis of petro-magic with the literary mode of magical realism is to consider how reality and representation converge: I coined *petro-magic-realism* in order to show how writers like Okri and Ifowodo imagine the pressures of a particular political ecology within a particular literary idiom. They draw attention to the devastating material effects and unimaginable disproportionalities of petro-magic, at the same time that they invoke literary precursors like Rushdie, Tutuola, and Fagunwa. Theirs is a particular kind of reimagining that encompasses petro-magic's castles in the air and its shacks in the swamp. Petro-magic offers the illusion of wealth without work; petro-magic-realism pierces such illusions, evoking a recognizably devastated, if also recognizably fantastic, landscape. As with the slippages in magical realism, here the geographical and aesthetic senses of *landscape* converge. Petro-magic-realism reveals the rust-inducing acid rain that follows the "rustling shower of money." Ifowodo and Okri are angels of history, surveying the wreckage of petroleum's progress from the treetops, the vantage of the fabled palmwine tapster.

These questions of mimesis are complicated by pressures of reception—the politics of reading within the global literary circuits that the new world literature takes as its purview. Magical realism has become as much a reading practice or marketing shorthand as a descriptor of realities or representational strategies existing *out there* in the world or in literary texts.[28] What I find most productive in petro-magic-realism is its potential to complicate and historicize the empty globalism of the label (or even "brand") *magical realism*, in which *magical* denotes anything unfamiliar to a European or American reader. Magical realism can function as something like the literary equivalent of the *resource-curse* hypothesis, which holds that bad governance and violent conflict are the inevitable lot of nation-states unlucky enough to be so well endowed with natural resources coveted by industrial capitalism that they can run their entire economies on them. The problem is perceived to be inherent to the abundant resource itself, or a perceived absence of traditions of democratic accountability in the places (disproportionately in the Global South) seen to be accursed, rather than resulting from the particular ways in which resource-rich states have been imbricated within global capitalism. Like oil itself, the resource-curse diagnosis hijacks the imagination: it is so appealing in First World analyses precisely because it lets the consumers of such resources, their governments, and the multinational corporations that supply them, off the hook. Similarly, an uncritical turn to magical realism as an interpretive rubric can allow readers to erect a comforting *cordon sanitaire* of "difference" and "distance" in narratives that might be read otherwise as tracing more complex geographies and complicities.

In a landmark essay, Stephen Slemon acknowledges that magical realism "threatens to become a monumentalizing category" by offering a "single locus upon which the massive problem of *difference* in literary expression can be managed into recognizable meaning in one swift pass," thereby "justifying an ignorance of the local histories behind specific textual practices."[29] Magical realism is often invoked when ethnographic interpretation proves inadequate for making sense of a text. In this mode of reading magical realism, the local literary genealogies that inform Okri's story (including the antelope folktale that Fagunwa textualizes) become a different kind of "curious irrelevance," or irrelevant curiosity.

The problem with the magic in magical realism is broader than the sanctioned ignorance of metropolitan readers of world literature, however. The relationship between realism and magic tends to be read as a binary opposition between the West and the rest, between a singular (but European) modernity and multifarious worldviews variously described as premodern, prescientific, pre-Enlightenment, non-Western, traditional, indigenous—or, in more recent and more subtly patronizing formulations, alternatively or globally modern. In his metacritique of magical realism, Michael Valdez

Moses notes that "if the paternity of the magical realist novel is everywhere the same" (in the European realist novel and its attendant ideology), then "in each locale where the magical realist novel is born, its mother appears to be different, distinct, and as it were, native to the region."[30] This analogy casts magical realism and its many mothers as the product of cross-cultural, imperialist polygamy. The cumulative effect of such strangely binary readings of magical realism—in which one term is always the same, the other always different—is the consolidation of the West as a single entity confronting innumerable local traditions.

This reification of the West seriously undermines claims for magical realism's subversive, antihegemonic, or decolonizing thrust. In this context, Zamora and Faris's playful definition of magical realism as a "return on capitalism's hegemonic investment in its . . . now achieving a compensatory extension of its market worldwide" is telling.[31] Read in terms of reception and the reduction of difference, magical realism tends to move smoothly along the traffic lines of culture, power, and profit established by colonialism, rather than to obstruct or reroute them. Such magical realist habits of reading (some of which, in turn, congeal in the production and circulation of texts) tend to reduce the fantastic or excessive into a reified and unexamined "cultural difference," rather than performing the critical acts of reimagining that I have in mind with petro-magic-realism's exposure of petro-magic's tricks.

The relationships among reality, representation, and the politics of reception in magical realism have important implications for narratives of modernization. Ifowodo's juxtaposition of incommensurable local and global markets, as well as Okri's tale of the tapster's nightmarish experience in Delta Oil Company territory, thematizes the conflict between established and emergent modes of production (and, I would add, modes of combustion and resource extraction) that Fredric Jameson posits as constitutive of magical realism.[32] In these landscapes of petro-nature, the texts reimagine a decidedly modern reality while drawing upon fantastic idioms with venerable literary histories that include both local narrative traditions and world literature. In other words, these texts insist that petro-magic is no vestige of precapitalism, nor a simple clash between tradition and modernity, or local content and foreign form. Petro-magic's enchantment (and eventual disenchantment) is that of petro-modernity, in a world-ecological sense: at once intensely local and shaped by global capitalism.[33] In both texts, the ecological ravages of petromodernity effect a forcible disenchantment by "driv[ing] away spirits from the forest," at the same time that petro-magic effects its own modern enchantments, creating and obscuring its own reality. In turn, petro-magic-realism stages the illusions of petro-magic only to puncture them, revealing the phantasmagoric effects of petro-violence. The modernity of petro-magic-realism obstructs the consumption of magical realist texts as nostalgic encounters with an exotic yet vanishing world. It rejects the quarantine of the imagination that underwrites the resource-curse diagnosis.

As Valdez Moses points out, the production and consumption of magical realist texts by "those *who would like to believe* in the marvelous" but who do not actually believe involves a tacit assumption that a disenchanted "modern world . . . is the only one with a historical future" (106). It is this kind of assumption about history as a narrative of progress into a future perfect of secular modernization that makes the dialectical and contradictory effects of underdevelopment and unimagining so difficult to grasp, particularly in extractive economies that generate wealth and poverty simultaneously. This assumption of linear progress allows Ifowodo's Major Kitemo to invoke the fearful prospect of rust as a reason for continued exploitation of the Niger Delta without seeing the damage that rust has already done. It allows Michael Watts to describe the juxtaposition of "ultramodern," capital-intensive oil installations with extreme poverty as a "horror" that aims to evoke surprise at an economic contrast that implies historical anachronism. When read with a critical awareness of these kinds of assumptions, petro-magic-realism can uncover both the lies of petro-magic and the temporal contradictions of underdevelopment. It can implicate metropolitan consumers of magical realist texts and petroleum products not in modernization's inevitable disenchantment of vestigial tradition, but rather in petro-modernity's phantasmagoric ravagements of societies and lifeworlds. This ravagement too is historical and dynamic, not easily contained within static boundaries or binaries between the West and the rest, the comforting assurances of a resource curse. Petro-magic's effects might be reimagined to include the "man-camps" and gas flares visible from space that are among the most extreme effects of North Dakota's recent oil boom. Across the United States, the rapid spread of the technologies of horizontal drilling and hydraulic fracturing is outpacing effective regulation and posing difficult choices for landowners and communities—particularly those in rural and Rust Belt regions long out of work, now promised easy wealth. Petro-magic: it's the future.

NOTES

1. Wendy Griswold, *Bearing Witness: Readers, Writers, and the Novel in Nigeria* (Princeton: Princeton University Press, 2000), 37–38.

2. Ukoha Ukiwo, "Empire of Commodities," in *Curse of the Black Gold: 50 Years of Oil in the Niger Delta*, ed. Ed Kashi and Michael Watts (Brooklyn, N.Y.: PowerHouse Books, 2008), 73.

3. Michael Watts, "Sweet and Sour," in *Curse of the Black Gold*, ed. Kashi and Watts, 44.

4. Michael Watts, "Violent Environments: Petroleum Conflict and the Political Ecology of Rule in the Niger Delta," in *Liberation Ecologies: Environment, Development, and Social Movements*, ed. Richard Peet and Michael Watts (New York: Routledge, 2002), 256.

5. Ryszard Kapuściński, *Shah of Shahs*, trans. William R. Brand and Katarzyna Mroczkowska-Brand (San Diego: Harcourt Brace Jovanovich, 1985), 35.

6. Jennifer Wenzel, "Petro-Magic-Realism: Towards a Political Ecology of Nigerian Literature," *Postcolonial Studies* 9, no. 4 (2006). Political ecology seeks to understand the "convergences of culture, power, and political economy" that inform conflicts over "defining, controlling, and managing nature" (Nancy Peluso and Michael Watts, "Violent Environments," in *Violent Environments*, ed. Peluso and Watts [Ithaca: Cornell University Press, 2001], 25). I'm interested in how questions of knowledge and cultural representation intersect with questions of nature and political representation: a political ecology of literature would be concerned not only with thinking about literature *as if* it were a commodifiable resource like oil, but also with the multivalent relationships between literary production and conflicts involving natural resources.

7. Watts, "Sweet and Sour," 47.

8. Rob Nixon uses "unimagined communities" to describe those who are not included when Andersonian national community is being imagined: these subnational communities are rendered invisible, in the name of national development. Rob Nixon, *Slow Violence and the Environmentalism of the Poor* (Cambridge: Harvard University Press, 2011), 150–74. I intend something else with *unimagining*: the relationships among transitive processes, within communities and outside them, through which a polity as a whole is *unmade* or a crisis is made to exceed the capacity to be imagined.

9. I'm paraphrasing Imre Szeman's pithy formulation of oil's paradoxically divergent economic, infrastructural, and environmental implications. See Imre Szeman, "Subjects of Oil? Energopolitics, Materialism, and Agency," *Cultural Critique* (forthcoming).

10. Fernando Coronil, *The Magical State: Nature, Money, and Modernity in Venezuela* (Chicago: University of Chicago Press, 1997), 4.

11. "Petro-magic" is Michael Watts's felicitous paraphrase of Coronil's argument about the magical aspects of the oil state; see Watts, *Petro-Violence: Some Thoughts on Community, Extraction, and Political Ecology* (Berkeley: Berkeley Workshop on Environmental Politics, Institute of International Studies, 1999), 7.

12. Andrew H. Apter, *The Pan-African Nation: Oil and the Spectacle of Culture in Nigeria* (Chicago: University of Chicago Press, 2005), 14, 201. Apter links oil to occult practices of "money magic" in southern Nigeria, whereby human blood and body parts are illicitly transmuted into currency (249). Apter draws on Karin Barber's classic essay, "Popular Reactions to the Petro-Naira," *Journal of Modern African Studies* 20, no. 3 (1982).

13. In *Petro-Violence*, Michael Watts theorizes the complex of ecological, economic, social, and political modes of violence associated with oil extraction.

14. Ogaga Ifowodo, *The Oil Lamp* (Trenton, N.J.: Africa World Press, 2005), 38. Major Kitemo is based on Paul Olusola Okuntimo, commander of the Rivers State Internal Security Task Force who coordinated with Shell in the Ogoni region in the 1990s.

15. Salman Rushdie, *Midnight's Children* (New York: Avon, 1982), 4.

16. Ifowodo depicts what Arturo Escobar distinguishes as organic, capitalist, and techno-nature in Arturo Escobar, "After Nature: Steps to an Antiessentialist Political Ecology," *Current Anthropology* 40, no. 1 (February 1999): 1–30.

17. The short-term rhythms of capitalist production disregard slower, longer organic cycles of growth and renewal, thereby "robbing the soil" of its fertility and making sustainable forestry "an industry of little attraction to private and therefore capitalist enterprise" (Karl Marx, *Capital*,

ed. Frederick Engels, trans. Samuel Moore and Edward Aveling [New York: International Publishers, 1967], 1:505–6, 2:244).

18. Environmental historian Jason Moore theorizes "world-ecology" as a corollary to world-system; it refers not (necessarily) to "the ecology of the [entire] world" but rather to how capitalism "progressively deepens the world-historical character of microlevel socio-ecologies in the interests of the ceaseless accumulation of capital.... With the rise of capitalism, ... varied and heretofore largely isolated local and regional socio-ecological relations were incorporated into—and at the same moment became constituting agents of—a capitalist world-ecology." Jason Moore, "Capitalism as World-Ecology," *Organization & Environment* 16, no. 4 (December 2003): 447.

19. These rust flakes also link Ifowodo's infernal landscape to the "flakes of fire" in Dante's *Inferno*, in a passage that serves (with *Midnight's Children*) as epigraph to *The Oil Lamp*.

20. The African oil palm, *Elaeis guineensis*, yields wine, oil, and kernels. Unlike palm wine, which spoils quickly and does not travel well, palm oil and kernel were drawn into global circuits of exchange. The European palm trade in West Africa dates as far back as the 1480s and was worth a million pounds by 1840. In the colonial era, Nigeria's Oil Rivers region was named for palm oil, perhaps as indispensable for nineteenth-century industry as petroleum later became. Palm oil (known as "Lagos oil") was used as an industrial lubricant, an edible oil, and in the making of soap, tin, and candles. Beyond their significance as exportable commodities, palm oil and kernel have been used locally for food and illumination; the tree yields materials for building, roofing, and other household uses. Within the riverine economy of the Oil Rivers region, jars of palm oil functioned as currency. See Martin Lynn, *Commerce and Economic Change in West Africa: The Palm Oil Trade in the Nineteenth Century* (New York: Cambridge University Press, 2002), 1–3.

21. Ben Okri, "What the Tapster Saw," in *Stars of the New Curfew* (New York: King Penguin, 1988), 187, 185.

22. Amos Tutuola, *The Palm-Wine Drinkard* (New York: Grove Weidenfeld, 1984), 7.

23. For further discussion of Tutuola's novel and its parody in Karen King-Aribisala's novel *Kicking Tongues*, see Wenzel, "Petro-Magic-Realism."

24. D. O. Fagunwa, *Forest of a Thousand Daemons*, trans. Wole Soyinka (New York: Random House, 1982), 12–13. Ato Quayson argues that, since Okri is Urhobo, his deployment of Yoruba narrative and cosmological traditions reflects the "development of a broadly Nigerian consciousness" and the possibility of a national literary tradition produced through "strategic filiation with a specific discursive field irrespective of ethnic identity." In this view, intertextuality becomes a mode of reimagining national community. See Ato Quayson, *Strategic Transformations in Nigerian Writing* (Bloomington: Indiana University Press, 1997), 101–2.

25. See, for example, Wendy B. Faris, *Ordinary Enchantments: Magical Realism and the Remystification of Narrative* (Nashville: Vanderbilt University Press, 2004); and Brenda Cooper, *Magical Realism in West African Fiction: Seeing with a Third Eye* (New York: Routledge, 1998). For a more expansive view, see Gerald Gaylard, *After Colonialism: African Postmodernism and Magical Realism* (Johannesburg: University of Witwatersrand Press, 2006).

26. Alejo Carpentier, *The Lost Steps*, trans H. de Onis (Harmondsworth: Penguin, 1968), 252. See also Carpentier, "On the Marvelous Real in America," in *Magical Realism: Theory, History, Community*, ed. Lois Parkinson Zamora and Wendy Faris (Durham: Duke University Press, 1995).

27. Franz Roh, "Magical Realism: Post-Expressionism," in *Magical Realism,* ed. Zamora and Faris.

28. Reading the sociological circulation of postcolonial literature in terms of its marketing of cultural difference, Graham Huggan and Sarah Brouillette suggest that commercial and other editorial Pures not only determine what gets published but also shape what gets written in the first place. In other words, representation and reception cannot be isolated from each other. Graham Huggan, *The Postcolonial Exotic: Marketing the Margins* (New York: Routledge, 2001); Sarah Brouillette, *Postcolonial Authors in the Global Literary Marketplace* (New York: Palgrave, 2007).

29. Stephen Slemon, "Magical Realism as Postcolonial Discourse," in *Magical Realism,* ed. Zamora and Faris, 409, 422.

30. Michael Valdez Moses, "Magical Realism at World's End," *Literary Imagination* 3, no. 1 (2001): 115, 110.

31. Lois Parkinson Zamora and Wendy B. Faris, "Introduction," in *Magical Realism,* ed. Zamora and Faris, 2.

32. Fredric Jameson, "On Magical Realism in Film," *Critical Inquiry* 12, no. 2 (1986): 311.

33. On world-ecology, see note 18.

12

Refined Politics

Petroleum Products, Neoliberalism, and the Ecology of Entrepreneurial Life

MATTHEW T. HUBER

In his recent book, journalist Peter Maass reflects on the apparent powers of oil: "Across the world, oil is invoked as a machine of destiny. Oil will make you rich, oil will make you poor, oil will bring war, oil will deliver peace."[1] In discourses like these, oil is invoked as a *singular force* capable of producing singular effects–oil wars, oil addictions, and oil states. In one sense, this is a preeminent example of what Karl Marx called *fetishism*—that is, according material "things" a kind of autonomous power divorced from the social relations that make such "things" possible.[2] One confusion (of many) is that oil is invoked as a singular commodity. Yet a single commodity must have a specific material use value, and crude oil—in itself relatively *useless*[3]—is valuable precisely because it is the antecedent of a multiplicity of use values. The complex hydrocarbon assemblage of crude oil is only commodified through the refining process, which by its very nature creates numerous petroleum products.

Indeed, much like commodity fetishism, oil fetishism obscures the complex socio-ecological processes through which crude oil *becomes* commodified. Refineries are an often invisible but massively consequential node of socio-ecological transformation and waste production. Refining inevitably leads to leaks, spills, and the flaring of greenhouse gases (and other air pollutants).[4] The process of crude transformation deploys highly flammable materials through intense amounts of heat and pressure, making deadly explosions and fires a necessary evil of operations.[5] The products and

"Refined Politics: Petroleum Products, Neoliberalism, and the Ecology of Entrepreneurial Life" by Matthew Huber. *Journal of American Studies* 46, special issue 2 (May 2012): 295–312. Copyright 2012 Cambridge University Press; reprinted with permission.

wastes of the refinery include known carcinogens such as benzene and arsenic that lead to severe burns, skin irritation, chronic lung disease, psychosis, and elevated cancer risks among workers and nearby communities.[6] The 150-mile stretch of refineries and chemical plants along the Gulf Coast, also known as "Cancer Alley," has become the epicenter of the environmental justice movement, where communities struggle to prove scientifically that the concentrated levels of cancer and death all around them are a direct result of oil and chemical pollution.[7] Refineries are also tremendously energy- and water-intensive. An estimate using United Nations data suggests that a single refinery in the United States consumes the electricity equivalent of 30,633 households.[8] For cooling towers and steam generation, it takes an estimated 1.53 gallons of water for each gallon of crude processed.[9]

Despite these many problems, refineries remain a necessary site through which crude oil becomes useful in modern economic life. Singularizing oil distracts from one of the most potent narratives through which oil's embeddedness within everyday life is naturalized. From the beginning, the petroleum industry has consistently attempted to remind the public that their lives are *saturated* with not just one but a multiplicity of petroleum products. In this essay, I link the refining process with this particular cultural politics of "life" made possible through petroleum products. While it is increasingly commonplace to claim oil is central to "the American way of life," I further particularize this well-worn phrase as a *neoliberal* imaginary of "life" as purely a product of atomized private energies and choices. While the refinery process itself produces a distinct set of "fractions" distilled and cracked from a barrel of crude oil, petroleum products provide the supplementary materiality for the appearance of *segmented lives* each tidily controlled within the private spaces of the home, automobile, and workplace.

This essay proceeds in three sections. First, deploying a Gramscian-Foucauldian approach, I articulate my own particular interventions within debates over "neoliberalism." Second, I briefly review the basics of the refining process and suggest that the very nature of the process itself *ensures* multiple products. Third, I examine a specific cultural object produced by the petroleum industry that actively constructs an imaginary of petroleum-dependent life: an "educational" film for sixth- to eighth-graders prepared by the American Petroleum Institute titled *Fuel-Less* (a parody of the hit 1995 film *Clueless*).

ENERGIZING NEOLIBERALISM

We often think too much about the politics of energy—geopolitics, petrostates, oil-spill regulation—and not enough about how energized practices prefigure particular forms of politics.[10] If the textbook definition of energy is the ability to do material work, I pose a different question: Can energy do *political work*? More specifically, I aim to interrogate the role of petroleum products in both powering and provisioning neoliberal forms of common sense.

Neoliberalism can be seen as a specific hegemonic political formation. Antonio Gramsci's conception of hegemony calls attention to the ideological aspects of social power that produce forms of "common sense." As Gramsci describes, "Common sense is not something rigid and immobile, but is continually transforming itself, enriching itself with scientific ideas and with philosophical opinions which have entered ordinary life."[11] Thus, consent is secured only through a fractious struggle to produce commonsense sentiments that serve to reinforce existing power relations as natural and just. Yet it is important not to succumb to an *idealist* theory of hegemony. As Raymond Williams makes clear, hegemony is best theorized as a material *lived process*: "[Hegemony] is a whole body of practices and expectations, over the whole of living: our senses and assignments of energy, our shaping perceptions of ourselves and our world. It is a lived system of meanings and values."[12] Of course, energy is central to any understanding of "ordinary life" or the "whole of living." Energy is the stuff of material life—the food, the fuel, the muscles, and the fire, soot, and smog emblematic of the fossil age. On the one hand, the geographies of life itself must be materially produced out of particular relations with energy—relations with food, heating fuel, transportation fuel, etc. On the other hand, these historically sedimented and energized geographies themselves produce a particular cultural politics of "life."[13] The cultural politics of life focuses on how wider narratives make normative claims about particular modes of living as a universal model. The materiality and cultural politics of life always invokes historically specific forms of spatial practice entangled with normative visions of what constitutes "the good life."

In this essay, I aim to link the ecology of petroleum products to a specifically neoliberal cultural politics of life. Neoliberalism is understood as a coherent set of practices, policies, and ideas, including free market ideology, deregulation, and the cutback of social services.[14] I offer three interventions to these debates. First, there has been a proliferation of accounts detailing the "neoliberalization of nature"—showing how neoliberal policies have, successfully, and unsuccessfully, commodified, privatized, and marketized various realms of biophysical nature, such as fisheries, wetlands, and forests.[15] Yet few have asked how nature–society relations are internal to the process of "neoliberalization" itself; a socio-*ecological* process entangled within particular regimes of resource, energy, and waste production. Thus, I aim to shift attention away from the neoliberal politics of ecology (or nature) to a framework that considers the *ecology of neoliberal politics.*[16]

Second, despite detailed accounts of neoliberalism's intellectual lineages, policy outcomes, and resistance,[17] there is still little work explaining why neoliberalism succeeded as a *popular political project.* As David Harvey puts it, "Neoliberalism increasingly defines the common sense way many of us interpret, live in, and understand the world. We are, often without knowing it, all neoliberals now."[18] As Wendy Larner reminds us, the "complex appeal" of neoliberal tropes such as "freedom" and "choice" are not simply handed down by intellectual elites, but need to be understood as grounded in the daily practices that animate neoliberal subjectivities.[19]

Third, although many "periodize" neoliberalism as emerging in the 1970s and 1980s, I contend that U.S. neoliberalism is rooted in the restructuring of capitalism—and everyday life—in the postwar period (itself a product of crisis and labor struggles during the 1930s).[20] I argue that a specifically *neoliberal* cultural politics of "life" *grew* during the postwar period only to become hegemonic during the 1970s and beyond to the present. As postwar accumulation was materialized through the construction of vast sprawling suburban housing tracts, Keynesian ideas of government intervention and the social safety net were slowly transformed into an increasing politics of privatism. As many suburban historians have shown,[21] the political victories of the Right in the United States (i.e. neoliberalization) depended upon the mobilization of a petty-bourgeois stratum of white suburban homeowners increasingly distrustful of government handouts, high taxes, and the redistribution of wealth. Underlying the suburban geography of private homeownership is what Evan McKenzie refers to as an "ideology of hostile privatism."[22] The hostility itself emerges from what Edsall and Edsall call "conservative egalitarianism,"[23] which posits that everyone has an equal opportunity to work hard and succeed in life.

Here, as many have recognized,[24] Foucault's ideas provide a richer micropolitical lens through which to view the macrostructural concept of hegemony. This suburban politics of life shows considerable overlap with Michel Foucault's 1978–79 lectures on neoliberalism recently published as *The Birth of Biopolitics*.[25] These lectures, given during the infancy of neoliberal hegemony, hold tremendous insight into the micropolitics of neoliberal subjectivity. According to Foucault, what distinguishes classical liberalism from neoliberalism is the latter's concentration on competition and the former's focus on market exchange. In an ideal neoliberal society governed by competition, Foucault suggests that the "enterprise form" will dominate the social body. According to postwar German strands of neoliberal thought, the materialization of this enterprise form is assured through private property: "First, to enable as far as possible everyone to have access to private property . . . the decentralization of the places of residence, production, and management."[26] This requires constructing a particular "politics of life," or *Vitalpolitik* as Alexander Rüstow coined it, which means "constructing a social fabric in which precisely the basic units would have the form of the enterprise, for what is private property if not an enterprise? What is the house if not an enterprise? . . . I think this multiplication of the 'enterprise' form within the social body is what is at stake in neo-liberal policy."[27] Thus, private homeowners run their house like a business. So-called responsible homeowners construct a family budget tracking spending against revenue, make investments with savings and pensions, and maintain a healthy long-term relation with credit markets. Thus, it is up to the individual to make the right choices in the context of a competitive society. Individual responsibility fuses with an entrepreneurial outlook on the whole of living: "The individual's life itself—with his [sic] relationships to his private property, for example, with his family, his household, insurance, and retirement—must make him into a sort of permanent and multiple enterprise."[28] Thus, the construction of a propertied mass of homeowners—an "ownership society," as George W. Bush called it—creates a situation where your *own very life* is seen as a product of your entrepreneurial choices. Your investment in education, your hard work, your competitive tenacity, all combine to make a life—to *make* a living—for yourself. As is becoming more and more common, we hear that we are "the CEO of our lives."[29] And the product of a "successful" life is expressed through a set of material prerequisites—a home, a car, a family.

As an aside, this cultural politics of life differs markedly from Marx's vision of *proletariat* life. For Marx, the proletariat was defined by his or her propertylessness—"the proletarians have nothing to lose but their chains."[30] The question for a propertyless proletariat is: How will I live? The answer was of course to desperately sell your labor power in exchange for a wage. In contrast, the question for the propertied mass of workers/entrepreneurs is: What will I do with my life? Of course, this question assumes that your life itself is purely a product of atomized choices and individualized efforts.

In what follows, I link petroleum to this view of entrepreneurial life. It was during the postwar period that petroleum became the critical material and energetic basis of everyday life centered on single-family homeownership, automobility, and the nuclear family.[31] And oil's imbrication within a vision of entrepreneurial life is not singular. It not only provided the gasoline to propel masses of atomized individuals through the dispersed geographies of social reproduction (home, work, school),[32] but it also provided the material for much of this sociospatial infrastructure—asphalt for roads, vinyl siding for homes, countless plastic commodities to fill the home. The petroleum industry has consistently crafted a narrative through which "life" is ultimately dependent upon not just one, but multiple, products. In order to understand crude oil itself as an assemblage of multiple potentialities, it is important to understand first the basics of the refining process.

REFINED LIVES

Crude oil is of course the product of millions of years of "fossilized sunshine" expressed in unoxidized marine plant life.[33] In addition to nitrogen and sulfur, crude oil is a complex assemblage of different kinds of hydrocarbon molecules that vary according to the type of crude extracted in a particular region. Refineries can be seen as particular expressions of the historically specific relations between petroleum and society. Key to the contemporary use of petroleum is the refining process of fractional distillation. Fractional distillation allows producers to segment a given amount of crude into a variety of hydrocarbon *fractions* from light gases with lower boiling points and fewer carbon molecules to heavy tarlike substances with extremely high boiling points and more carbon molecules.[34] Although the process of distillation has been traced back to ancient Egypt and China for lamp oil,[35] petroleum refining in the nineteenth century coalesced with modern chemistry to produce a certain kind of knowledge of distillation as a molecular process of chemical transformation. A Yale University chemist named Benjamin Silliman Jr. is credited with explaining in molecular terms how refining techniques could be applied to petroleum to produce high yields of the illuminant kerosene.[36]

The majority of refiners set up rudimentary distillation towers to transform crude into kerosene as quickly and haphazardly as possible.[37] With kerosene representing a middle fraction in a given barrel of oil, the lighter fractions (gases, gasoline, and naphthas) were simply flared off into the atmosphere, and the heavier tarlike materials such as petroleum coke were disposed of in nearby water systems.[38] As the petroleum boom proceeded on a mountain of waste, chemists and engineers began to imagine a given barrel of crude as not simply a profitable means to kerosene, but rather as a vertical hierarchy of different hydrocarbon molecules that each could be transformed into marketable products (Figure 12.1).[39] Indeed, Silliman claimed that the residual or waste products of the refinery process should not be ignored. He emphasized, "The crude product contained several distinct oils all with different boiling points. . . . My experiments prove that nearly the *whole* of the raw product may be manufactured without waste."[40] While kerosene—and later gasoline—was still the most profitable product, a given barrel of oil began to be imagined as containing not one, but hundreds, of petroleum products. As early as the late 1870s, heavier fractions were being marketed as lubricants, waxes, petroleum jelly, and even chewing gum.[41]

Fractional distillation, however, only took the petroleum industry so far. Such a process was ultimately dependent upon the particular biophysical quality of the crude oil itself. Lighter (and sweeter) crude oil produced higher yields of gasoline and heavier crudes produced less. Frustration with low yields for lower-quality crudes led to the development of "cracking" technologies that could break apart heavier hydrocarbon molecules into smaller, lighter ones more amendable to producing gasoline. During the 1920s alone, refiners in the United States increased gasoline yields from 25 percent to 39 percent using thermal cracking.[42] The 1930s saw the first widespread

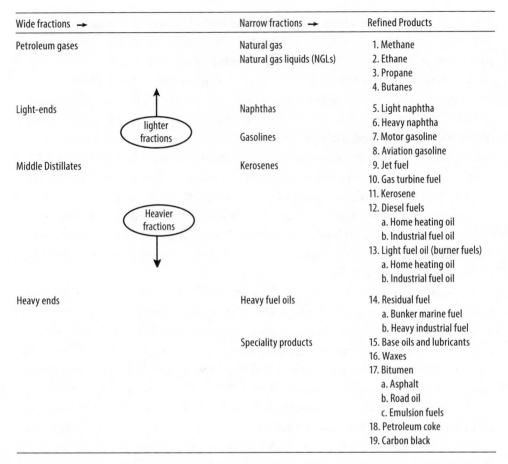

Wide fractions →	Narrow fractions →	Refined Products
Petroleum gases	Natural gas Natural gas liquids (NGLs)	1. Methane 2. Ethane 3. Propane 4. Butanes
Light-ends	Naphthas Gasolines	5. Light naphtha 6. Heavy naphtha 7. Motor gasoline 8. Aviation gasoline

Middle Distillates	Kerosenes	9. Jet fuel
		10. Gas turbine fuel
		11. Kerosene
		12. Diesel fuels
		   a. Home heating oil
		   b. Industrial fuel oil
		13. Light fuel oil (burner fuels)
		   a. Home heating oil
		   b. Industrial fuel oil

Heavy ends	Heavy fuel oils
	
	Speciality products
		14. Residual fuel
		   a. Bunker marine fuel
		   b. Heavy industrial fuel
		15. Base oils and lubricants
		16. Waxes
		17. Bitumen
		   a. Asphalt
		   b. Road oil
		   c. Emulsion fuels
		18. Petroleum coke
		19. Carbon black

(diagram: arrow pointing up labeled "lighter fractions"; arrow pointing down labeled "Heavier fractions")

FIGURE 12.1. Crude as a vertical hierarchy of different hydrocarbon molecules that each could be transformed into marketable products. Adapted from Morgan Patrick Downey, *Oil 101* (New York: Wooden Table Press, 2009).

development of the use of chemical catalysts (such as aluminum chloride) to allow hydrocarbon molecules to break up more quickly at lower temperatures (catalytic cracking).

Since catalytic cracking of heavier hydrocarbon molecules into lighter ones actually increased the octane rating of fuels, the federal government underwrote a massive 527 percent expansion of catalytic cracking capacity between 1941 and 1945 to expand the production of high-octane jet-engine fuel for the war effort.[43] One of the major by-products of cracking are olefins (e.g., ethylene and propylene), which do not occur in nature and became the vital feedstock for the production of petrochemicals and plastics.[44] Thus, alongside the ramped-up production of high-octane jet fuel came the rise of the petrochemical industry and the sprawling multiplicity of use values from plastics to pesticides and synthetic fibers.[45] It bears remembering that "war" sets the material conditions for a period called "postwar," and, indeed, much of the postwar petroleum economy was built upon the war economy.

Much of the postwar vision of life, leisure, and freedom depended upon the construction of geographies of everyday social reproduction outside the workplace. In the realm of life as opposed to work, individuals could construct their own privatized spaces of freedom. The petroleum industry was keen on tying oil to an overwhelming set of basic everyday products—food, plastics, medicine, clothing, and energy— to create a sense of the unavoidability of oil within that space of freedom. During the late 1940s and 1950s, the petroleum industry began to promulgate a narrative best summed up in *Esso*'s (later *Exxon*) tagline of the early 1950s—"petroleum helps to build a better life."[46] Life itself—always imagined as the life of white nuclear families in the suburbs—was seen as comprising multiple building blocks in which petroleum products were implicated at every stage. Indeed, the idea of an American standard of living came to be *equated* with petroleum consumption. A particular ad offered this fun fact: "Did you know—that a nation's progress (and their standard of living) can be measured pretty well by its consumption of petroleum?"[47] Petroleum, called a "chemical wonderbox" on a national television celebration of the seventy-fifth anniversary of Standard Oil in 1957,[48] became constitutive of a whole set of lived practices and visions of the good life.

The reproduction of everyday life itself implies the repetition of practices. The material *anchor* of this vision of everyday life was centered upon the home and the automobile. Figure 12.2 reveals a 1950s ad campaign from Shell entitled "From A to Z—An Alphabet of Good Things about Petroleum." Shell provided an exhaustive list of letter-specific things and practices linked to petroleum. One of the key aspects of this imagery is the *succession* of practices depicted with the implication that each one is tied to some petroleum product. In the advertisement for H, H was for Hydrocarbon (we consume seventeen pounds a day!), Heart medicine, Heat, Horsepower, and the offshore frontier of H_2O (a necessary frontier to satiate increasing demand, of course). The central message was that H was for "Home"—"Oil research helps with quick drying paints, no polish floors, durable plastic table tops, and weather defying asphalt-shingled roofs. In many ways, your home is a house that oil built." If your home is the product of your own entrepreneurial efforts, it is *also* a product of petroleum products.

The vision of "life" underwritten by the petroleum industry was contradictory. Life was constructed as a privatized (white) realm of social reproduction that was made possible through free competition and the individual (male) breadwinner's own entrepreneurialism. Yet this individual "life" was also perilously *dependent* upon

FIGURE 12.2.   1956 Shell Advertisement.

not only petroleum but also the petroleum industry. These ad campaigns actively sought to remind consumers that their lives—as singular, heroic, entrepreneurial projects—were only made possible through petroleum products.

In the wake of the 1969 Santa Barbara oil spill, the rise of the modern environmental movement, and the oil shocks, the dependence of life upon petroleum products was increasingly problematized. The crisis of stagflation and profitability slowdowns in the 1970s created the conditions for new policy consensuses based upon free markets, low taxes, and the demonization of government and the redistribution of wealth.[49] Richard Nixon and Ronald Reagan perfected the southern strategy of harnessing votes from middle-upper-income strata in the Sunbelt suburbs (among other suburbanized geographies) who found neoliberal discourses resonant and appealing.

The entrepreneurial logic of suburban life was fueled with and through petroleum products. Petroleum both powered and provisioned a particular lived geography—a "structure of feeling"[50]—that allows for an appearance of privatized command over space and life, or petro-privatism. Individuals are propelled from private homes in private automobiles to privatized workplaces and consumption locations. Notwithstanding the immense public investment that makes it possible, the lived geographies of oil consumption allows for a construction of a realm of "the public" as irrelevant and burdensome. Government programs were seen as "unfair" handouts to individuals who simply had not made the right choices "in life." The language of "family values" also centered life upon the home and family as a privileged site for cultural cultivation and refinement against a hostile and degraded external culture. Contrary to some arguments,[51] family values centered on the home were not uneconomic or purely "cultural" values, but rather were absolutely fused with economic concerns with one's own entrepreneurial capacities to make a living, keep a job, and be responsible.

The petroleum industry benefited greatly from the neoliberal turn toward deregulation and free market ideology. In 1981, Reagan immediately set up a task force to query industry about the most "burdensome" regulations, and environmental regulation of the chemical industry was atop the list.[52] The challenge for the petroleum industry was to remind consumers of their inextricable dependence on the industry, but also harness the logic of entrepreneurial life to reconcile the ecological contradictions of oil-fueled life. By the 1990s, oil companies began to construct a sophisticated "greenwashing" campaign that situated the solutions to environmental problems firmly in the hands of privatized individually responsible actors. Next, I will examine a specific cultural object produced by the American Petroleum Institute (API) that attempted to reconcile the tensions between ecology and an oil-fueled life.

"ALMOST EVERYTHING I HAVE THAT'S REALLY COOL COMES FROM OIL"

The focus of the petroleum industry on the multiplicity of products constituting "life" has continued into the present. Yet it is important to understand how the politics of oil-based life have shifted into our current neoliberal era. As the perils of oil-dependence have intensified since the 1970s, the petroleum industry has had to more carefully construct a positive narrative of oil-saturated life *and* environmental responsibility. The American Petroleum Institute's educational film *Fuel-Less* (a parody of the hit film *Clueless*) attempts to reconcile these tensions not only through an emphasis on the unavoidability of petroleum products but also by situating the appreciation of petroleum products as the basis of an entrepreneurial capacity to make contributions toward environmental sustainability.

The 1995 film *Clueless* could be seen as a kind of critique of the superficiality of neoliberal celebrations of the consumer. The story tracks the redemption of the main character, Cher, whose wealth and privilege shields her from recognizing the importance of friendship and social service. However, the film—released at the zenith of "end of history" neoliberal hegemony—reproduces the core of neoliberal visions of entrepreneurial life discussed above. Cher's low point is when she is denied the key material conduit of petro-privatism—she fails her driving test. Moreover, Cher's redemption is negotiated through an individualist vision of life—wealth and privilege had prevented Cher from *making something of herself*. Thus, the film's happy ending presents a new Cher (fresh off a shopping binge) committed to a life of authentic relationships and volunteerist service to the community (she runs her school's disaster relief effort).

The API's parody tracks the same plotline from privilege to redemption. The film begins with Crystal's narration amid images of a materially comfortable suburban bedroom, "You're absolutely not going to believe this, but almost everything I have that's really cool comes from oil." Immediately, Crystal discovers to her horror (she literally screams in terror) that all her petroleum-based products have disappeared. Deprived of makeup, clothes (she is forced to wear a potato sack), and a car, she is compelled to walk, "a whole six blocks to school." At the outset, the film establishes a specific material vision of life and standard of living undergirded by a multiplicity of petroleum products.

Crystal has taken petroleum for granted. Her redemption is guided by her "extremely cool science teacher Ms. Watkins." After hearing Crystal's story, Ms. Watkins easily concludes, "Crystal, it sounds to me like everything in your life that comes from oil is gone." The film then proceeds to take Crystal on a journey through the petroleum commodity chain to discover the wonders of oil. The first lesson—at an oil well—is to establish oil's foundation *in nature*. Ms. Watkins teaches, "The oil coming out of this oil well comes from plant and animal remains which were buried beneath the earth's surface millions of years ago." Crystal initially finds this "a bit gross," but Ms. Watkins counters, "Gross? Are you kidding? Nature is elegant. You see these remains broke down into chemical compounds of hydrogen and carbon . . . hydrocarbons." Crystal replies, "So oil is like natural?" Ms. Watkins then traces hydrocarbons to plastic shopping bags. Crystal replies, "Gee, I never thought of plastic as natural before." The lesson serves to literally naturalize the presence of petroleum products in everyday life and deflect particular critiques of the toxicity and artificiality of plastics and other synthetic petrochemicals.

Crude oil is imagined as the ingredient not only in the material comforts of life but also in the multiplicity of medical technologies preventing death. "Today, relatively few Americans die of these ailments [tuberculosis, diphtheria, diarrhea], because of petroleum-based medicines such as sulfa drugs which helped to conquer these diseases. The vitamins you take and many of the medicines you use contain chemical constituents of oil." One conclusion is that those who oppose the petroleum industry implicitly oppose the saving of lives through modern medicine.

But Crystal is impatient and eager to get her belongings back. Ms. Watkins reassures Crystal that her reunification with her oil-based belongings would be simple, "With a little bit of understanding and appreciation you'll get your belongings back." Here is the API's prime message—although oil gets a bad reputation for causing environmental disasters, wars, and terrorism, consumers need to *appreciate* all the good things that oil brings to life. Only then will the public recognize that all the good things we associate with life—mobility, home, family—are inextricably tied to petroleum products.

The next lesson takes Crystal and Ms. Watkins to the critical site of hydrocarbon transformation discussed above—a refinery. Crystal is introduced to a mullet-headed petroleum engineer who educates Crystal about the refinery process:

> Crystal, the process we use to convert oil materials used in products like cosmetics is called fractional distillation. Essentially we take crude oil, and we place it into a large crock pot, technically called a fractionating column, and we cook it until it vaporizes and as the vapors cool, they separate out into gasoline, into heating oil, into lubricating oil, and into wax. This white glop—soon to become a football helmet—started out as oil. That's because when ethylene and propylene gases from the fractionating column are exposed to certain metals under high pressure and temperature, they turn into what we call polymers—you know them as plastics. . . . Cool chemical wizardry also makes oil part of the recipe for toothpaste, telephones, TV sets, skateboards, shampoos, computers, CDs, contact lenses, cars, credit cards, you name it.

Crude oil is useless without the "cool chemical wizardry" that makes oil-based hydrocarbons the key ingredient of countless commodities. Again, it is the overwhelming multiplicity of products that helps craft the API's narrative of the *unavoidability* of oil.

Once Crystal learns of oil's connection to her most prized commodities—clothes—her educatory transformation intensifies, "I'm thinking something must really be wrong with me, because I'm beginning to find this stuff sort of interesting." Yet her interest is still guarded with skepticism. Crystal questions the core paradox of oiled life—the relation between mass oil consumption and continued ecological crisis. Overwhelmed with the mountain of disposable oil-based products in the school cafeteria, she exclaims, "But, then look what happens—we just throw that stuff away." Ms. Watkins concedes the point—"You're right. We sure don't want to pollute our water and use up our landfills with products from oil that could easily be reused. That's why it's so important to recycle plastics like our lunchroom cups, and to recycle used motor oil from your car. . . . Fortunately, the earth isn't close to running out of oil anytime soon, but we need to be responsible about using what Mother Nature provides." Thus, it is upon the terrain of individual actions and choices through which the contradictory relation between oil and life can be reconciled. Channeling the most powerful logics of entrepreneurial life—individual responsibility—the oil industry

encourages us all to take individual control over our own private consumption decisions. As Ted Steinberg has recently argued, this form of "green liberalism" has more
in common with neoliberal ideology than is often recognized.[53]

Crystal is still worried about something consumers have little control over: "But
then I remembered those oil spills I used to see on the news. That's responsible?
What about oil spills?" In this case, Ms. Watkins assures Crystal that the industry
itself has it under control. "Well there's no failsafe guarantee against oil spills, but
there's a reason why you don't hear about them as much as you used to; because
stricter standards for oil tankers and the people who pilot them have been adopted.
At the same time, the oil industry is developing new ways to deal with oil spills." Of
course, if the API had simply waited three years (*Fuel-Less* is copyrighted in 2007),
the notion that "we don't hear about" oil spills anymore would seem laughable in the
face of the Deepwater Horizon disaster.

Finally, Crystal comes to the necessary epiphany, "You know, I think I get it now.
You're right—nature is just as elegant as Armani—who was a very important designer.
And oil is like 'neat.'" Immediately, her potato sack is transformed into an elegant business suit, and Ms. Watkins happily announces, "You've just gotten your things back
which means you truly understand and appreciate the value of oil and how useful it is."
The film ends one month later and, like *Clueless*, Crystal has produced a new life for
herself. She has taken charge of the school recycling program and committed herself to
environmental responsibility. Ms. Watkins is quite pleased because it is these kinds of
efforts—consumers and capital in tandem—that can reconcile oil-based life with ecology into the future, "Efforts like these enable us to continue doing the things we enjoy
thanks to the products that oil provides. . . . [A] large part of our future lies in the hands
of people like you who really understand and appreciate the issues, and are willing
to pitch in and help." In this case, the effort to "pitch in" is negotiated on privatized
terms—the choices of consumers and capital that aggregate into a sustainable future.

Crystal then neatly summarizes her profound transformation toward seeing her
life as a singular heroic entrepreneurial project. "I knew that I could make a difference." The film ends with classic imagery of suburban life: Crystal and some male
suitors hanging out in a large driveway with an SUV Ford Explorer and Volvo. The
irony of an environmental message behind an image of fuel-inefficient automobility
was apparently lost on the API.

Overall, the film reproduces the notion that *lives*—like specific petroleum fractions—
are singular, fragmented, and self-made entities. The API seeks to make clear that
lives are only made *through* petroleum products. Yet the embedded nature of petroleum within the American standard of living—a standard that consumes nearly 25
percent of petroleum with only 4 percent of global population[54]—is by no means

universal, and it takes political work to normalize this energy profligacy. Moreover, as it becomes clearer that it is precisely this profligacy that is driving continued geopolitical conflict, spectacular (and more everyday) pollution, and climate change, the oil industry has been forced to craft a narrative that reconciles oil-based profligacy with environmental responsibility. Clearly the producers of *Fuel-Less* were keen on projecting a vision of environmental politics that ultimately conforms with neoliberal forms of politics that have been such a great boon to the industry.

CONCLUSION

We often think our "addiction" to oil is purely a material relation—urban spatial form, disposable plastics—but we need to think more deeply about how our relation to petroleum also shapes the way we think and feel about politics. The oil industry has consistently attempted to *lodge* petroleum products within a very powerful vision of "the American way of life." In this essay, I have attempted to call attention to the refinery process and the multiplicity of by-products from a given barrel of crude as being a key way through which the industry projects the unavoidability of oil in everyday life. This allows the oil industry to equate opposition to oil as opposition to cherished ideals of home ownership, freedom, and family. Yet the "American way of life" saturated in petroleum products is not simply about material profligacy; it is about a specific vision of life as best negotiated through market forces in opposition to any notion of public or collective solidarity. The construction of suburban spatialities of property ownership seemed to reinforce what Perry Anderson considers the core of capitalist consent: "the fundamental form of belief by the masses that *they exercise an ultimate self-determination within the existing social order.*"[55] With all the "work" (or energy) accomplished through taken-for-granted hydrocarbons, individuals could imagine themselves as masters of their own lives. Without the popular support of an energized suburban populism, neoliberal policies—including the kinds of deregulation that laid the basis for the Deepwater Horizon—could not be constructed as "commonsense" alternatives to the inefficacy of the public sector.

As the perils of petroleum-based life have become all too apparent, the petroleum industry has been forced to craft a "solution" to these problems firmly within the neoliberal logics of entrepreneurial life. As the story of Crystal illustrates, *making a life for yourself* requires not only "appreciating" the role of petroleum but also taking responsibility to "pitch in" alongside the aggregated environmental efforts of privatized consumers and businesses. The underlying message is that petroleum-based life is sustainable through market forces, free choice, and individual responsibility.

Thus, our oil addiction is perhaps most problematic not because we drive too far to work, but because it supplements an insidious ideology of privatism. Rather than see our energy crisis as solved through private consumer choices and cap and trade carbon market schemes, we need to imagine new ways of life and living that can once again construct popular resonances around notions of public solidarity and the viability of collective management of environmental problems.

NOTES

1. Peter Maass, *Crude World: The Violent Twilight of Oil* (New York: Alfred A. Knopf, 2009).

2. See Karl Marx, *Capital, Vol. 1* (London: Penguin, 1976), 163–77.

3. Granted, some regions with high levels of crude production and little refining capacity do burn crude oil for electricity. Smil reports that burning oil in its crude state accounts for about .5 percent of crude output. See Vaclav Smil, *Oil: A Beginner's Guide* (Oxford: Oneworld, 2008), 146.

4. See Dara O'Rourke and Sarah Connolly, "Just Oil? The Distribution of Environmental and Social Impacts of Oil Production and Consumption," *Annual Review of Environment and Resources* 28 (2003): 603–7.

5. A visceral account of the precarious explosiveness of refineries is provided by Lisa Margonelli, *Oil on the Brain: Petroleum's Long Strange Trip to Your Tank* (New York: Broadway Books, 2007), 48–65.

6. O'Rourke and Connolly, "Just Oil?," 605.

7. For the role of the chemical industry in covering up the cancer risks for workers and communities in the production of polyvinyl chloride, see Gerald Markowitz and David Rosner, *Deceit and Denial: The Deadly Politics of Industrial Pollution* (Berkeley: University of California Press, 2002). For influential accounts of the environmental justice movement in relation to the chemical industry, see Robert D. Bullard, *Dumping in Dixie: Race, Class, and Environmental Quality*, 3rd ed. (Boulder: Westview Press, 2000); and Barbara L. Allen, *Uneasy Alchemy: Citizens and Experts in Louisiana's Chemical Corridor Disputes* (Cambridge: MIT Press, 2003).

8. United Nations Statistics Division, Energy Statistics Database. http://data.un.org/Data .aspx?d=EDATA&f=cmID%3AEL%3BtrID%3A0925.

9. M. Wu, M. Mintz, M. Wang, and S. Arora, Center for Transportation Research, Energy Systems Division, Argonne National Laboratory, *Consumptive Water Use in the Production of Ethanol and Petroleum Gasoline* (January 2009), http://www.transportation.anl.gov/pdfs/ AF/557.pdf.

10. For a recent novel attempt to link fossil fuel with a historically specific vision of "democracy," see Timothy Mitchell, "Carbon Democracy" *Economy and Society* 38 (2009): 399–432.

11. Antonio Gramsci, *Selections from the Prison Notebooks* (New York: International, 1971), 326.

12. Raymond Williams, *Marxism and Literature* (New York: Oxford University Press, 1977), 110.

13. The focus on "life" is influenced primarily by Marx and Engels's *The German Ideology*, where they situate "the real life-processes" as the central object of historical materialist inquiry. See Karl Marx and Friedrich Engels, *The German Ideology* (New York: Lawrence and Wisehart, 1970), 47. Kathi Weeks has also recently called for a renewed Marxist politics of "life" against work. See Kathi Weeks, "Life within and against Work: Affective Labor, Feminist Critique, and Post-Fordist Politics," *Ephemera* 7 (2007): 233–49. A renewed interest in the politics of life has

also emerged out of recent efforts to combine Foucault's insights on biopolitical subjectivities and Marx's critique of capital. The most famous of these is of course the work of Michael Hardt and Antonio Negri, *Empire* (Cambridge: Harvard University Press, 2000). See also Mauruzio Lazzarato, "The Concepts of Life and Living in the Societies of Control," in *Deleuze and the Social,* ed. Martin Fuglsang and Bent Meier Sorensen (Edinburgh: Edinburgh University Press, 2006), 171–90; and Tiziana Terranova, "Another life: The nature and political economy of Foucault's genealogy of Biopolitics" *Theory, Culture & Society* 26, no. 6 (2009): 234–62. The most sustained philosophical elaboration of these ideas is laid out in Jason Read, *The Micro-politics of Capital: Marx and a Pre-History of the Present* (Albany: SUNY Press, 2003), 103–52. Overall, this perspective on "life" should be distinguished from an emerging literature on what Nikolas Rose calls "The Politics of Life Itself." See Nikolas Rose, *The Politics of Life Itself: Biomedicine, Power, and Subjectivity in the Twenty-first Century* (Princeton: Princeton University Press, 2006).

14. See David Harvey, *A Brief History of Neoliberalism* (New York: Oxford, 2004); and Jamie Peck, *Constructions of Neoliberal Reason* (New York: Oxford, 2010).

15. See Nik Heynen, James McCarthy, Scott Prudham, and Paul Robbins, eds., *Neoliberal Environments: False Promises and Unnatural Consequences* (London: Routledge, 2007).

16. Similarly, Jason Moore has argued that we need to not only focus on capitalism's effects on the environment, but also theorize the ecology of capital: "Capitalism does not *have* an ecological regime; it *is* an ecological regime." See Jason Moore, "Transcending the Metabolic Rift: A Theory of Crisis in the Capitalist World Ecology," *Journal of Peasant Studies* 38 (2011): 34.

17. Helga Leitner, Jamie Peck, and Eric Sheppard, eds., *Contesting Neoliberalism: Urban Frontiers* (New York: Guilford Press, 2006); and Kim England and Kevin Ward, eds., *Neoliberalization: States, Networks, Peoples* (Oxford: Blackwell, 2007).

18. David Harvey, *Cosmopolitanism and the Geographies of Freedom* (New York: Columbia University Press, 2009), 57.

19. Wendy Larner, "Neoliberalism?" *Environment and Planning D* 21 (2003): 511.

20. To be sure, the intellectual ideological roots of neoliberalism are commonly located in the 1930s and postwar era (e.g., Hayek and the Mount Pelerin Society), but I am interested in the broader populist roots of neoliberalism.

21. Among the best studies of this phenomenon include Mike Davis, *Prisoners of the American Dream: Politics and Economy in the History of the American Working Class* (London: Verso, 1986); Thomas Edsall and Mary Edsall, *Chain Reaction: The Impact of Race, Rights, and Taxes on American Politics* (New York: W. W. Norton, 1991); Evan McKenzie, *Privatopia: Homeowners Associations and the Rise of the Rise of Residential Private Government* (New Haven: Yale University Press, 1994); Lisa McGirr, *Suburban Warriors: The Origins of the New American Right* (Princeton: Princeton University Press, 2001); and Matthew Lassiter, *The Silent Majority: Suburban Politics in the Sunbelt South* (Princeton: Princeton University Press, 2006).

22. McKenzie, *Privatopia,* 19.

23. Edsall and Edsall, *Chain Reaction,* 147.

24. See Donald Moore, *Suffering for Territory: Race, Place, and Power in Zimbabwe* (Durham: Duke University Press, 2005); and Mike Ekers and Alex Loftus, "The Power of Water: Developing Dialogues Between Foucault and Gramsci" *Environment and Planning D: Society and Space* 26 (2008): 698–718.

25. Michel Foucault, *The Birth of Biopolitics: Lectures at the College De France, 1978–1979,* trans. Graham Burcell (New York: Picador, 2008).

26. Ibid., 147.

27. Ibid., 148.

28. Ibid., 241.

29. A recent example: "Employment experts have some advice for the many Americans either looking for work or fearing they soon will be: Consider yourself an entrepreneur—of your own working life." James Flanigan, "Manage Your Career as a Business," *New York Times,* October 14, 2009.

30. Karl Marx and Friedrich Engels, "Manifesto of the Communist Party," *The Marx-Engels Reader,* ed. Robert Tucker (New York: W. W. Norton, 1978), 469–500.

31. Ian Rutledge, *Addicted to Oil: America's Relentless Drive for Energy Security* (London: I. B. Tauris, 2006). For an insightful discussion of the relations between oil and the biopolitics of security in the United States, see David Campbell, "The Biopolitics of Security: Oil, Empire, and the Sports Utility Vehicle," *American Quarterly* 57 (2005): 943–72.

32. Following Cindi Katz and others, I use the term "social reproduction" deliberately. See Katharyne Mitchell, Sallie A. Marston, and Cindi Katz, eds., *Life's Work: Geographies of Social Reproduction* (Oxford: Blackwell, 2004).

33. Alfred Crosby, *Children of the Sun: A History of Humanity's Unappeasable Appetite for Energy* (New York: W. W. Norton, 2006), 62.

34. The heaviness and viscosity of crude oil is generally paralleled by the number of carbon atoms. See William L. Leffler, *Petroleum Refining in Nontechnical Language,* 4th ed. (Tulsa: Penn Well, 2008), 25–40.

35. Morgan Patrick Downey, *Oil 101* (New York: Wooden Table Press, 2009).

36. Harold F. Williamson and Arnold R. Daum, *The American Petroleum Industry: The Age of Illumination, 1859–1899* (Evanston: Northwestern University Press, 1959), 69–72.

37. Ibid., 212.

38. Ibid.

39. Figure 1 is drawn by the author and adapted from Downey, Oil 101, 101.

40. Benjamin Silliman Jr., *Report on the Rock Oil, or Petroleum* (New Haven: J. H. Benham's Steam Power Press, 1855), 6, 20.

41. Williamson and Daum, *The American Petroleum Industry,* 232–51.

42. Ibid., 436.

43. Harold F. Williamson, *The American Petroleum Industry: The Age of Energy, 1899–1959* (Evanston: Northwestern University Press, 1963), 790.

44. Williamson, *The American Petroleum Industry,* 423; Leffler, *Petroleum Refining in Nontechnical Language,* 53–56.

45. Peter Spitz, *Petrochemicals: The Rise of an Industry* (New York: Wiley and Sons, 1988), 116–56.

46. See one example, the Esso advertisement "50 Centuries of This . . . 50 Years of This," at <http://users.adam.com.au/gasmaps/esschar.jpg>.

47. *Esso Oilways,* November 1950.

48. "Standard Oil: New Jersey 75th Anniversary Entertainment Special," October 13, 1957. The Paley Center for Media, New York.

49. Bruce Schulman and Julian Zelizer, eds., *Rightward Bound: Making America Conservative in the 1970s* (Cambridge: Harvard University Press, 2008).

50. Williams, *Marxism and Literature,* 128–35.

51. Thomas Frank, *What's the Matter with Kansas?: How Conservatives Won the Heart of America* (New York: Metropolitan Books, 2004).

52. See Daniel Faber, *Capitalizing on Environmental Injustice: The Polluter-Industrial Complex in the Age of Globalization* (Lanham, Md.: Rowman and Littlefield, 2008), 128.

53. Ted Steinberg, "Can Capitalism Save the Planet? On the Origins of 'Green Liberalism,'" *Radical History Review* 107 (2010): 7–24.

54. See Energy Information Agency, "International Energy Statistics," http://www.eia.gov/cfapps/ipdbproject/IEDIndex3.cfm?tid=5&pid=54&aid=2.

55. Perry Anderson, "The Antinomies of Antonio Gramsci," *New Left Review* 100 (November–December 1974): 5–78 (30).

13

Gendering Oil

Tracing Western Petrosexual Relations

SHEENA WILSON

By employing a feminist lens to "follow the oil" and trace "the webs of relations and cultural meanings through which oil is imagined as a 'vital' and 'strategic' resource,"[1] I wish to interrogate the relationship between human rights and gender and racial equality and the petro-discourses that are newly oriented around ecology in our contemporary moment.[2] As with many cultural transformations and their associated ideological turns, women's relationship to oil, to the environment, and to the petro-cultures of the twentieth and twenty-first centuries in the West is portrayed in the mainstream media in a limited number of largely superficial ways: first, through embedded feminism and women's rights as they intersect with human and ethno-cultural rights; second, through consumerism; and third, through the recuperation of the female body as a canvas on which to spectacularize politics—largely with explicit consumer aims.

The latter half of this article focuses on representations that either neutralize or trivialize women's political and economic relationships to oil, through an analysis of the Ethical Oil media campaign that began in 2010 in Canada, a Beyond Petroleum commercial of the early millennium, the *Vogue Italia*'s August 2010 fashion spread called "Water & Oil," and the general discourse about eco-fashion and greening beauty regimes. These instances demonstrate the way women's identities have been intentionally constructed to naturalize a particular relationship between Western women and oil. The narratives and gender constructs that inform many advertising and fashion images also function to undermine contemporary women's more serious engagements with the environment, and as such contribute to a broader array of cultural, rhetorical, and legal efforts currently under way to criminalize environmentalists and activists in legislation as well as in the popular media.[3] The first half of this essay discusses the media reception of two examples of contemporary women's environmental activism—the Idle No More protests in Canada and Chief Theresa Spence's subsequent hunger strike in December 2012 and January 2013—in order to establish the cultural and ideological context in which recent representations of women and oil intervene. By examining these forms of activism, I will show how Aboriginal women's political activism in the context of mainstream media is reconstructed to the detriment of the resistance movements and to other community members—namely, Aboriginal men and youth, who risk being constructed as terrorists in discourses of petro-violence.

Chief Theresa Spence and Idle No More: Petropolitics and the Construction of the Terrorist

In the current dominant discourses circulating in Canada and the Western world, environmental messages are acceptable when they are controlled and shaped by petro-invested governments, industry organizations, and corporations. These environmental messages articulate concerns about health and safety, environmental stewardship, and performance, all within the context of a neoliberal discourse of increasing expansion and exploitation of resources. Within this paradigm, environmentalism, environmentalists, environmental science, and scientists—especially women and minority citizens acting on behalf of environmental agendas—become the targets of media attack, perceived not only as potential obstacles to oil extraction but also as threats to the proliferation of capitalism itself, since oil and capitalism are imagined as symbiotic.

In our contemporary moment, environmental activists are increasingly constructed as environmental terrorists, both through counterterrorism units and associated mainstream discourses. In Canada, on February 9, 2012, the federal government responded to left-leaning environmental movements resisting oil sands extraction through a sliding semantics that identified these groups and their actors as terrorists. The government-authored report, "Building Resilience against Terrorism: Canada's Counter-Terrorism Strategy," contends that "although not of the same scope and scale faced by other countries, low-level violence by domestic issue-based groups remains a reality in Canada. Such extremism tends to be based on grievances—real or perceived—revolving around the promotion of various causes such as animal rights, white supremacy, environmentalism, and anti-capitalism."[4] By May 1, 2012, Canadian environment minister Peter Kent had also used criminalizing rhetoric to characterize the activities of Canadian environmental groups, claiming that these organizations had been used "to launder offshore foreign funds for inappropriate use against Canadian interest."[5] And, as Aboriginal communities have begun to mount more organized protests against the

infringement of their treaty rights that are linked to issues of environmental protection and that include barricades and other forms of demonstration, it is important
to remain cognizant of what feminist scholar Heather M. Turcotte has already identified as the slippages within "academic and state representations of petro-terrorists,
petro-gangs, and victims of gender violence . . . that produce the figure of the petro-
terrorist-gang-member" for public consumption and foreign policy."[6] It is imperative
to consider how these discourses use conflicting representations of women and of
feminism in the West that have ramifications not only for women in Western petro-
cultures but also for ethno-cultural communities and other marginalized groups.

Media and advertising tactics performed to minimize women's relationship to the
environment are in direct contrast to the very serious involvements of women actively
engaged in environmental movements in Canada and around the world. Robert R. M.
Verchick, for example, points out that many of the "most visible and effective environmental justice organizations are led by and consist mainly of women. . . . Thus, while
'environmental justice' describes an environmental movement and a civil rights movement, it also describes a *women's* movement . . . a *feminist* movement."[7] In Canada,
as around the world, a significant percentage of female environmental activists are
women of color or women from minority communities—especially Aboriginal women
who are disproportionately impacted by environmental changes.

In her work on the Niger Delta, for example, Heather Turcotte has found that
women in petropolitics are typically invoked only as objects of law who struggle against
violence in their communities. She theorizes that activist women are naturalized as
mothers and grandmothers in mainstream discourses—maternal protectors of the
environment "rationalized as unpolitical and external to the political economy."[8] She
argues that women's protests are rearticulated "in ways that omit deeper histories of
interconnected state violence and women's anti-imperialist engagements with state
power."[9] By contrast, racialized men are constructed as "terrorists" when active in these
same petro-resistance movements that are sometimes initiated or led by women.[10]

The Idle No More movement went public at a November 10, 2012, Saskatoon
teach-in organized by four women activists—Nina Wilson, Sheelah McLean, Sylvia
McAdam, and Jessica Gordon. These activists were concerned about the effects of a
Harper government omnibus bill, C-45, that infringed on Aboriginal rights.[11] Nonviolent political events then proliferated across the country and around the world. Chief
Theresa Spence of the Attawapiskat band in northern Ontario intensified public attention when she endured a forty-four-day hunger strike between December 11, 2012,
and January 24, 2013. These two actions, often conflated, led to broad public discussion of environmental and Aboriginal issues. Idle No More continues to gain momentum. Chief Spence has made headlines fighting for better educational opportunities

and living standards for her community since she became a chief in 2006. In spite of this, the media configures her in a formulaic manner—as an incompetent and possibly corrupt politician. Alternatively, as part of the recent petro-political resistance, she is for the most part a symbolic figure. Media attention tends to confine her principled political activism to health-related issues related to dietary regimes rather than to the long traditions of nonviolent civil disobedience and the hunger strike in particular. Twelve of the thirteen demands that she made to the Canadian government have been largely overwritten. Only one demand became highly visible: her desire to meet with Prime Minister Harper and Governor General David Johnston. This became the subject of much public debate and deflected focus from the issues at the heart of Spence's protest, positioning her as somewhat obstinate. Like the Idle No More movement, the remainder of Spence's concerns were centered on the omnibus bills C-38 and C-45, which undermine aspects of Section 35 of the Canadian Constitution, and Aboriginal treaty rights and ways of life in Canada. However, they were rarely, if ever, fully explained by the media.

Meanwhile, the mainstream media provided space for male Aboriginal leaders to declare that "it's time for the men to step up."[12] And the political discord within the Aboriginal movement and its leadership was given significant attention. This media focus rhetorically constructs Aboriginal struggles and debates as signs of the inability to organize and an ethno-cultural group turning in on itself—as opposed to the very common leadership conflicts any group faces when dealing with political issues. The Idle No More youth movement that Chief Spence inspired is currently being undermined through reports that its leaders "appear to have little control over the direction of the movement," and as of the end of January 2013 the movement itself is being increasingly linked to violence:[13] violence initiated by "aggressive elements within the existing [A]boriginal leadership structure" and violence against Aboriginals that rhetorically blames the victims for stirring up racist reaction.[14] Newspaper headlines read, "PM Harper Believes Idle No More Movement Creating 'Negative Public Reaction,' Say Confidential Notes."[15] The gang-rape of an Aboriginal woman in Thunder Bay, Ontario, on December 27, 2012, which initially received minimal attention from mainstream media, was finally reported on *The Current*, a major Canadian radio broadcast on January 25, but it was contextualized by an introduction that places the onus on the movement and not on the history of colonialism and racism: "Idle No More is inflaming long-standing tensions between Aboriginal and non-Aboriginal communities. In Thunder Bay, police investigate a possible hate crime and the mayor regrets that his plan to keep people safe has failed."[16] This rape is relevant to the Idle No More movement because, as it was reported, "During the attack the men allegedly told the victim it wasn't the first time they had committed this type of crime and 'it wouldn't be the last.' She [the victim] told police they [the attackers] also told her, 'You Indians deserve to lose your treaty rights,' making reference to the recent Idle No More events in Thunder Bay."[17] This attack and another alleged "starlight tour" reported in the CBC broadcast,[18] while part of a long history of violence against Aboriginal women and men, are now inspiring fear in Aboriginal communities and being used to suggest that youth might be safest distancing themselves from the activism of the Idle No More movement.[19] The violated female body, in association with the Idle No More movement, functions to spectacularize and market petro-violence to the media-consuming public as a reaction to the Idle No More resistance movement rather than as endemic to the legacies of colonial logic, human rights abuses, and gender-sexual violence on which Western petrostates are founded. This is but one manifestation of how the appropriation of women's political power can create dangerous outcomes for entire communities. The foundations for these practices are outlined in the following section, which analyzes how the frameworks of consumerism and embedded feminism have defined women's relationships to oil in such a way as to naturalize imperial and neoliberal agendas.

WESTERN PETROCULTURES, WOMEN, AND MINORITIES

The histories of feminism and oil are intertwined.[20] In the century and a half since oil's potential as a major energy source was first discovered and harnessed, a world oil industry has emerged to transform the distribution of world power between nations, the everyday lived reality of all citizens of the Western world, and increasingly the daily experience of people across the globe. The age of oil in the West is virtually synchronous with the women's rights movement:[21] after similar periods of development, both the oil industry and the Western women's rights movements had gained significant momentum by the early twentieth century. In this same moment, photography began to be used as a tool to construct new feminine identities for the public imaginary that have promoted certain concepts of beauty, domesticity and housewifery, motherhood, personal and family hygiene, and women's autonomy. Alternately affirming and contesting these concepts, women's rights movements and women's lives have transformed over the last century of the age of oil; many of these transformations can be attributed directly to industrialization and the petroleum-related innovations that came to define gender dynamics and gender roles in Western petroculture(s). The age of oil is rife with ironies that have resulted in both feminist advances as well as the reinforcement of long-standing patriarchal conceptualizations of woman as object and as property, popularized through the pervasiveness of the female image as it has been recuperated by capitalist, consumerist, neoliberal discourses of the late twentieth and twenty-first centuries.

There are a number of visual and rhetorical tropes that position women's relationships to oil—particularly Western, white, middle- and upper-class women's relationships—in consumer terms. These tropes build on historical practices that have linked women to social and cultural developments by targeting them as consumer and as commodity—the "consumer consumed." Women have long been identified as major consumers of petroleum products—fashion products being one example. "Ecofeminism" in mainstream popular media has come to be signified by reductive and trivial issues. Magazine and new-media headlines ask questions such as "Are You a Green Beauty?" "What's Your Clothing's Footprint?" and "Are You an Eco-Fashionista?"[22] Advice is given about "How to Go Green: Fashion and Beauty."[23] In the Western cultural practice of reducing women's social engagement to consumerism and the neoliberal practice of Starbucks logic, as Slavoj Žižek has called it, consumerism and social justice are collapsed into one act through "products that contain the claim of being politically progressive acts in and of themselves . . . [and in which] political action and consumption become fully merged."[24] And this logic—Starbucks logic—expresses itself in gendered ways. Specifically in relationship to petroleum and oil, this logic reinforces patriarchal social, political, and economic norms.

This consumerist logic is neither new nor restricted to a surface resistance of oil and the petroleum industry. Similar instances might include anti-animal testing labels on cosmetic products, or the grunge movement's ironic or distorted cultural refraction of the global antipoverty movements of the last several decades, whereby designers sell a spectacle of scarcity. These trends disguise consumption as political awareness, or even activism, and elide discussion of the more systemic and infrastructural processes that imbricate oil, our capitalist economy, and our culture in ways that are much more complex to untangle than simply purchasing petroleum-free mascara.

This practice of both objectifying women, particularly middle- and upper-class white women, as consumable products and supposedly empowering them as consumers themselves, through fashion and beauty products, translates into a performance of political engagement both as the eco-fashionista with a *green* wardrobe and *green* cosmetic bag and as the often dehumanized object onto which social resistance is draped, projected, and performed as a commercial strategy in the guise of political resistance. Western women's relationships to petroleum have been constructed in the social imaginary as a site of spectacle through which resistance to petroculture is signified yet undercut.

CANADA AND THE ETHICAL OIL BILLBOARD CAMPAIGN: GENDER, SEXUAL RELATIONS, AND COLONIALISM AS PETRO-DISCOURSES

Ezra Levant, a conservative Canadian media personality, published the book *Ethical Oil: The Case for Canada's Oil Sands* in 2010.[25] This book promotes a right-wing, pro-Canadian-oil-sands vision for the country. In fact, Levant was at one time, around the new millennium, a political candidate in the same party as Stephen Harper— current prime minister of Canada.[26] Furthermore, Levant uses polemical and preju- diced vocabulary to characterize foreign interests and the nationals of multiple coun- tries. In fact, Levant has a history of publishing pejorative material.[27] Furthermore, the book targets certain ethnic groups within Canada, including Aboriginals. The rhetoric first promoted by this book has been sustained through a website, EthicalOil .org, started by Alykhan Velshi. In 2011, Ethical Oil also ran a media campaign that included billboards and television commercials (also made available online). The campaign explicitly rebranded Canadian oil/tar sands oil as an ethical source of oil.[28] This is relevant at this particular historical moment because it, at least temporarily, reoriented the public debate in Canada away from a discussion of the environment and toward a discussion of oil's foundational and integral role in Canadian national identity, by fetishizing oil as a national resource linked to Canadian pioneerism and innovation, ingenuity, and integrity, as well as Canada's international reputation as a liberal, democratic, peacekeeping nation with an excellent human rights record.[29]

The campaign links what are apparently disparate issues in ways that rely on the preexisting sociocultural fetishization of oil as a thing, with powers that make it capa- ble of changing the nation. In this case, the Ethical Oil campaign imbues the oil sands with the potential to redirect power toward good and ethical ends—the propagation of multiculturalism, for example, or the advancement of the Canadian dream, a ubiq- uitous myth linking economic expansion with personal and financial fulfillment. Playing on historical identity tropes and the myth of the Canadian nation, Levant simply affirms what Canadians wish to believe about themselves, and he does so by weaving oil into the social, political, and economic fabric of the nation.

These popular beliefs about multiculturalism and racial and gender equality became the focal point of the Ethical Oil billboard and television campaign of 2011, which employs images meant to represent foreign female nationals, as well as white and nonwhite Canadian women, as a strategy to reinscribe oil into the nation myth. The concept of saving foreign women in this campaign invokes a power-discourse that rhetorically positions and visually reinforces for all Canadians the narrative myth of nation whereby Canadians are benevolent, tolerant, democratic protectors of human rights. One of the billboards juxtaposes the images of two women. The image on the left of the billboard appears under a red-banner heading that reads "Conflict Oil" with a secondary red-banner message, "Conflict Oil Countries: Women Stoned to Death," superimposed over the black-and-white image of a burka-clad woman who is being buried alive in preparation for stoning. The image on the right half of the billboard is a color photo of the graduation-gown-clad female mayor of Fort McMurray, the largest urban center in the oil sands region of Northern Alberta.[30] Her picture appears under the green-banner message "Ethical Oil," and the second green-banner message superimposed over the middle of the image reads "Canada's Oil Sands: Woman Elected Mayor." Her identity is emphasized in the fine print in the bottom right-hand corner of the graduation photo that reads "Mayor Melissa Blake." By contrast, the woman in the left-hand photo is not similarly identified. According to various sources, this image dates back to either the 1970s or 1980s, and the female subject is an Iranian woman. However, the billboard itself provides no such context.[31] This burka-clad woman simply stands in as a synecdoche for the perceived oppression of women in Muslim areas of the world. The visual rhetoric of this billboard not only situates foreign women in a position of victimization whereby they must be rescued—[32] a form of embedded feminism whereby women's rights are used to justify foreign policy, often in the form of political or military intervention by Western nations into the affairs of Eastern nations—but it also validates the status of Western women. In this paradigm, women's liberation from traditional private-sphere roles becomes the only evidence required or necessary to demonstrate the superior status and civilization of Western nations, despite the ongoing feminist struggles of women in the twenty-first-century West.

This billboard promotes a variation on the themes of embedded feminism and consumerism as social justice. At the bottom of the billboard, another green-banner message reads "Ethical Oil. A Choice We Have to Make." The color codes of red and green make evident that viewers are being encouraged to choose—or rather that they have no choice but to choose—Canadian oil from oil sands for ethical reasons that link Canadian Oil, by proxy, to women's rights. The billboard is not a direct call for war or invasion in the name of women's rights, although one could easily argue that it is a tentative step toward that, as political-media attention has increasingly highlighted tensions with Iran over its nuclear program and "the threat posed by Iran to Middle Eastern oil supplies";[33] however, as a reaction to women's oppression, it calls for embargo, even if only at the level of individual consumer choice. The use of the Iranian woman's image is an example of what Turcotte theorizes as the "uncritical representation of gender violence in other geopolitical locales [that suppresses] the state's simultaneous and daily consumption and violation of women for its own nation-building practices."[34] The use of the foreign burka-clad woman is part of a larger rhetorical practice that fails to read and understand exceptional moments of gender violence in Other contexts as such, and instead invokes these images of violence against women because they sustain the Western narratives of foreign-woman-as-victim. As Turcotte argues in regard to the United States, locating violence against women in other places "obscures and denies the dismal histories of gender and sexual violence endemic to the United States"—or, as I would argue, in the West in general—and the "consequent 'rescue narratives' demand victimized 'third world women' . . . must be saved from 'ethnic' perpetrators."[35] In the Canadian context, the Ethical Oil campaign is written into an accepted history of petropolitics that justified the suppression of Other peoples (domestically and internationally) and the invasion of foreign nations—for the purpose of gaining control of petroleum resources—by invoking the national myth of Canada as a defender of human rights. This billboard implies, through its false logic, that Canadians can choose continued security and freedom not only for Canadian women but also for foreign women by supporting Canadian oil. Resisting the oil sands industry, within this visual rhetorical frame, becomes akin to supporting female repression in other regimes, to supporting terrorism as it has been loosely defined in the post-9/11 era, and to forsaking the advances of Western feminist movements. The consumer "choice" provided by the artificially constructed parameters of this campaign justifies the perpetuation of Western oil-consuming lifestyles driven by current petro-economies as a strategy to shift the discussion away from environmentally focused critiques of the oil sands industry and its planned expansion.

Considering this billboard campaign within the context established by the book *Ethical Oil: The Case for Canada's Oil Sands*, with its blatantly racist, imperialist, and sexist language, adds another layer of meaning to the visuals. Take, for example, the billboard that juxtaposes the image of a beautiful and smiling Aboriginal woman dressed in oil-rig garb: her head tilted and looking beatifically upward to the left from under an Esso hardhat, protective industrial eyewear, earmuffs, and overalls. The green-banner messages read "Ethical Oil" and "Canada's Oil Sands: Aboriginals Employed." This is juxtaposed with the red-banner messages of "Conflict Oil" and "Sudan's Oil Fields: Indigenous Peoples Killed" superimposed on the image of militiamen walking through the desert with a human skull in the forefront of the image. The Aboriginal woman's presence in the billboard campaign appropriates her identity at two levels. Her sexuality and ethnicity are employed simultaneously as a tool of erasure in at least two ways: by preempting potential criticism of the rampant gender inequality in the oil industry by creating a simplified visual claim that women do benefit financially from oil-field employment, and by undermining the ongoing collective criticism and resistance against oil industry expansionism coming from Aboriginal communities in Northern Alberta and elsewhere in Canada. This image visually reinforces Levant's scathing criticism of Aboriginal (First Nations, Métis, and Inuit) resistance to the oil sands as hypocritical and unjustified because he claims that the oil industry provides jobs and fiscal return for Aboriginal communities.[36] This image of the young Aboriginal woman reinvents the Canadian oil industry, and by proxy

Canada itself, as a site of gender equality, in contrast to the alleged misogyny and sexism found in other countries. The image also aims to create doubts about the pervasiveness of anti–oil sands attitudes among Aboriginal peoples, and to raise questions around the legitimacy of these communities to resist treaty breeches that disrupt traditional practices on Aboriginal lands. However, by this same economic logic, all citizens of the Global West, if not the entire world, can be silenced into complicity due to the way petroleum has shaped our daily lives and the global economy. This is not the case, nor should it be, in light of increasing evidence of environmental impact that has inspired a demand for change from across the political spectrum. Nevertheless, this oversimplified visual and rhetorical strategy attempts to weaken public sympathy for Aboriginal resistance against the environmental impacts of the oil industry.

Prior to and during the period of the Ethical Oil billboard campaign, a number of Aboriginal communities publicly resisted industrial encroachment on their land and exposed the impacts of air, water, and land pollution in the area of Northern Alberta. There were also a number of environmental groups and activists making headlines with their anti–tar sands protests both in Canada and around the world at the time of the campaign. Rather than address these issues that involve both human rights and environmental concerns within Canada, the Ethical Oil campaign resituates petro-violence and human rights abuses outside the Canadian context. Situating the Aboriginal body—both the beautiful female in the "Aboriginals Employed" billboard as well as the ethnic male, potentially Aboriginal, in the "Good Jobs" billboard discussed below—within the neoliberal infrastructure of the petroleum industry serves to erase the invisibilized Indigenous body that resists the oil sands and the oil industry. In doing so, petro-protests that demand social and eco-justice can be redefined as petro-violence. The activists in these movements are also easily reinscribed as terrorists.[37]

The racist underpinnings of the appropriation of the Aboriginal female figure and Aboriginal identities in general become even more explicit within the context of the other stereotypical and reductive billboard messages in the campaign. First is a billboard that, on the left, declares "Conflict Oil" and "Dictatorship" over a collage of the flags of Saudi Arabia and Iran, with faint superimposed images of Ayatollah Khomeini and President Ahmadinejad overlaid on the flag of Iran, all contrasted against the message on the right side of the billboard that features the Canadian flag flying behind green-banner messages of "Ethical Oil" and "Democracy." Another billboard represents the dichotomy of the red-banner messages, "Conflict Oil" and "Forced Labour," with an image of Hugo Chavez, versus the green-banner messages of "Ethical Oil" and "Good Jobs" visually supported by the image of a young pleasant-looking man, quite possibly Aboriginal. Yet another billboard presents the idea of "Conflict Oil" and "Degradation" illustrated with an image of a black man walking through a muddy or oily field gesturing toward a large fire burning in the background, likely in the Niger Delta, which is then contrasted with the green-banner messages of "Ethical Oil" and "Reforestation" and the image of someone standing in a wooded area—possibly meant to represent one of the people responsible for the Syncrude reforestation project in Northern Alberta. Yet another set of red-banner messages reads "Conflict Oil" and "Funds Terrorism," juxtaposing a photograph of the back of someone's head, clad in a Saudi Arabian keffiyeh and looking into the distance at an oil rig, with the adjacent green-banner messages "Ethical Oil" and "Funds Peacekeeping" superimposed on the image of a Canadian peacekeeping monument. Photographed from below so as to heighten its majestic appearance, this sculpture is silhouetted against a blue sky and accompanied by a Canadian flag blowing in the wind. A final billboard pairs "Conflict Oil" and "Persecution" substantiated by the image of two blindfolded men being prepared for hanging, on the left, paralleled by the green-banner message on the right-hand side that reads "Ethical Oil" and "Pride" visually substantiated by two interlocking male hands, clasped together, each with a rainbow bracelet at the wrist.

In these simplistic binary arguments, the Ethical Oil campaign reveals the gendered and racialized messages that have been naturalized as part of Canadian, and even Western, petrocultural narratives. Read together, the various billboard images situate in bas-relief the identity-based fantasies of the entire Ethical Oil campaign, whereby foreign women of color are figured as victims of horrific violence, Canadian Aboriginal women are recuperated as symbols of Western gender and ethnic equality and representatives of the progressive employment practices of the oil industry, and white Canadian women are celebrated as civic leaders and symbols of democracy. These images obfuscate the historical, ongoing systemic and cultural racism against First Nations peoples in Canada and overwrite the very low percentage of female politicians.[38] By focusing on gender-sexual violence and inequality elsewhere, the campaign accomplishes what Elizabeth Swanson Goldberg has identified as the practice of creating markets of violence for consumption that erases these issues within the nation-state.[39]

BP TELEVISION COMMERCIAL:
TRIVIALIZING WOMEN'S RELATIONSHIP TO OIL

The use of women's images in the Canadian Ethical Oil campaign is one manifestation of a larger petrosexual discourse that recuperates the images and roles of Western women in reductive ways that trivialize women's relationship to both oil and the environment. The ongoing cultural narrative defines women as either consumers or as bodies onto which politics can be projected and performed. Therefore, the popular media inundates women with a barrage of petroleum consumer products and services that they should wear, buy, or surgically implant; more recently, this also includes media messages around what products women should not consume, thereby defining ecofeminism not as a political stance that requires serious engagement and action, but as a consumer choice.

One such example is a BP television commercial of the early millennium that revises a series of identity tropes to promote the fantasy of the "new BP," no longer "British Petroleum," but "Beyond Petroleum." This media campaign was designed to reinvent BP's image as a good corporate citizen, part of the global environmental solution, and part of a team of social and technological innovators. The commercial both begins and ends with slight modifications on the phrase "beyond darkness there is light," represented through a light bulb in the opening sequence and by placing the viewer in the position of a speeding train rushing through a tunnel and bursting into daylight, suggesting invention, modernity, and progress. However, within this visual and narrative frame, the viewer is also presented with a number of problematic racial and gender stereotypes. Innovation and environmental citizenship are constructed in relationship to women and people of color in anything but new or forward-thinking terms. The commercial's initial image is of a young woman, in black alternative-style dress, wearing black lipstick, her hair dyed black, and with multiple piercings and tattoos. The narrator explains that "behind a thorn, there is a rose." Through a rapid time-lapse sequence we see this same woman stripped down and transformed to her natural self: a young, smiling, blond beauty, whose pouty glossed lips and subtle head movement seem to suggest that she is flirting with the camera. Both identities are stereotypical, but the one that resists conformity is ironically reinvented according to classic (read: white, youthful, innocent) female beauty standards. Another time-lapse sequence illustrates an increasingly pregnant woman who eventually holds a child, looking down on it in the classic mother–child embrace—like Mary gazed on Jesus. This is paired with the message "Beyond patience, fulfillment," which visually establishes motherhood as the ultimate goal for all women. Furthermore, this mother is a woman of color depicted in the nude—symbolic of a proximity to nature, a common but stereotypical representation of women of color in advertising. This, in direct contrast to the white woman, in the "Beyond Pain, Joy" sequence, who embodies artifice and represents a higher class. For the white woman, "pain" is represented by breast-implant surgery, and "joy" by standing in a low-cut black cocktail dress—objectified as the camera zooms in, focusing on her cleavage. But realistically speaking, implants are not the "joy" of the woman. The woman herself will almost certainly have no sensation in her nipples after surgery, thereby making her an object and not a subject of the experience of erotic pleasure, if "joy" refers to sex. If it does not, then in this heterosexual visual economy of male spectatorship, "joy" must refer to her presumed enjoyment at being admired as an object of beauty.

Race in this commercial is represented in equally problematic terms. The graceful dancer in the commercial appears to be a nonwhite male. The fastest runner is represented by a black male. All the faces flashed in sequence during the word "disease" are Asian, as is the young male doctor featured during the word "cure." This commercial would have been produced in the first few years of the new millennium, when the predicted Asian flu pandemic of 2000 that never materialized was worldwide news for months. By contrast, white men in the commercial have either trivial problems—one is out of toilet paper and the other has locked his keys in the car—or they are visually signified as important, located as part of the military industrial complex, through terms of "power" and "responsibility," and visually associated with a nuclear explosion.

This commercial, which claims innovation and progress as part of BP's branding campaign, in fact makes recourse to the accepted status quo, particularly as it pertains to gender roles and stereotypical ethnic and racial identities within the neoliberal complex. Innovation is not required; merely the implication thereof is enough narratively and visually to invent an identity linked to innovation—Beyond Petroleum.[40] These rhetorical moves function to bolster the mythical value of oil, to fetishize oil, by crediting it more generally with the wonders of life, both those that are natural manifestations (motherhood, for example), as well as those attributed to science and innovation (nuclear energy). This discourse reinforces the value of BP's role as an oil and energy provider: sustaining life as we know it and life in all its infinite possibilities. In claiming that it is providing energy, the very stuff of life, to us in groundbreaking new ways, BP insinuates itself into multiple discourses that suggest the possibilities of innovation, including environmentalism (à la corporations). This rhetorical strategy allows corporate interests to redirect and co-opt the zeal of eco-movements in ways that satisfy consumer guilt but do little to disrupt business as usual.

VOGUE ITALIA'S "WATER & OIL"

Much as visions of women in professional roles are marketed as proof of "ethical oil" and women's bodies are represented as objects (not subjects) of petrocultural pleasure, so too are women's bodies used to market fashion by staging a spectacle of political engagement against oil. The August 2010 issue of *Vogue Italia* contained a photo spread shot by Steven Meisel, featuring the model Kristen McMenamy.[41] The photo spread was a reaction to the BP oil spill that flowed for three months in the Gulf of Mexico in 2010. In this fashion editorial, the female body is recuperated through the guise of political activism to promote consumerism.

Kristen McMenamy, a white female model, is represented in some instances as an oiled and dead bird, and in others as a woman covered in oil. In one particular photo the model's gloved and feathered hand is transformed into an oiled bird. The female figure is dehumanized not merely through the compartmentalized focus on one part of her body, but also through her transformation into an animal in other shots. This fashion spread, as a bid to suggest political engagement, builds on long-standing fashion and advertising usages of the segmentation of women's bodies, women depicted as animals (especially women of color), the stalked or raped woman, the murdered female corpse, and the dehumanization of women to sell high fashion. In only one image is the model explicitly human. However, where the model is alive, she eerily resembles a zombie or a recently traumatized civilian victim of war—namely, an Eastern woman, signaled by a headscarf, quite possibly that of a Muslim woman. This photo editorial makes no call to action but merely aestheticizes female death and female victims of violence—albeit this time glamorizing big oil as the aggressor. Furthermore, representing women's disempowerment in relationship to oil via a white supermodel in a *Vogue Italia* exposé, a high-fashion magazine targeted at middle- and upper-class women, contextualizes oil and environmentalism as First World issues with First World solutions.

Intermingling deep-rooted icons of misogynistic violence (the female corpse), environmental devastation (the oiled bird), and gothic horror (the nonhuman zombie), the *Vogue Italia* photo spread contributes to a much broader cultural discourse of disaster circulating through Western popular culture that reimagines the oil spill as a monstrous exception. As Alain Badiou has argued, "We live within an Aristotelian arrangement: there is nature, and beside it right, which tries as much as possible to correct, if needs be, the excesses of nature. What is dreaded, what must be foreclosed, is what is neither natural nor amendable by right alone. In short, what is *monstrous*."[42] It has become acceptable to respond to oil spills as moments of exception and to the associated symbols—the bird, for example—with expressions of empathy and outrage. This carefully constructed response ensures that the broader oil assemblage goes unquestioned: like the representation of women as victims of violence and civilian casualties of war, like the disregard for Aboriginal treaty rights and traditional ways of life, oil has been entirely naturalized. Despite the fact that the air and water pollution from daily industrial activity around the oil sands in Northern Alberta over the course of a year is equivalent to that from a major oil spill every year, this activity has been naturalized. By contrast, the monstrous oil spill is perceived as something that needs to be brought under control and disciplined.[43] Therefore, it becomes culturally acceptable to protest the impacts of oil only at the moment of crisis. And, of course, the irony is that this protest, just like the *Vogue Italia* fashion spread, is socially condoned and therefore not a resistance against the system but part of its *natural* cycle of cannibalizing women's bodies in a bid to promote capitalism and the petro-economy.

CONCLUSION

Understanding the role of gendered petro-relations as they have been historically constructed and as they continue to be perpetuated by Ethical Oil, the *Vogue Italia* fashion spread, related eco-fashionista discourses, and mainstream media representations of resistance movements such as Idle No More reveals the degree to which Western neoliberal petro-discourses are invested in promoting specific female identities and definitions of Western feminism. These iterations justify our current oil-consuming lifestyles as an issue of women's rights through basic rhetorical strategies that reinforce women's relationship to petroleum products in consumer terms and that recuperate the female body as a canvas on which to spectacularize and perform politics. Furthermore, the export of these specific female identities brings with it the promise of new female consumer markets. Therefore, empowered feminist identities outside those sanctioned by the mainstream neoliberal petro-discourses are depoliticized and renegotiated in the public sphere. Women activists—especially those of color—who refuse to conform to the eco-fashionista consumerist identities marketed to them, and whose agendas to protect the environment will potentially disrupt industrial development and business as usual, pose a particular threat to established forms of hegemonic power and therefore risk becoming the target of violence,[44] just as the men that align themselves with these women also risk being framed by discourses of petro-violence and marked as terrorists. It is not merely the threat of alternative energies that is being resisted. Reliance on wind and solar energy will not, in and of themselves, reconfigure power relations, nor eliminate racism, sexism, and class disparity, any more than oil in and of itself causes wealth, poverty, war, or militarism.[45] Therefore, as we make the necessary moves toward alternative energy sources in the twenty-first century, it is valuable to consider how we might be simply fetishizing these energies, as we have done with oil, in ways that perpetuate the capitalist status quo and the web of associated relationships and power dynamics of gender, race, ethnicity, and class. Rather than succumb to the fears of what the end of cheap oil might mean for the future in general, or for feminism in particular,[46] it may be more productive to reevaluate current discourses around oil and alternative energy sources for their potential to disrupt rather than to simply reproduce intersectional social inequalities.[47]

NOTES

1. Matthew T. Huber, "Oil, Life, and the Fetishism of Geopolitics," *Capitalism Nature Socialism* 22, no. 3 (2011): 32-48.

2. Ibid., 36.

3. In 2012, a counterterrorism unit was established in Alberta to protect the energy sector. See "Counter-Terrorism Unit to Protect Alberta Energy Industry," *Canadian Broadcasting Corporation*, June 6, 2012, http://www.cbc.ca/news/business/story/2012/06/06/rcmp-counter-ter rorism-oil.html. And over the summer of 2012, the media reported on concerns of environmental terrorism. An article published on July 29, 2012, in *The Canadian Press* cites a "newly declassified intelligence report" stating that there is a "growing radicalized environmentalist faction" in Canada that is opposed to the country's energy-sector policies: "The RCMP criminal intelligence assessment, focusing on Canadian waters, cites potential dangers from environmental activists to offshore oil platforms and hazardous marine shipments, representing perhaps the starkest assessment of such threats by the Canadian security community to date." See "Radical Environmentalism on the Rise, RCMP Report Says," *Canadian Broadcasting Association*, July 29, 2012, http://www.cbc.ca/news/politics/story/2012/07/29/pol-radical-environmentalism-gr owing-intelligence-report-warns.html.

4. "Building Resilience against Terrorism: Canada's Counter-Terrorism Strategy," *Public Safety Canada*, last modified December 4, 2012, http://www.publicsafety.gc.ca/prg/ns/2012 -cts-eng.aspx#s3.

5. Quoted in "Environmental Charities 'Laundering' Foreign Funds, Kent says," *Canadian Broadcasting Corporation*, May 1, 2012, http://www.cbc.ca/news/politics/story/2012/05/01/ pol-peter-kent-environmental-charities-laundering.html.

6. Heather M. Turcotte, *Petro-Sexual Politics: Global Oil, Legitimate Violence, and Transnational Justice* (Charleston, S.C.: BiblioBazaar, 2011), 213.

7. Robert R. M. Verchick, "Feminist Theory and Environmental Justice," in *New Perspectives on Environmental Justice: Gender, Sexuality, and Activism*, ed. Rachel Stein (New Brunswick: Rutgers University Press, 2004), 63.

8. Turcotte, *Petro-Sexual Politics*, 208.

9. Ibid.

10. Ibid.

11. "Idle No More: How It Began," *Province*, January 7, 2013, http://www.theprovince.com/ news/bc/Idle+More+began/7780924/story.html.

12. Bruce Campion-Smith, "Idle No More: Spence Urged by Fellow Chiefs to Abandon Her Fast," *Toronto Star*, January 18, 2013, http://www.thestar.com/news/canada/2013/01/18/ idle_no_more_spence_urged_by_fellow_chiefs_to_abandon_her_fast.html.

13. Deveryn Ross, "Idle No More's Real Challenge," *Winnipeg Free Press*, January 24, 2013, http://www.winnipegfreeP.com/opinion/westview/idle-no-mores-real-challenge-188173011 .html.

14. Ross, "Idle No More's Real Challenge," *Winnipeg Free Press*, January 24, 2013.

15. Jorge Barrera. "PM Harper Believes Idle No More Movement Creating 'Negative Public Reaction,' Say Confidential Notes," *APTN National Notes*, January 25, 2013, http://aptn.ca/ pages/news/2013/01/25/pm-harper-believes-idle-no-more-movement-creating-negative -public-reaction-say-confidential-notes/.

16. Duncan McCue, "Idle No More and Tensions in Thunder Bay," Special edition of *The Current*, CBC Radio, January 25, 2013, http://www.cbc.ca/player/News/Canada/Audio/ ID/2329105939/.

17. Valerie Taliman, "Rape, Kidnapping Being Investigated as Hate Crime in Thunder Bay," Indian Country Today Media Network, January 7, 2013, http://indiancountrytodaymedianet work.com/article/rape-kidnapping-being-investigated-hate-crime-thunder-bay-146797.

18. "Starlight tours" refers to the unofficial police practice of dropping people, often Aboriginals or other marginalized individuals, outside of city limits in what are often intemperate and life-threatening Canadian weather conditions, forcing them to walk back to the city or home. On January 7, 2013, CBC Radio ran a report about a local man who was driven to the edge of town by police in Thunder Bay, Ontario. The report speaks of the history of this practice. See "Starlight Tours," *Superior Morning, CBC Radio*, January 7, 2013, http://www.cbc.ca/superiormorning/episodes/2013/01/07/starlight-tours/.

On January 21, 2013, a young First Nations man in Saskatoon, Saskatchewan, was reportedly driven outside the city limits and told by officers to "Idle No More." See David Giles, "Saskatoon Police Investigating Alleged 'Starlight Tour,'" *Global News*, January 22, 2013, http://www.globalnews.ca/saskatoon+police+investigating+alleged+starlight+tour/6442793549/story.html.

19. "No More Stolen Sisters: Justice for the Missing and Murdered," *Amnesty International Canada*, n.d., http://www.amnesty.ca/our-work/issues/indigenous-peoples/no-more-stolen-sisters.

20. The Seneca Falls women's rights convention was held in New York in 1848. See "Women's Rights Convention in Seneca Falls, NY," *The Susan B. Anthony Center for Women's Leadership*, n.d., http://www.rochester.edu/sba/suffragewomensrights.html. Ten years later, in 1858, the first commercial oil well was established in Oil Springs, Ontario. See "Black Gold: Canada's Oil Heritage," *County of Lambton Libraries Museums Galleries*, n.d., http://www.lclmg.org/lclmg/MuseumofCanada/BlackGold2/OilHeritage/OilSprings/tabid/208/Default.aspx. A year later, an oil rig in Pennsylvania struck oil. See "The Story of Oil in Pennsylvania," *Petroleum Education: History of Oil*, The Paleontological Research Institution, n.d. http://www.priweb.org/ed/pgws/history/pennsylvania/pennsylvania.html.

21. It is important to note that there have been other class and race/ethnic struggles during the petroleum era, which this essay also references, and cannot escape referencing, because of the manner in which popularly naturalized discourses around oil and new energies are situated in relationship to women, women of color, and other marginalized groups.

22. "Quizzes: Fashion & Beauty," *Planetgreen.com*, n.d., http://planetgreen.discovery.com/games-quizzes/quizzes-fashion-beauty.html.

23. "How to Go Green: Fashion & Beauty," *Planetgreen.com*, n.d., http://planetgreen.discovery.com/go-green/green-index/fashion-beauty-guides.html.

24. Slavoj Žižek, "Censorship Today: Violence, or Ecology as a New Opium for the Masses, part 1," *Lacan.com*, n.d., http://www.lacan.com/zizecology1.htm.

25. This section borrows and builds on ideas previously published in Sheena Wilson, "Ethical Oil: The Case for Canada's Oil Sands, Review," *American Book Review* 33, no. 3 (2012): 8–9.

26. In 2002, Levant was nominated as the Canadian Alliance candidate in the Calgary Southwest riding, but he stepped down so that Stephen Harper, who had been elected party leader, could run in that riding. This led to the eventual election of Harper, leader of the Canadian Conservative Party, as Prime Minister of Canada. See Jane Taber, "Meet Harper's Oil-Sands Muse," *Globe and Mail*, September 10, 2012, http://m.theglobeandmail.com/news/politics/ottawa-notebook/meet-harpers-oil-sands-muse/article1871340/.

27. Levant has had various complaints brought against him: both human rights complaints for the publication of the Danish cartoons of the Prophet Muhammad in 2006 and for making false claims about George Soros, to name two examples. And again in March 2014, yet another

lawsuit was filed again Levant, this time for $100,000, related to blog posts he made that allegedly damaged the reputation of a Saskatchewan lawyer by labeling him a "jihadist and a liar." http://www.winnepegfree press.com/canada/sun-news-host-ezra-levant-sued-for-libel-lawyer -seeking-100k-in-damages-248219811.html.

28. The Ethical Oil organization misappropriates terms and ideas, such as "grassroots" and "fair trade," that are typically associated with left-leaning social justice agendas, for the purpose of supporting a right-wing neoliberal agenda focused on promoting the expansion of the oil industry in Alberta. For more details on the book, see Wilson, "Ethical Oil," 8–9.

29. This redirection in public debate was certainly evident in Alberta and in Canada starting at the time of the book's release in the fall of 2010, and for the subsequent two years, 2011 through 2012. The book and its author received significant attention at the time of the book's release, and the ethical oil message was further perpetuated by the associated EthicalOil.org website. The site's "About" page explains that "EthicalOil.org began as a blog created by Alykhan Velshi to promote the ideas in Ezra Levant's bestselling book Ethical Oil: The Case for Canada's Oil Sands." See http://www.ethicaloil.org/about/. Furthermore, in mid-2011, the Ethical Oil organization released a billboard campaign and, later that same year, a thirty-second television advertisement that, by early September, had attracted international attention and a response from the Saudi government for the way that it juxtaposed women's rights in the two countries and, through this faulty logic, thereby asserted the superiority of Canadian-produced oil as ethical, stigmatizing Saudi oil as conflict oil.

The billboard images can be viewed online. See "In Pictures: Ethical Oil Ad Campaign," *Globe and Mail*, July 28, 2011, http://www.theglobeandmail.com/news/politics/ethical-oil-ad -campaign/article637242/. The television ad has been uploaded on YouTube. See "Ethical Oil TV Ad, Saudi Arabia (The ad Saudi Arabia doesn't want you to see)," YouTube video, 0:46, posted by "Ethicaloildotorg," August 26, 2011, http://www.youtube.com/watch?v=1SjZl qbDudI.

30. Fort McMurray is not officially a city but part of the Regional Municipality of Wood Buffalo in Northern Alberta.

31. According to a July 20, 2008, news article published by Radio Free Europe/Radio Liberty online, the article is from the 1980s. "Nine Iranians Sentenced to Death by Stoning," *Radio Free Europe/Radio Liberty*, July 20, 2008, http://www.unhcr.org/refworld/country,,,,IRN,,4889 d086c,0.html. However, a major Canadian newspaper, *Globe and Mail*, claimed the image was from the 1970s. See "In Pictures: Ethical Oil Ad Campaign," *Globe and Mail*, July 28, 2011.

32. For a more specific discussion and definition of "embedded feminism," albeit in a different context, see the following: Helmut W. Ganser, "'Embedded Feminism: Women's Rights as Justification for Military Intervention?" (presentation at "Coping with Crises, Ending Armed Conflict: Peace Promoting Strategies of Women and Men," the international conference presented by the Gunda Werner Institute of the Heinrich Böll Foundation, Berlin, November 15, 2011), Gunda Werner Institute, n.d., http://www.gwi-boell.de/web/un-resolutions-helmut -ganser-embedded-feminism-presentation-2953.html.

33. Oliver Wright, "Britain Meets Gulf Allies over Growing Tensions in Iran," *The Independent*, January 24, 2013, http://www.independent.co.uk/news/uk/home-news/britain-meets-gulf-allies-over-growing-tensions-in-iran-8278101.html.

34. Heather M. Turcotte, "Contextualizing Petro-Sexual Politics," *Alternatives: Global, Local, Political* 36, no. 3 (2011): 204.

35. Ibid.

36. Levant provides an inflected critique of Chief Al Lameman, and by proxy all other Aboriginal people who contest the oil sands, in the book *Ethical Oil*. Levant writes: "Of course, oil sands firms are careful to make sure they stay on good terms with Aboriginals in the area, even if a few noisy malcontents like Beaver Lake Cree chief Al Lameman figure they'd rather sue the industry. . . . They've unilaterally declared themselves to be the sole 'keepers' of an enormous swath of land that crosses the border between Saskatchewan and Alberta. But Lameman and his reserve are not keepers of that massive swath of land in any meaningful way; they don't tend to it; they don't look after it or protect it; they don't improve it or develop it, and they certainly don't work it. They just claim it for themselves. Keepers is right. If Al Lameman has his way, he'll end up keeping thousands of Aboriginals from improving their lives, getting an education, escaping poverty. All of those jobs, all that education, all those opportunities ended, all for one band" (213–15).

37. Turcotte discusses at length how, in the context of the Niger Delta, the reporting of and even the academic analysis of "ethnoracial, national, and gender-sexual inequalities within and between Nigeria–U.K.–U.S. relations" have misidentified petro-protests as terrorist activities, in part because there has been a failure to recognize the degree to which these movements "developed out of strategies of community justice within women's organizations. These were neither 'terrorist acts' nor the community 'turning in on itself' but, rather, expressions of justice often supported by male youth groups, which created spaces through which community engagement could address inequalities." "Contextualizing Petro-Sexual Politics," 208.

38. According to the database for "Women in national parliaments" compiled by the Inter-Parliamentary Union based on information from 190 countries, last updated October 31, 2012, Canada ranks 47th in the world for female participation in the "lower or single house" with a percentage of only 24.7. Afghanistan, by contrast, ranks significantly higher in 37th place, with 27.7 percent participation. *Inter-Parliamentary Union*, http://www.ipu.org/wmn-e/classif.htm.

39. See Elizabeth Swanson Goldberg, *Beyond Terror: Gender, Narrative, Human Rights* (New Brunswick: Rutgers University Press, 2007).

40. Peter Hitchcock, in his article "Everything's Gone Green: The Environment of BP's Narrative," explains in detail how Lord John Browne, CEO of British Petroleum, during the Beyond Petroleum rebranding campaign "reads corporate responsibility as 'showing willing' about fighting climate change," with little transformation to the corporate practices. Special issue, *Imaginations: Journal of Cross-Cultural Image Studies* 3, no. 2 (2012): 110. http://www.csj.ualberta.ca/imaginations/wp-content/uploads/2012/09/001_Full_V2_3.2.pdf.

41. "Steven Meisel," *art + commerce*, n.d., http://www.artandcommerce.com.

42. Alain Badiou, *The Century* (Cambridge: Polity Press, 2007), 176–77.

43. "In 2010, a pair of studies from University of Alberta ecologist David Schindler concluded that [the oil sands] industry [in Northern Alberta] was releasing heavy metals and hydrocarbons at levels that were, in some cases, already toxic to fish and equivalent to a major oil spill every year. Schindler's work was later backed up by a provincially appointed review panel." Bob Weber, "Government-Funded Study Concludes Toxic Hydrocarbons from Oilsands Pollute Lakes," *Global Calgary*, January 7, 2013, http://www.globaltvcalgary.com/government-funded+study+concludes+toxic+hydrocarbons+from+oilsands+pollute+lakes/6442783562/story.html.

44. "Women Human Rights Defenders," *Association for Women's Rights in Development*, n.d., http://www.awid.org/Our-Initiatives/Women-Human-Rights-Defenders; Inmaculada Barcia and Analía Penchaszadeh, "Ten Insights to Strengthen Responses for Women Human Rights

Defenders at Risk," *Association for Women's Rights in Development*, 2012, http://www.defending
women-defendingrights.org/pdf/WHRD%20Ten%20Insights%20ENG.pdf.

45. Huber, "Oil, Life, and the Fetishism of Geopolitics."

46. Sharon Astyk has written and blogged about how "the women's movement has never
fully acknowledged the degree to which women's social roles have changed not just due to activism, but due to energy resources. This comparative blind spot means that we have also failed
to grasp how vulnerable those gains are." While Astyk rightly indicates that women's lives have
been transformed by the petroleum-derived energy sources, the oversimplified cause-and-effect
relationship drawn between oil and feminism fetishizes the power of oil and fails to acknowledge the many other socioeconomic and political power relations that have played a role in the
feminist advances of the last two centuries. See "Peak Oil Is Still a Women's Issue and Other
Reflections on Sex, Gender, and the Long Emergency" (blog post), *Casaubon's Book*, Science-
blogs.com, January 31, 2010, http://scienceblogs.com/casaubonsbook/2010/01/31/peak-oil
-is-still-a-womens-iss/.

47. For more information on the notion of intersectionality, see Sheena Wilson, "Petro-
Intersectionalities: Oil, Race, Gender, and Class," in *Fueling Culture: Politics, History, Energy*, ed.
Imre Szeman, Jennifer Wenzel, and Patricia Yaeger (forthcoming).

PART IV

Exhibiting Oil

14

Mixing Oil and Water

Naturalizing Offshore Oil Platforms in
American Aquariums

DOLLY JØRGENSEN

On June 26, 2010, the brand-new Gulf of Mexico exhibit at the National Mississippi River Museum & Aquarium in Dubuque, Iowa, opened devoid of life. The tanks were purposefully left empty, rather than showing the vibrant aquatic life of the Gulf, to highlight the oil spill associated with BP's Deepwater Horizon offshore drilling incident earlier in 2010. According to the museum's press release, the museum wanted "to open a Gulf exhibit recognizing the crisis that is happening on the Gulf Coast. . . . The exhibit, without fish, now has the opportunity to make a bold statement related to the oil spill in the Gulf Coast by asking Museum & Aquarium visitors to imagine a lifeless Gulf."[1]

While the Deepwater Horizon incident raised the American public's awareness of the environmental risks of offshore oil, environmental damage has by no means been the primary discourse about U.S. offshore oil over the last twenty-five years. Oil has been seen as an intruder in American oceans when the nation has confronted a large spill, most prominently the Exxon Valdez in 1989, but at other times the resource goes relatively unnoticed. Although Californians have a history of fighting against offshore oil drilling, in the Gulf of Mexico region offshore oil is a way of life.[2]

The National Mississippi River Museum & Aquarium's dramatic representation of an empty Gulf was a radical departure from the typical modern aquarium repre-sentation of the Gulf of Mexico. Rather than place the oil industry in opposition to the Gulf's ecosystem, aquarium displays have integrated the oil industry into it. Offshore oil structures are presented to the American public as an integral—and even necessary—part of nature in aquarium displays. Just as Stephanie LeMenager has argued that the La Brea Tar Pits in Los Angeles give visitors "the opportunity to think of oil as nature and/or art," these displays make oil technology natural.[3]

To examine the integration of oil and ocean, this essay analyzes depictions of offshore ecosystems in American aquarium displays owned by a variety of organizations, ranging from commercial restaurant chains to nonprofit environmentally focused organizations. The essay will argue that much of the public presentation of the offshore ecosystem is tied up with the development of Rigs-to-Reefs programs, which allow the conversion of offshore oil structures into artificial reefs after they are decommissioned from active use.[4] The Rigs-to-Reefs concept developed as part of broad strategies to improve U.S. fishing grounds and to make offshore oil production environmentally friendly, and Louisiana began the first state Rigs-to-Reefs program in 1986. In the wake of the Rigs-to-Reefs developments, aquariums located in the Gulf Coast chose to display Gulf of Mexico tanks highlighting the contribution of the oil industry to the Gulf's ecosystem. After discussing these Gulf exhibits and their context, the essay turns to an exhibit in California, which was introduced during a protracted legislative conflict about permitting Rigs-to-Reefs in that state. By placing these displays into the specific social and political context of the Rigs-to-Reefs program, we can see why aquariums developed hybrid schemes for the representation of marine life in the region that mixed oil and water.

Aquariums as Places of Learning

Before delving into the waters at aquariums, we need to understand a little about their function as places of learning. Only an extremely small portion of the U.S. population has ever seen an offshore oil platform, much less actually been to one. In addition, the vast majority of people have never been diving in the ocean to see the underwater life there. Aquariums, museums, and educational material play a key role in shaping how offshore structures and environments are perceived culturally.

Aquariums serve as a place of conservation learning. A significant three-year study, which surveyed more than 5,500 visitors of accredited institutions of the Association of Zoos and Aquariums, revealed that nearly half of the individuals surveyed believed their visit elevated their awareness of conservation and 42 percent of the individuals commented on the role of zoos and aquariums in education. Follow-up interviews with participants seven to eleven months after their visit showed that 61 percent of them were still able to talk about what they had learned from their previous visit.[5] A study of visitors to the National Aquarium in Baltimore has shown that visitors retained the conservation message of the aquarium in follow-up interviews six to eight weeks after their visit.[6]

Visiting aquariums is part of "free-choice science learning," recently defined as "the learning that individuals engage in throughout their lives when they have the opportunity to choose what, where, when and with whom, to learn."[7] As such, the displays contribute directly to the public understanding of science. Of course, not all visitors get the same things out of their visits to scientific institutions. A study of aquarium visitors found that individuals with limited knowledge but moderate to high interest gained the most conservation knowledge from their visit.[8] We can conclude, then, that the messages stressed in exhibits can make a lasting impression on the average visitor.

Aquarium exhibits, and other environmental educational material for that matter, are rarely neutral. They have a defined message—particularly a conservation message—that they want to impart. Contents may be influenced by controversies or recent events, such as the Gulf of Mexico exhibit in Dubuque.[9] In addition, corporate sponsorships of exhibits, which are a necessity for many nonprofit institutions to exist, can affect the contents of displays in various ways. Exhibit curators may prioritize certain elements within a display and de-emphasize others, or even redefine their own goals in order to appeal to their sponsors.[10] Exhibits are designed within a cultural and historical context that affects the information imparted to visitors.

Hybrid Exhibits

In the largest Gulf of Mexico tank in the world (Figure 14.1), which is located at Aquarium of the Americas in New Orleans, grouper, amberjack, and tarpon swim around and through a downscaled version of a steel offshore oil jacket (the understructure of an oil production platform). Stringrays and sand tiger sharks circle the four-hundred-thousand-gallon tank, while nurse sharks sit on the bottom among oilfield spare parts and piping. The legs of the replica platform are integrated seamlessly into the tank, and a replica platform itself sits above the water line. The exhibit highlights "Louisiana's Offshore Riches." A model offshore oil platform and text about the oil and gas industry is fitted alongside text and photographs about artificial reefs (Figure 14.2). The text on a board titled "Fishes and Rigs" lets the visitor know why the legs of an oil platform are appropriate in the exhibit: "You're looking at a scaled down model of an offshore production platform. Structures like this support marine ecosystems in the Gulf of Mexico and enhance populations of fish, shrimp, birds and other wildlife that feed on aquatic life." The message to the visitor is that the structure not only serves as marine habitat, but that it actually enhances life in the Gulf waters.

FIGURE 14.1. The largest Gulf of Mexico tank in the United States, with a prominent scaled-down version of an offshore oil jacket and platform. Aquarium of the Americas, New Orleans. 2009. Photograph by the author.

FIGURE 14.2. Louisiana's Offshore Riches display adjacent to the Gulf of Mexico tank. Aquarium of the Americas, New Orleans. 2009. Photograph by the author.

The giant steel structures provide the kind of habitat that the barren mud-sand bottom of much of the Gulf of Mexico within U.S. jurisdiction cannot.

The 125,000-gallon Islands of Steel exhibit at the nonprofit Texas State Aquarium in Corpus Christi shows visitors a similar view of Gulf aquatic life (Figure 14.3). Opened in 1990, the same year as its New Orleans counterpart, the exhibit features the steel legs of a jacket extending through the vertical space of the tank and numerous offshore species, including nurse sharks, turtles, amberjack, tarpon, and grouper, while the visitor is engulfed in barnacle-encrusted platform leg replicas.

One text on artificial reefs explains, "Production platforms, like natural reefs, help to increase the numbers and distribution of marine life in the Gulf. A typical platform provides about 4,000 square meters of surface area on which plants and animals can grow." All text is provided in both English and Spanish to facilitate understanding by more visitors. Accompanying photographs show close-ups of the small flora and fauna that grow on the steel legs—barnacles, oysters, hydroids, algae, and sponges all make an appearance in full color. The sea-life photos appear side by side with images of production platforms above water and dramatic views of the legs from below, both of which serve to connect the oil industry with bountiful nature.

Interpretive texts on the wall opposite the tank insist on this point, framing the "islands of steel" as "Aquariums without walls" that are vital to the Gulf of Mexico ecosystem: "Naturally occurring reefs and rocky areas are rare in the Gulf of Mexico. Instead, sea life dependent upon hard surfaces settle on oil and gas platforms. Small fishes, attracted by food, find shelter amid the 'legs.' Smaller fish attract larger predators, such as migratory fishes, which use the structures as feeding stations." The Gulf sea life is described here as dependent on these structures; these man-made steel homes make up for the lack of natural habitats. To further underscore this message, the Texas State Aquarium featured an exhibit hall that screened the film *Aquarium Without Walls* in the 1990s.[11] Made in 1989 and sponsored by Exxon Corporation, the film highlighted the life around the standing platforms, where "Mother Nature and man's technology meet, and a miracle of life is acted out." To illustrate the argument that "oil rigs have created homes for sea life unknown off Texas and Louisiana just a few decades ago," the video showed images of a dazzling array of life around the platforms: sponges, sea squirts, corals, blennies, bristle worms, angelfish, butterfly fish, sheepshead, red snappers, groupers, eels, sculpins, and many more make an appearance, affirming the argument made by the nearby "Islands of Steel" exhibit.[12]

While highlighting the benefit of the oil structures, oil is not left uncritiqued at the Texas State Aquarium. Before 2010, one small panel mentioned the 1979 Ixtoc I disaster, which spilled 134 million gallons of oil in the Gulf, reminding the visitor that "we must be cautious in our rush to harvest the ocean's riches," but the overall exhibit

FIGURE 14.3. Visitors are engulfed within an offshore structure that extends both overhead and in the tank where large fish, sharks, and turtles swim. Texas State Aquarium, Corpus Christi. 2012. Photograph by the author.

stresses the environmental benefits of oil platforms. Visitors picked up on this—a feature review of the newly opened facility in 1990 noted that while the one placard mentioned the Ixtoc I spill, the point of the Islands of Steel tank is that "the 3,500 oil and gas rigs in the Gulf do add desirable habitat for sea life."[13] The panel with the Ixtoc I spill has recently been replaced by one referring to the BP Deepwater Horizon accident in 2010 and an aerial photograph of the spill. In this new panel's text, the oil structures "add habitat," but "safety and response technology need to keep pace" with offshore exploration to avoid disastrous spills.

The sheer size of the tanks at the Aquarium of the Americas and the Texas State Aquarium allows the inclusion of relatively large replica oil structures, but even smaller aquariums built in the region after these two have chosen to show petroleum infrastructure as part of the habitat. Sea Center Texas, which is owned by the Texas Parks & Wildlife Department, has a five-thousand-gallon tank dedicated to petroleum structures as artificial reefs. The oil platform clearly has a place in the ecosystem according to the graphic displaying the habitat zones that appears at the bottom of many of the displays (Figure 14.4). The Downtown Aquarium in Houston, owned by Landry's Restaurants, likewise features the bottom of an oil structure in the Gulf of Mexico tank (Figure 14.5). At the ABQ Biopark Aquarium owned by the City of Albuquerque, New Mexico, instead of incorporating the structure into the standard large fish and shark tank, the designers chose to make a separate display using a replica jacket section.[14] The visitor can walk through the legs, and one tank of tropical fish is attached to a leg at eye level to show the colorful life around the platform legs (Figure 14.6).

All of these displays integrate the steel oil-jacket structure into the Gulf of Mexico ecosystem and highlight the ecosystem benefits brought by the structures' function as artificial reefs.[15] The visitor to these aquariums then is confronted with a hybrid

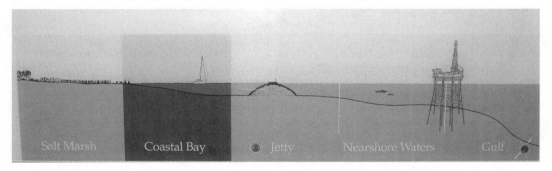

FIGURE 14.4. Bottom section of the interpretative display showing the various habitat zones off the coast of Texas. An offshore oil facility is prominently placed at the line between the Nearshore Waters and Gulf zones. Sea Center Texas, Lake Jackson. 2010. Photograph by the author.

FIGURE 14.5. Tank with replica of steel jacket, Downtown Aquarium, Houston. 2010.
Photograph by the author.

space—one that displays the built environment and technological artifacts alongside
of and integrated with nature. The Gulf of Mexico is a vast ecosystem, much of it open
water rather than dominated by such structures, yet these aquariums have chosen to
focus on this particular hybrid part of the ecosystem. Such a choice was not inevitable: a few aquariums outside the Gulf Coast region also have Gulf Coast tanks, for
example, the Tennessee Aquarium in Chattanooga and the National Mississippi River
Museum in Iowa, and these do not use oil structures as part of the habitat. They show
instead fish swimming around in tanks with sandy bottoms and a few rocky outcrops,
which is the dominant Gulf habitat. So why was a hybrid technological-environmental
space chosen as the ecosystem to display in the Gulf Coast?

FIGURE 14.6. Rigs-to-Reefs display with large structural components and a small fish tank. ABQ BioPark Aquarium, Albuquerque. 2011. Photograph by the author.

The Gulf of Mexico as a Man-Made Ecosystem

Specific historical circumstances affected the presentation of Gulf waters as an oil industry space. The hybrid spaces depicted in the aquarium exhibits have their genesis in understandings of the ecosystem of the Gulf of Mexico that came to the fore in the 1970s and 1980s.

By the late 1970s, thousands of offshore oil and gas structures dotted the map of the western Gulf of Mexico. The offshore oil business in the Gulf had started humbly in 1946 with Magnolia Petroleum Company's first operating platform five miles offshore of Morgan, Louisiana; by 1980, there were approximately four thousand production-related structures up to 1,000 feet deep and 130 miles offshore.[16] The bottom of the Gulf of Mexico where these oil wells stand is characterized as a gently sloping sand-mud flat, although some mud lumps and salt domes break up the smooth service. Most of this landscape is sparsely populated, with only a few areas with larger concentrations of organisms. The only natural coral reef in the western Gulf is Flower Garden Banks off the coast of Texas.

The understructure of production platforms rises from the seafloor to the surface in a crisscross pattern. These steel jackets support the weight of the above-water tanks, equipment, and platform housing; but under the water, they came to support sea life as well. Barnacles, spiny oysters, corals, and other colonizers attach themselves to the steel. Because the jackets extend vertically hundreds of feet through the water, various habitats are created with differing light and temperature conditions. The colonizers attract small fish that find food and hiding places in the new ecosystem, as well as larger fish that come for the prey and because the platforms are large visual markers.

By the 1970s, recreational fishermen had become regular visitors to offshore structures because of the larger sport fish that could be caught there. Photographs from the Louisiana State Archives from the late 1970s show Louisiana sportsmen engaged in the increasingly common practice of tying up boats to operating oil platforms in order to fish. A study titled *Recreational Fishing Use of Artificial Reefs on the Texas Coast* prepared for the Texas Coastal and Marine Council in 1978 showed that almost all offshore fishing-boat captains had previously fished off oil and gas structures in Texas and that half of all fishing trips visited them.[17] Thus, by the late 1970s, local knowledge defined the standing oil platforms as functional artificial reef habitats with significant fish populations.

In 1978, the Bureau of Land Management (BLM), which at the time was the major agency responsible for offshore oil leases, was required to prepare environmental impact statements for oil and gas leases in the wake of the 1976 Federal Land Policy and Management Act (FLPMA; 43 USC 1701–82), which made imperative an increased understanding of the effect of oil and gas operations on the Gulf ecosystem. The BLM contracted the Southwest Research Institute to prepare a large study of the off-shore ecology of the Louisiana Outer Continental Shelf. The study had two areas of inquiry: pollutant fate and effects, and the artificial reef characteristics of standing platforms.[18] When the study was finally issued in 1981, it described Louisiana's off-shore platforms as "artificial reefs which have apparently expanded the available habi-tat for numerous fish and invertebrate species that are dependent on hard banks as habitat."[19] The study identified several species that had not been previously docu-mented in the central Gulf waters or were thought to occur only on natural coral reefs, leading to the conclusion that "the platforms are contributing to the overall diversity of the OCS."[20] In this study and others issued in the early 1980s, standing jackets are described as beneficial reef habitat.[21]

Popular scientific and conservation magazines based in the Gulf carried stories highlighting the biological diversity of standing platforms. The earliest, an article titled "From Rigs to Reefs," published in 1975 in *Louisiana Conservationist,* touted the transformation of steel legs to artificial reefs supporting game and food fish. The arti-cle is richly illustrated with underwater photographs featuring underwater biologists inspecting algae growing on the structure, schools of spadefish, barracuda lurking among the dark crossbeams, and impressively sized catches made with rod-and-reel and underwater spears.[22] *Texas Parks & Wildlife* published a similar article in 1982 about the fishing and diving possibilities around offshore production platforms. This article likewise has full-color close-up underwater photos of the fish and coral resi-dents of platform jackets and a schematic showing the various habitat zones of the steel reefs. The article contrasts the vast "sediment-covered plain" of the Gulf of Mexico, which has "a lack of habitat diversity," with the abundant life around the platform.[23]

Within this context of portrayals of the Gulf ecosystem around standing platforms, it is no wonder that the exhibits that opened in 1990 at the Aquarium of the Americas and the Texas State Aquarium include scaled-down versions of offshore jackets as a major component of the tank habitat, with explanatory text reinforcing the integral nature of the steel structures. The displays contrast the natural mud bottom of the Gulf with the vibrant, colorful life on and around the structures just as the scientific studies and popular articles had done.

RIGS-TO-REEFS ON DISPLAY IN THE GULF

The context of offshore oil production in these aquariums goes beyond the ad hoc function of standing platforms as artificial reefs into the deliberate reuse of structures as reefs. The exhibits not only show the oil structures as part of the habitat—they also have accompanying text highlighting legislation that allows platforms to be preserved as reefs after being decommissioned under various programs collectively known as Rigs-to-Reefs. For example, a display titled "Artificial Reefs" on one side of the large Gulf of Mexico tank at the Aquarium of the Americas explains to visitors the benefits of the Rigs-to-Reefs program: "Production platforms were never meant to be permanent structures. Laws require they be taken away when their wells run dry. But platforms create underwater habitats for marine life—and good fishing—that are lost when platforms are removed. In 1987 Louisiana started the Artificial Reef Program, which allows some platforms to be sunk in selected spots where they continue to support marine life communities." The legislative development referred to in this exhibit grew out of the recognition in the 1980s that standing platforms added habitat and, thus, if those platforms were removed, the new habitat would be lost.

Artificial reefs created to increase recreational fish catches had become more and more popular in the United States after World War II. In 1961, the Sports Fishing Institute estimated that the number of anglers was growing at a rate of 270,000 persons per year, and the annual catch was around 300 million edible fish. The increasing numbers of anglers led sportfishermen and governmental agencies to pursue artificial offshore reef development in the late 1950s and 1960s in earnest. A vast array of materials— old car bodies and trolley cars, decommissioned ships, concrete rubble, stone fragments, and old tires, among other things—were sunk in offshore waters to create reefs. By late 1971, approximately 150 artificial reefs and reef complexes had been created off U.S. coastlines, most targeting commercial and recreational fish.[24] Texas and Louisiana had not been particularly active in the artificial reef boom, mainly because they already had lots of artificial reefs—their standing oil platforms. When proposals for a national artificial reef program came before the U.S. Congress beginning in 1978, the function of the platforms as reefs took a central position in the debates.

Two bills aimed at marine artificial reef development came before the House Committee on Merchant Marine and Fisheries in 1981.[25] The Subcommittee on Fisheries and Wildlife Conservation and the Environment held a one-day hearing to consider the question of a national program for artificial reef development in response to these bills. The bills themselves made no mention of offshore oil structures, but the chairman of the hearing, John B. Breaux of Louisiana, had personal experience with fishing at oil structures: "Having participated in many enjoyable fishing trips offshore Louisiana, within sight of many such structures, I can attest to the ability of these giant reefs to provide orientation, shelter, and food to fish throughout the water column."[26] Breaux's personal fishing experiences led him to automatically associate a coordinated reef program with offshore structures. Much of the testimony, both written and oral, also focused on offshore oil and gas structures, even though they were not overtly included in the bills.

Although nothing came of the 1981 bills, a later attempt would come to fruition as the National Fishing Enhancement Act of 1984, which mandated that the Department of Commerce write a plan for artificial reef development.[27] The resultant National Artificial Reef Plan of 1985, a general thirty-nine-page guidance document, focuses on improving recreational fishing. The section on potential materials of opportunity listed ships, concrete, tires, and oil and gas structures.[28]

In the wake of the National Artificial Reef Plan, both Louisiana (in 1986) and Texas (in 1989) established state-level artificial reef plans focused on Rigs-to-Reefs to allow the conversion of obsolete offshore oil structures into artificial reefs. In both state plans, oil structures were touted as vital fish habitat.[29] The Louisiana Artificial Reef Plan highlighted the unintentional yet critical nature of habitats on offshore structures:

> The development of the oil and gas industry in the Gulf of Mexico resulted in the creation of this country's most extensive artificial reef system. . . . For over 40 years, Louisiana fishermen have benefited from the increased biological activity associated with this unintentional artificial reef habitat. Since these platforms are so commonplace off the Louisiana coast, many citizens and management groups believe that they are permanent and will always be available for fishing. This is, however, not the case. . . . It was, therefore, imperative that Louisiana recognize this potential loss of habitat and plan to offset it by either creating new artificial reefs or preserving existing structures.[30]

The Texas plan also stressed the ecological enhancement effect of offshore structures, going so far as to recommend that "the [Texas Parks and Wildlife] Department should actively pursue acquiring offshore platforms for use as artificial reefs in the Gulf of Mexico, in deference to other structures."[31] This language shows that by 1990, the link between offshore oil and gas structures and improved fish habitat was clear. I should also note that these plans were written by leading fishery scientists, referencing up-to-date scientific publications that basically came to the same conclusion that petroleum structures converted into artificial reefs make good fish habitat.

As this legislation was under way, scientists continued to study the relationship between the Gulf environment and offshore oil structures through the 1980s, focusing on reusing obsolete structures deliberately as artificial reefs. The Gulf of Mexico Information Transfer meetings became a regular venue for both scientists and industry representatives to advocate making artificial reefs out of structures. In the papers given at these meetings, the Gulf ecosystem was clearly one in which steel structures and fish coexist; the only question was how to best implement artificial reef programs to take advantage of the standing de facto reefs when the time came to remove the structures.[32] Jackets turned into artificial reefs were the natural choice.

These developments factored into the way that the Gulf of Mexico aquarium tanks are interpreted. Just as the Aquarium of the Americas interpretative text mentions the Louisiana Rigs-to-Reefs program, a similar educational display at the Texas State Aquarium discusses Rigs-to-Reefs (without naming the program) as a way to create long-lasting habitats: "In the past, rigs and platforms were dismantled and removed at the end of their production lives causing the destruction of established reef communities. Today, the petroleum industry offers the structures to coastal states that place them in select locations in the Gulf of Mexico. The habitats may last as long as one hundred years before they are destroyed by corrosion." The Downtown Aquarium in Houston likewise has a placard next to its Gulf tank that explains, "Over the past twenty years, more than 200 platforms no longer in use have given rise to rich communities that have boosted the sport fishing industry in the Gulf as well as provided areas for continued research in the field of aquaculture." These texts place the exhibit design within the context of a specific regulatory framework that was developed for the Gulf of Mexico—one that emphasizes the beneficial effect of offshore structures both during and after their life span as production facilities. The benefit to sportfishing is specifically included, which makes sense considering that sportfishermen were the first and loudest advocates of Rigs-to-Reefs.

The artificial reef tank at Sea Center Texas, which opened in 1996, is the most explicit display about the regulatory framework of the exhibit. In this case, the interpretive text puts the underwater jacket into the context of other artificial reef-creation projects in which "water and the flat seafloor get a little help," noting that "since the 1940s, the Texas Parks and Wildlife Department has been placing artificial reefs in nearshore waters to attract plants and animals that normally can't live here. These reefs are built of materials like stone rubble, trees, concrete, old ships and oil rigs." A bronze commemorative plaque is placed prominently on the wall next to the tank in honor of the Texas Artificial Reef Program and its original sponsor, Senator J. E. "Buster" Brown of Lake Jackson (Figure 14.7). The plaque has a particular context: the facility is run by Texas Wildlife and Fisheries, which is responsible for the Texas Artificial Reef Program and is located in Lake Jackson, the home of the author of the bill that created the program. Given the interconnections joining aquarium, environmental agency, and legislation, it seems entirely unsurprising that the institution decided to highlight the Artificial Reef Program next to a tank with steel oil-jacket legs in the habitat.

The bronze plaque, however, hints at another important issue—sponsorship of exhibits. In this case, the Texas Artificial Reef Program itself had partially funded the aquarium. Aquariums rely to a great extent on corporate sponsorships, and not unexpectedly, oil companies are primary sponsors of Gulf of Mexico tanks. In 1988, Tenneco Inc. and two of its operating divisions announced a $250,000 contribution

FIGURE 14.7. A bronze plaque next to the Rigs-to-Reefs tank honors the enabling state
legislation and its author, Senator J. E. "Buster" Brown. Sea Center Texas, Lake Jackson.
2010. Photograph by the author.

to the Aquarium of the Americas to sponsor the Gulf of Mexico exhibit. Bob Taylor,
vice present and general manager for Tenneco Oil Exploration and Production explained the sponsorship rationale: "The Petroleum Wing of this facility will help
demonstrate the positive impact of oil and gas platforms on marine life, thus providing further opportunity to explain the relationship between the petroleum industry
and the abundant aquatic life of the Gulf of Mexico."[33] In this statement, Taylor even
refers to the section of the aquarium as "The Petroleum Wing," which, although not
the official name (it is named "Gulf of Mexico" on all facility maps), is indicative of
how visible the oil industry is in the exhibit. The corporate sponsors are honored
with their company logos on one of the posts separating the tank glass sections; in
2009, the sponsors included a who's who list of major oil and gas companies operating in the area—BP, Shell, ExxonMobil, Tenneco, and Chevron. As noted above,
Exxon created the film that was shown at the Texas State Aquarium; the company is

also still listed as the sponsor of the panel "Aquarium without walls." These sponsorships do not mean that the oil companies had a direct hand in the exhibit design, but it does show that the exhibits reflect an understanding of the ecology of the Gulf that oil companies would agree with—that offshore structures provide key habitat in the Gulf of Mexico, making them environmentally beneficial.[34]

CONTROVERSY IN CALIFORNIA

While the Gulf of Mexico Rigs-to-Reefs programs moved forward in the 1980s with little resistance, California entered a protracted battle over Rigs-to-Reefs legislation in the late 1990s.[35] Many Californians have vehemently opposed offshore oil developments, particularly after the Santa Barbara oil spill of 1969, which many people consider one of the significant catalysts of the American environmental movement.[36] In spite of heavy debate and extensive objections from environmentalist groups and trawlers, a bill (SB1 of 2000–2001) that would have permitted conversion of obsolete structures into artificial reefs eventually passed the California legislature. The governor, however, refused to sign the bill and there were not enough votes to override the veto. After the bill's defeat, the proponents of Rigs-to-Reefs continued to advocate the program. The sportfisherman group United Anglers of Southern California (UASC) and California Artificial Reef Enhancement (CARE), a nonprofit organization supported by Chevron, held media events, including an underwater platform tour in 2003 and a Rigs-to-Reefs conference in March 2007. The continued political pressure eventually paid off with the passage of Assembly Bill 2503 (2009–10), which was signed into law in June 2010 and permits reef conversions.

The debate was most heavily centered in Santa Barbara—home of the spill, the Environmental Defense Center (EDC), which most vocally opposed the legislation, and the University of California, Santa Barbara, where Dr. Milton Love, a marine biologist with extensive research showing the positive effects of California's offshore platforms as fish habitat, worked. Santa Barbara's waters became the focal point of a contention over what was natural and what was not.

Within this context, in 2008 the Ty Warner Sea Center, an extension of the Santa Barbara Museum of Natural History located on a wharf in town, offered visitors several encounters with the life around standing offshore structures through images. A documentary film titled "Life Under the Platforms" produced by the CARE organization in 2007, was shown in two places in the center. First, in the section highlighting research at the University of California, Davis, several video monitors continuously play some of the film's underwater recordings from around the Santa Barbara platforms (Figure 14.8). Second, the film in its entirety was screened in the center's small theater room every half hour.

FIGURE 14.8. One of two screens showing the film "Life Under the Platforms" as part of the Santa Barbara Channel ecosystem exhibits. Ty Warner Sea Center, Santa Barbara. 2008. Photograph by the author.

The film offers a view of platforms strikingly similar to the Gulf of Mexico exhibits. The disembodied narrator begins with the contrast of seen and unseen nature: "Man-made structures would seem to contribute little to life in the ocean. But beneath the twenty-seven oil platforms dotting the California coastline, it's a different story." The film focuses in on Dr. Love's work, which included extensive surveys of several structures from 1995 to 2001. Videos of the colorful underwater biodiversity are interspersed with interview clips with Love, who stresses the multitude of organisms living on and around the structures. The structures are naturalized in the film, just as in the Gulf aquarium exhibits, becoming "homes for starfish, mussels, barnacles, anemones, and scallops."

The center's showing of "Life Under the Platforms" in the midst of political controversy over a potential Rigs-to-Reefs program in California highlights the choices that are made in aquarium exhibits. If the Rigs-to-Reefs opponent discourse had been adopted, the structures would have been portrayed as polluters and only the near-shore natural reef systems of the Santa Barbara Channel would be visible to visitors.[37] Instead, the proponent discourse was adopted, perhaps because of the availability of a beautifully produced film with colorful and fascinating underwater footage. While the proponent position has scientific merit, the center's staff still made a choice to integrate the film as a significant part of its presentation of the Santa Barbara Channel ecosystem. As Dr. Love says in one interview segment of the film, "Whatever you think about platforms and the oil industry, the platforms themselves are really lovely underwater."

SHOW AND TELL

These American aquarium exhibits show visitors a particular part of the ocean's eco-system and tell the story of a beneficial relationship between offshore oil and ocean life. The inclusion of offshore structures in these exhibits comes from a particular scientific, social, and political context. Gulf fishermen's experiences with standing offshore platforms encouraged many scientific studies of the structures as habitat in the Gulf of Mexico. Anecdotal experience from fishing at offshore rigs combined with scientific findings bolstered arguments for creating Rigs-to-Reefs programs that would turn obsolete jackets into permanent artificial reefs. The major Gulf of Mexico aquarium exhibits at the Aquarium of the Americas and Texas State Aquarium both opened in 1990 on the heels of Rigs-to-Reefs programs being established in Louisiana and Texas. The later Sea Center Texas and Houston's Downtown Aquarium opened after the Rigs-to-Reefs programs were well established. The choice then of incorpo-rating offshore structures in aquatic displays of the Gulf should come as no surprise. The thinking is clear: fish swim around structures in the Gulf because they provide habitat, therefore the structures should be integrated in the tanks as habitat. As the visitor experiences the Gulf of Mexico display, amberjack, grouper, and sharks swim rhythmically through the water, darting around and through the scaled-down version of an oil jacket. The steel structure becomes a place teeming with life.

There is no arguing that vibrant communities of sea life do indeed crawl on and swim through offshore oil structures and artificial reefs constructed from old jackets, but the interpretation of these ecosystems and the choice to use oil structures in the aquarium displays is not a given; rather, it was a product of context. The showing of the film "Life Under the Platforms" at the Ty Warner Sea Center could not have happened at another point in time. The film's creation was a direct product of the political context of the fight over Rigs-to-Reefs in California. By integrating the film into the center's exhibits, the center becomes aligned with the proponents of the contested legislation. These exhibits are an accurate depiction of one subsection of the Gulf and Santa Barbara Channel ecosystems, but they are just that—only one specific sub-ecosystem. The choice by the aquariums to show this particular view of ocean life reveals the social, political, and scientific context in which the exhibits were developed—that American oceans are a harmonious meeting place of oil and water.

NOTES

1. National Mississippi River Museum & Aquarium, "Imagine a Lifeless Gulf," Press release, June 18, 2010.

2. See Robert Gramling and William Freudenburg, "Attitudes Toward Offshore Oil Development: A Summary of Current Evidence," *Ocean & Coastal Management* 49 (2006): 442–61; and William Freudenburg and Robert Gramling, *Oil in Troubled Waters: Perceptions, Politics, and the Battle over Offshore Drilling* (Albany: SUNY Press, 1994).

3. See Stephanie LeMenager, "Fossil, Fuel: Manifesto for the Post-Oil Museum," chapter 16 in this volume.

4. Although offshore platforms are technically not rigs (the former are used in the production phase of oil and gas extraction, while the latter is used during the drilling phase), the phrase Rigs-to-Reefs refers to the repurposing of any offshore oil and gas structure as an artificial reef. Locals in the Gulf Coast often use "rig" as a synonym for a standing offshore oil and gas platform when they refer to fishing around it.

5. John H. Falk, Eric M. Reinhard, Cynthia L. Vernon, Kerry Bronnenkant, Nora L. Deans, and Joe E. Heimlich, *Why Zoos & Aquariums Matter: Assessing the Impact of a Visit* (Silver Spring, Md.: Association of Zoos & Aquariums, 2007).

6. Leslie Adelman, John H. Falk, and Sylvia James, "Impact of National Aquarium in Baltimore on Visitors' Conservation Attitudes, Behavior, and Knowledge," *Curator* 43 (2000): 33–66.

7. John H. Falk, Martin Storksdieck, and Lynn D. Dierking, "Investigating Public Science Interest and Understanding: Evidence for the Importance of Free-Choice Learning," *Public Understanding of Science* 16 (2007): 455–69.

8. John H. Falk and Leslie M. Adelman, "Investigating the Impact of Prior Knowledge and Interest on Aquarium Visitor Learning," *Journal of Research in Science Teaching* 40 (2003): 163–76. As a corollary, the authors note that "most experts do not find museum-like settings ideal for dramatically furthering their knowledge" (172).

9. For a discussion of the challenges of updating an exhibit to address a new public controversy, see Susan Macdonald and Roger Silverstone, "Science on Display: The Representation of Scientific Controversy in Museum Exhibitions," *Public Understanding of Science* 1 (1992): 69–87.

10. Garry C. Gray and Victoria Bishop Zendzia, "Organizational Self-Censorship: Corporate Sponsorship, Nonprofit Funding, and the Educational Experience," *Canadian Sociological Association / La Société canadienne de sociologie* 46 (2009): 161–77.

11. Sally Hoke, Texas State Aquarium, personal communication. The author viewed the film at the Louisiana State University library, which holds a copy in its Education Resource Center.

12. *Aquarium Without Walls*, written, produced, and directed by Paul K. Driessen, funding provided by Exxon Corporation, 1989.

13. Joe Nick Patoski, "Go Fish," *Texas Monthly*, October 1990, 180.

14. ABQ Biopark Aquarium follows the ecosystems of the Rio Grande from its headwaters along the Texas–Mexico border to the Gulf of Mexico, so even though New Mexico is not considered in the Gulf region, the aquarium design follows the patterns of the Gulf region.

15. Aquarium design from the mid-1980s onward has focused on habitat simulation, stressing the animal in its ecological context, rather than the older method of highlighting the animal as artistic object on display. See Dennid Doordan, "Simulated Seas: Exhibition Design in Contemporary Aquariums," *Design Issues* 11, no. 2 (1995): 3–10, for an overview of large-scale habitat simulation as an aquarium exhibition design strategy.

16. Committee on Disposition of Offshore Platforms, *Disposal of Offshore Platforms*, prepared for the National Research Council (Washington, D.C.: National Academy Press, 1985), 9.

17. Robert B. Ditton and Alan R. Graefe, *Recreational Fishing Use of Artificial Reefs on the Texas Coast*, contract report (Austin: Texas Coastal and Marine Council, 1978).

18. C. A. Bedinger, ed., *Ecological Investigations of Petroleum Production Platforms in the Central Gulf of Mexico*, 3 vols., MMS 1981-16 (San Antonio: Southwest Research Institute for the Bureau of Land Management, 1981).

19. Ibid., 2:89.

20. Ibid., 3:16.

21. A study commissioned by the Fish and Wildlife Service stated that platforms have a "reef" effect because of the biofouling organisms that can attach to the structures and create food and shelter for fish: Benny J. Gallaway, *An Ecosystem Analysis of Oil and Gas Development on the Texas-Louisiana Continental Shelf* , FWS/OBS-81/27 (Washington, D.C.: U.S. Fish and Wildlife Service, Office of Biological Services, 1981). Another study, issued in 1982, came to the same conclusion that petroleum structures increased the abundance and diversity of reef fish in the Gulf: Continental Shelf Associates, *Study of the Effect of Oil and Gas Activities on Reef Fish Populations in the Gulf of Mexico OCS Area*, MMS 1982-40 (Washington, D.C.: Bureau of Land Management, 1982).

22. McFadden Duffy, "From Rigs to Reefs," *Louisiana Conservationist* 27 (1975): 18–21. The magazine is the official publication of the Louisiana Department of Wildlife and Fisheries.

23. Maury Osborn Ferguson, "Underwater Communities," *Texas Parks & Wildlife* 40 (1982): 2–7. The magazine is the official publication of the Texas Parks and Wildlife Department.

24. Ann Weeks, "Fish Cities: A New School of Design," *NOAA* 2 (April 1972): 2–7.

25. H.R. 1041 and H.R. 1897. Earlier attempts at legislation had also been made (S. 3094 of 1978 and H.R. 4413, H.R. 4714, and S. 325 of 1979), but no hearings were held.

26. Hearings before the Subcommittee on Fisheries and Wildlife Conservation and the Environment of the Committee on Merchant Marine and Fisheries, House of Representatives, Ninety-Seventh Congress on Establishment of a National Artificial Reef Policy–H.R. 1041, H.R. 1897, September 11, 1981, Serial No. 97-35 (Washington, D.C.: U.S. Government Printing Office, 1982), 1–2.

27. There were several failed legislative attempts to get a national artificial reef plan, but I will not discuss them here.

28. Richard B. Stone, *National Artificial Reef Plan*, NOAA Technical Memorandum NMFS OF-6 (Washington, D.C.: U.S. Department of Commerce, 1985).

29. See Dolly Jørgensen, "An Oasis in a Watery Desert? Discourses on an Industrial Eco-system in the Gulf of Mexico Rigs-to-Reefs Program," *History and Technology* 25 (2009): 343–64, for a full discussion of the discourses at work in the Gulf of Mexico Rigs-to-Reefs concept.

30. Charles A. Wilson, Virginia R. Van Sickle, and David L. Pope, *Louisiana Artificial Reef Plan*, Technical Bulletin No. 41 (Baton Rouge: Louisiana Department of Wildlife and Fisheries, 1987), vii.

31. C. Dianne Stephan et al., *Texas Artificial Reef Fishery Management Plan*, Fishery Management Plan Series No. 3 (Austin: Texas Parks and Wildlife Department, 1990), 1.

32. This is obvious in the contents of the papers in the sessions on Rigs-to-Reefs in *Proceedings: Fourth Annual Gulf of Mexico Information Transfer Meeting*, November 15–17, 1983, New Orleans, OCS Study MMS 84-0026 (New Orleans: Minerals Management Service, 1984); and *Proceedings: Fifth Annual Gulf of Mexico Information Transfer Meeting*, November 27–29, 1984, New Orleans, OCS Study MMS 85-0008 (New Orleans: Minerals Management Service, 1985). Major studies later include Villere Reggio, *Rigs-to-Reefs: The Use of Obsolete Structures as Artificial Reefs*, OCS Report MMS 87-0015 (New Orleans: Minerals Management Service, 1987); and Reggio, compiler, *Petroleum Structures as Artificial Reefs: A Compendium*, OCS Study MMS 89-0021 (New Orleans: Minerals Management Service, 1989).

33. Southwest Newswire, "Tenneco Inc. Announces $250,000 Contribution to Aquarium of the Americas in New Orleans," May 3, 1988.

34. I want to note that while oil companies generally supported the Rigs-to-Reefs concept, the programs were primarily pushed by recreational fishing and diving interests. Since the inception of programs in Louisiana and Texas, only about 10 percent of the obsolete structures have been donated by the oil companies as artificial reefs. In many cases, there is actually very little cost savings and more paperwork to donate the structure, thus the industry may be more interested in it as an environmental goodwill activity than as a money-saving activity.

35. For more details about the California debate, see Dolly Jørgensen, "Environmentalists on Both Sides: Enactments in the California Rigs-to-Reefs Debate," in *New Natures: Joining Environmental History with Science and Technology Studies*, ed. Dolly Jørgensen, Finn Arne Jørgensen, and Sara B. Pritchard (Pittsburgh: University of Pittsburgh Press, 2013); Donna Schroeder and Milton Love, "Ecological and Political Issues Surrounding Decommissioning

of Offshore Oil Facilities in the Southern California Bight," *Ocean and Coastal Management* 47, no. 1 (2004): 21–48; and Dan Rothbach, "Rigs-to-Reefs: Refocusing the Debate in California," *Duke Environmental Law and Policy Forum* 17 (2006): 283–95.

36. Teresa Sabol Spezio has analyzed the Santa Barbara oil spill's influence on American environmental policy in "Rising Tide: The Santa Barbara Oil Spill and Its Aftermath" (PhD diss., University of California, Davis, 2011).

37. See Jørgensen, "Environmentalists on Both Sides," for a discussion of this discourse.

15

Petroaesthetics and Landscape Photography

New Topographics, *Edward Burtynsky, and the Culture of Peak Oil*

CATHERINE ZUROMSKIS

> If a landscape, as we say, "draws us in" with its seductive beauty, this movement is inseparable from a retreat to a broader, safer perspective, an aestheticizing distance, a kind of resistance to whatever practical or moral claim the scene might make on us. Raymond William's remark still holds true: "A working country is hardly ever a landscape." The invitation to look at a view is thus a suggestion to look at nothing— or more precisely, to look at looking itself—to engage in a kind of conscious apperception of space as it unfolds itself in a particular place.
>
> —W. J. T. Mitchell, *Landscape and Power*

Landscape is a cultural formation. It is not simply what is out there (geographically, ecologically, spatially), but how we figure it ideologically, politically, economically, and aesthetically through representation. Coming into its own as an aesthetic subject in the sixteenth century, the landscape tradition has taken shape in a variety of media and contexts from the earthy naturalism of seventeenth-century Dutch landscape painting to the imagined Arcadian idylls of Nicholas Poussin, and from the bombast of Alfred Bierstadt's views of the Rocky Mountains to the ephemeral instantaneity of John Constable's watercolor cloud studies. Consistent throughout the landscape genre, however, is a tacit understanding that through framing and manipulation, views of the land and its features (both natural and man-made) become not only expressions of the artist's taste but also vital tools in the construction of social and national identities and, indeed, a culture's perceptions of the world. Landscape photography is no exception. Despite the medium's indexical faithfulness to the reality of what it represents, the history of landscape photography furnishes myriad examples of highly motivated, even contrived, representations that figure the land and its resources in very particular terms. The project of British colonialism in India, for example, was reinforced through the landscape views of Samuel Bourne, who struggled (and sometimes failed) to represent Indian geography as domesticated and familiar to British pastoral tastes.[1] Westward expansion in the postbellum United States was captured on film by photographers like William Henry Jackson, Timothy H. O'Sullivan, and Carleton Watkins, who married the sublimity of the "untouched" natural landscape to the technologies (the railroad and the photograph itself) that would make the land and its resources more accessible.[2] And American landscape photographer par excellence Ansel Adams photographed the natural expanses of Yosemite, Yellowstone, and the Grand Tetons in order to celebrate the spiritually transformative power of nature at its most pristine and to advocate for the preservation of wilderness through National Parks. In each case, the natural world, framed and documented by the photograph, becomes a vehicle for both a particular ideology and a set of feelings specific to that historical moment.

This essay will explore a somewhat different kind of photographic landscape, but one that is equally bound up in the ideologies and sensations of its history. In the late twentieth and early twenty-first centuries, the land, both in the United States and globally, has become increasingly marked by the industry and culture of petroleum. Not only is petroleum obtained from the land in the form of fossil fuels and central to the ubiquitous automobile culture that has dramatically altered the land and our relation to it in the modern and contemporary age, but oil has also come to be synonymous with almost every facet of modernity, from transportation, urban design, global manufacturing, and national defense to consumption in all its forms: the food we eat, the clothes we wear, the toys we give our children. Thus to understand the effect of the petroleum industry on landscape photography in late twentieth-century American culture is not just to document the proliferation of oil wells and automobiles, though that is certainly a part of it. One must also take into account the way that petroleum refigures culture in its entirety. In her essay "The Aesthetics of Petroleum, After *Oil!*," Stephanie LeMenager suggests that culture (in her study, literature and film) help us to negotiate on a sensorial level the effects of "petromodernity," which she defines as "a modern life based on the cheap energy systems long made possible by petroleum."[3] The landscape photography I will address here depicts this shift to petromodernity by increasingly turning away from the wilderness as metaphor (either for a pristine and transcendental natural world or a blank canvas for conquest and development) and toward a depiction of the land that treats man-made structures from housing developments and highways to industrial and commercial architectures as now inseparable from the geography they inhabit. Like all landscapes, these images are in part politically motivated and fuse aesthetics with the real-world contingencies of land use, population, resource allocation, and ecology. Yet they do so in a way that challenges conventional landscape aesthetics, refusing the poetics of the picturesque and sublimity of nature in favor of a more conflicted and ambiguous aestheticization of the oil industry itself and the modern convenience culture it makes possible. In so doing, they highlight both the physical and the psychological terrains of petromodernity in Western culture.[4]

Following LeMenager, this essay will examine what landscape photography does to frame both the industry of oil and the pervasive petroculture it supports on both a political and an affective level. What do the photographic landscapes of petromodernity stand to tell us about petroleum as not only the most powerful and influential industry of the twentieth century but also a critical element of modern culture and our American social imaginings? Moreover, what can we make of Raymond Williams's contention, quoted in the epigraph, that "a working country is hardly ever a landscape"?[5] While the frank appraisal of landscapes of petromodernity, glutted in some parts of the world with tract houses, office parks, freeways, gas stations, and the billions of vehicles they serve, and in others with refineries, oil fields, and new mountains and seas of industrial waste, would seem to offer a political call to arms, the overt aestheticization of petroculture in its many forms would seem to undercut those politics. As W. J. T. Mitchell suggests, "The invitation to look . . . is thus a suggestion to look at nothing—or more precisely, to look at looking itself."[6] What kinds of possibilities, then, does this introspective self-appraisal offer for, if not imagining a world less dependent on petroleum, then at least better understanding the ways that oil captivates contemporary culture? To answer these questions, I will examine two episodes from the history of North American landscape photography, each concurrent with a critical moment in the history of petromodernity. The first is the 1975 exhibition *New Topographics: Photographs of a Man-Altered Landscape*. Bringing together an initially somewhat haphazard selection of photographers, the relatively small exhibition looms large in histories of photography for the way that its participants turned their lenses away from the natural world and toward a postwar American landscape increasingly colonized by housing developments, highways, office parks, and gas stations in all of its banal familiarity. Though curator William Jenkins insisted on the directness of the work in the show, pointedly eschewing any connection to the photoconceptualism that was so central to the American art of 1960s, more recent scholarship has painted a more complex, and more contingent, view of this moment in photo history. I will argue that the photographs of *New Topographics* created a vision of a postwar America that is both formally lyrical and politically ambiguous, demonstrating an unresolved tension about the changing backdrop to American life and the cultural effects of petromodernity at precisely the moment when the crisis of peak oil first emerges in the United States. The second is Edward Burtynsky's dramatic and revealing photographic exhibition from 2009 titled *Burtynsky: Oil* (also a book of the same name). Equally a product of its era, Burtynsky's *Oil* documents the ecological nightmare of oil production and consumption in the postmillennial (and what is often erroneously perceived in the First World as postindustrial) moment. Though on its face, Burtynsky's work seems to offer an exposé of the oil industry, the aestheticization of his subjects complicates understandings of his photography as an activist or journalistic project. Pairing sublime scale and detail with a cultivated political quietude, Burtynsky's landscapes of petro-postmodernity reflect both the problematic excesses of a petroleum industry out of control and the problem of a world that cannot comprehend itself without the comforts of that same industry and the culture it creates.

NEW TOPOGRAPHICS AND THE AESTHETICS OF AMBIGUITY

The group exhibition *New Topographics: Photographs of a Man-Altered Landscape* was never intended to cause the kind of impact upon the history of photography that it did. The 1975 exhibition at the George Eastman House International Museum of Photography and Film in Rochester, New York, was the kind of exhibition museums often put together: a group show highlighting a particular thread of contemporary art practice.[7] The photographers in the show—Robert Adams, Lewis Baltz, Bernd and Hilla Becher, Joe Deal, Frank Gohlke, Nicholas Nixon, John Schott, Stephen Shore, and Henry Wessel Jr.—all shared a fascination with the cultural landscape of petromodernity, from its coursing highways dotted with gas stations and fast-food restaurants to its rapid commercial and residential expansion, particularly in the American West. Yet while there was clearly a coherence of subject matter in the works of the various photographers, curator William Jenkins identified a rather different thread of commonality connecting the *New Topographics* photographers: style. For Jenkins, what the ten photographers brought to the fore was a frankness and simplicity of presentation, one, he argued, that continued the stylistic work of Ed Ruscha in the 1960s, epitomized in his self-explanatorily titled photo books such as *Twenty Six Gasoline Stations* (1962) and *Every Building on the Sunset Strip* (1966). Unlike, Rucha, however, Jenkins did not see the work of the *New Topographics* photographers as conceptual. While Ruscha's photographs of gas stations were about something else ("aesthetic issues," as Jenkins put it), Jenkins argued that the photographers in his exhibition were, quite simply, making photographs of gas stations. And therein lay their signature style. While Jenkins is quick to acknowledge that the images in question were "far richer in meaning," it was primarily their straightforward photographic approach—making pictures "stripped of any artistic frills and reduced to an essentially topographic state"—that defined his exhibition.[8] Contrary to what the title of the exhibition might suggest, then, the "topographics" of the exhibition's title are more an aesthetic metaphor than a particular social, historical, or geographical reality. The photographers of *New Topographics* in Jenkins's description emerge almost as photographic cartographers, at a technical remove from any emotional, social, political, or ecological issues surrounding the spaces and places their photographs represent.

This visual frankness and flatness of affect is significant, both in art historically (it can still be seen in the work of myriad postmodern and contemporary photographers) and for the purposes of my argument here. It is critical to what I identify as the ambiguity of these landscapes as representations of everyday life in the United States at a point (shortly after the 1973 oil crisis) when the West's dependence on petroleum was first being called into question. Yet as growing scholarship on this exhibition demonstrates, Jenkins's overall assessment of the show as purely about style was both modest and reductive. First, and perhaps most striking about the assortment of photographers included in the exhibition, the range of attitudes displayed toward the landscapes of petromodernity suggests more engagement with content than Jenkins seems ready to admit. While Lewis Baltz and Robert Adams highlight the utter plainness, monotony, and isolation of urban sprawl in the American West (a theme that Robert Adams made particularly resonant in his book *The New West* from the previous year), the sole color photographer in the show, Stephen Shore, offered a far more accepting and familiar (if not quite celebratory) vision of car culture and contemporary small-town life. Where Nicholas Nixon made sweepingly majestic cityscapes of Boston from a bird's-eye view, Frank Gohlke kept his camera low, in the weeds, as it were, capturing the very surface of the man-altered landscape itself in expanses of asphalt, irrigation canals littered with detritus, and sad little patches of overlooked urban greenery. Moreover, contrary to Jenkins's contention that the *New Topographics* photographers embrace the form but not the content of Ed Ruscha's conceptualism, the Bechers' serial photographs of industrial architecture seem to be more about repetition and difference as qualities of both photography as a medium and industrial modernity as a historical phenomenon than about style or modernist photographic aesthetics. Recent scholarship has emphasized this variety of perspectives as central to the meaning of the exhibition as a whole. Toby Jurovics argues, for example, that far from cultivating aesthetic distance from their photographic subjects, the *New Topographics* photographers were struggling to represent and "reengage," in very different ways, a familiar landscape that was growing increasingly strange.[9] Even more pointed is Finis Dunaway's examination of the way that the photographers of *New Topographics* used photography to cultivate a new mode of ecological citizenship.[10] In contrast to Ansel Adams and other Sierra Club photographers who saw ecological activism in transcendentalist terms and highlighted the transformative power of a pristine, untouched, and problematically inaccessible wilderness, the photographers of *New Topographics* seem to advocate for what Dunaway sees as a more practical relationship to the natural world grounded in proximity and the incorporation of natural elements into the day-to-day lives of urban and suburban dwellers. Ultimately, then, despite Jenkins's attention to style and form over content, *New Topographics* can be seen to offer a subtle and varied, but nevertheless substantive, commentary on the social and cultural experience of petromodernity.

One of the challenges of conceptualizing *New Topographics* as a photographic movement (as opposed to a collection of individual artists and images) is the aesthetic and geographical range of the photographs it includes. Yet if we follow Jurovics's and Dunaway's suggestion and understand these photographers as socially engaged with the ecological and sociological topographies they represent, perhaps the most dominant theme to emerge is a sense of ambiguity, a struggle to negotiate the comforts and conveniences of postwar America with the changes it has brought to the American landscape, and a desire to reconcile the real and imagined landscapes of the American dream. Some of the more critical visions of this postwar American landscape come from Robert Adams and Lewis Baltz. Shortly before his inclusion in *New Topographics*, Adams gained recognition for his monograph *The New West* (1974), which arguably marked the paradigm shift in American landscape photography that would be more broadly fleshed out in the *New Topographics* exhibition the following year. The photographs in *The New West* (some of which were featured in *New Topographics*) depict suburban developments in and around the cities of Denver, Boulder, and Colorado Springs. Shot in black and white and often at a palpable distance from the houses, billboards, gas stations, and RVs they represent, Adams's photographs are very much landscapes. Unlike the street photography that defined the previous generation of American art photographers, Adams's photographs are rarely populated, and when they are, the people in them (usually a single person) are dramatically isolated—dwarfed, both physically and emotionally, by the cold expanse of the man-altered terrain around them. Yet these photographs are all the more unsettling for the way they also capture the postwar optimism of American culture, one fueled (metaphorically and literally) by the products of the petroleum industry.

Adams's photograph *Longmont, Colorado* (1973), for example, is one of a number of images of the expanse of suburban housing developments in Colorado springing up in the decades following World War II (Figure 15.1). The image depicts a split-level suburban tract house shot from the back. Located at the end of a cul-de-sac, the house appears to be a relatively comfortable one with a large chimney and a spacious back-yard with a roomy concrete patio. The backyard, which occupies the foreground of the photograph, is well appointed with cushioned lawn furniture, a wooden picnic table, and a built-in charcoal grill. Its newness is evident both in the clean, unweathered exterior of the house itself, and in a tuft of long grass at the edge of the back lawn to the right of the image indicating the presence of an adjacent vacant lot yet to be developed in this emerging bedroom community. And with its verdant and well-manicured lawn and clear blue sky, the house and its backyard seem an inviting domestic setting, the perfect spot for a family cookout, a game of catch, or simply lounging in the afternoon sun with a magazine and a glass of iced tea. Yet the photo is oddly devoid of any human presence. Not only are there no people (this is actually a false statement—careful scrutiny reveals a single woman striding across the driveway of another house toward her car in the far distance), but with the exception of a few small details that indicate the owners of this house have a small child—a diminutively sized lawn chair on the back patio, a stuffed toy glimpsed in the upper right-hand

FIGURE 15.1. Robert Adams, *Longmont, Colorado*, 1973. Copyright Robert Adams. Courtesy of the Fraenkel Gallery, San Francisco, and Matthew Marks Gallery, New York.

window—there are few signs that this house is even lived in. Rather than offering a portrait of a home and all of the messy particularities that implies, *Longmont, Colorado* is instead an image of a carefully maintained space ready to be used. Like the photograph in a real estate listing, this image is less a trace of history than a stage, a suggestion of potential. Though what that potential might be is left artfully ambiguous. Further accentuating this ambiguity is Adams's viewpoint. Despite the dramatically flat expanse of the landscape in the image, Adams is somehow situated above it. He is perhaps on a dramatic rise behind the house or shooting from the upper story of another house adjacent, but the effect is one of distance and disengagement from the scene. However pleasant this little house may be, the image suggests a sense of unease and withdrawal, a refusal to commit to either a celebration or a rejection of the suburban domesticity on display here.

Lewis Baltz's work has a similarly cool affect and clean formalist style, yet while Adams often concentrates on residential landscapes, the Baltz photographs included in *New Topographics* were exclusively of commercial spaces—various warehouses and office parks—in the city of Irvine, California. The choice of Irvine was surely not accidental; Irvine is a company town. Built on the site of a former ranch, most of the city sprang up in little more than a decade between 1959 (when the University of California began to develop its campus there) and 1971, when it was formally incorporated. Baltz's singular photographs of this strangely ahistorical urban landscape are immediately identifiable for their radical frontality. Of the twenty photographs included in the exhibition, all but three are shot directly perpendicular to the building facade they depict (the other three are at a precise 45-degree angle). Equally striking is the almost complete lack of signage or corporate branding of any kind.[11] Instead, the photographs depict spare facades marked only by the geometry of windows, doors, and the occasional drainpipe, and horizontal bands of asphalt and grass (there are almost no trees in these landscapes). As with Adams's work, indeed more so, these spaces are strikingly devoid of human presence. With the exception of parking spaces for cars and doorways for people to enter and exit these various structures, Baltz's photographs exclude anything that even gestures to the human body. Thus, as coolly observed as Baltz's photographs are, there is an insidiousness to these empty spaces of labor and commerce, an ever so slight postapocalyptic feel that gestures toward the profligacy of late capitalist commerce.

A particularly striking example is *South Wall, Mazda Motors, 2121 East Main Street, Irvine* (1974) (Figure 15.2). Like the majority of the Baltz photographs in the exhibition, *South Wall, Mazda Motors* depicts an expanse of wall sandwiched between narrow strips of grass and sky. The building extends beyond the edge of the frame on both right and left, transforming a three-dimensional structure into a dramatically

FIGURE 15.2. Lewis Baltz, *South Wall, Mazda Motors, 2121 East Main Street, Irvine, 1974*, from *The New Industrial Parks Near Irvine, California*, gelatin silver print, 20.3 × 25.4 cm. Copyright Lewis Baltz. Courtesy of the Galerie Thomas Zander, Cologne.

two-dimensional facade. The only sense of depth comes from a shallow overhang at the top of the facade and a sparsely ornamental rock garden underneath it. The wall itself is composed largely of a bank of mirrored plate-glass windows that neatly reflect the facing landscape back at us: some scraggly trees, a set of power lines, a street with a few cars, and, in the distance, more homogenous commercial architecture. Baltz preserves the ironically unpopulated emptiness of this man-made landscape by positioning himself opposite a concrete building support that at once perfectly bisects the expanse of windows and blocks his own reflection from being included in the image.

In contrast to Adams's anticipatory domestic landscape, Baltz offers a landscape utterly unfit for and disinterested in human habitation. It is particularly fitting here that the title identifies this structure as Mazda Motors (though its unclear if we are looking at a corporate office or a dealership). The mirrored windows and the token detail of the cars in the reflected landscape evoke the atomizing and alienating social effect of car culture and urban sprawl. One presumes that whatever function this building has, people work there, yet the mirrored windows bar us from seeing them. Like the Bonaventure hotel in Fredric Jameson's famous essay on postmodernism, the windows reflect the surrounding city and in so doing achieve a "peculiar and placeless dissociation" from its local environs, one that Baltz echoes in his severe and sparse formalist approach.[12] The effect again is one of unease and ambiguity. Baltz depicts a real space, but this knowledge makes it all the more unsettling. The subtly comic gesture of identifying the utterly generic facade by its street address points to the homogeneousness of these structures while reminding the viewer that this facade is more than a metaphor; it is a concrete place, albeit one barely discernable from countless other such spaces.

Among all the photographers included in *New Topographics,* Stephen Shore is perhaps the most distinctive, both for his optimism and for his vivid use of color photography (two qualities in his work that are not unrelated). Stephen Shore's *Beverly Boulevard and La Brea Avenue, Los Angeles, California, June 21, 1975* was actually not included in the *New Topographics* exhibition, but it was shot by Shore that same year and it is one of the more explicit engagements with petroculture in the oeuvres of any of the *New Topographics* photographers (Figure 15.3). The image is a street view of the 7000 West block of Beverly Boulevard in Los Angeles on a sunny day. In contrast to his fellow *New Topographics* photographers, Shore's work overall (and this work in particular) has less of a sense of desolation. Though one has to squint to pick out any human beings in this street scene (they are there—an auto mechanic is working on a car at the Texaco station, and some people are waiting for a bus on Beverly, just past the intersection at La Brea), the scene nevertheless has a sense of activity and dynamism to it. It is a street scene where "business as usual" is going on around us, even if we cannot quite see it. Cars zip along the street, and the vivid color gives the image a sense of familiarity. This is not the real transformed into artful black and white, but life at its most frankly pedestrian. Indeed the scene is so familiar it seems almost contemporary—we see an iconic L.A. thoroughfare, a billboard for McDonald's, two gas stations, a Chevron and a Texaco, across the street from each other, offering that all-American quality of choice in all things. The persistence of these familiar icons of petromodernity is telling, yet to the keen eye details emerge that date the image to a particular historical moment: an old faux-wood-paneled station wagon idles at the light on La Brea, the gas pumps at the Chevron station have rotating number gauges rather than digital screens and card readers, what might be a McDonald's Big Mac is advertised for a mere eighty-five cents. The models of the cars and the gas pumps are perhaps familiar from those iconic images of the 1973 oil shortage, where similar cars waited in lines at similar pumps, indeed, probably at these very gas pumps. And by contemporary standards, the car culture on Beverly Boulevard in 1975 seems a bit too modest, even quaint. Of course, Shore could not have known exactly how photographs like these would read thirty-five or forty years later. Yet embedded in his fixation with the plain, the everyday, and the familiar is an understanding that these

FIGURE 15.3. Stephen Shore, *Beverly Boulevard and La Brea Avenue, Los Angeles, California, June 21, 1975*, 1975–2004, C-print, 20 × 24 inches. Courtesy of the 303 Gallery, New York.

minutiae have the potential to evoke a world and a realm of affect associated with it. And while Shore's work has a warmth to it that is missing (or rather, stridently foreclosed upon) in the work of peers like Adams, Baltz, Gohlke, and others, there is nevertheless a subtle ambiguity here as well. *Beverly Boulevard and La Brea Avenue* evokes a culture of casual, comfortable petroleum consumption, but one that requires blocking out the emerging notion that such consumption may not always be possible.

Formally groundbreaking as the *New Topographics* may be, then, it is also sociologically and economically prescient in the way it depicts the man-altered landscape. Photographers like Baltz, Adams, and even Shore seem to stand hesitantly on the edge of a precipice. On the one hand, they highlight the utter normalcy of the postwar landscape, a banality that would seem to make it almost invisible were they not to turn their cameras squarely upon it. (Even then, as accounts of the exhibition suggest,

the subject matter gave many museum visitors pause. As one remarked to an inter-
viewer, "I don't like them—they're dull and flat. There's no people, no involvement,
nothing.")[13] On the other hand, and "dull" as they may seem to some, these photographs figure a contemporary national culture (mostly) blind to the very effects and
conditions of its own existence, a social emptiness that forecasts the growing obsolescence of American industrial labor, the rampant spread of commerce, and the eclips-
ing of subsistence on a human, rather than a corporate, scale. Emerging concomitant
to American culture's first taste of the finitude of oil and petroculture (the 1973/74
OAPEC oil embargo that led to shortages and gas rationing in the United States was
still fresh in people's minds when the exhibition was mounted), the photographic
landscapes of *New Topographics* located parallel currents of optimism and pessimism
about American futurity, and found in the everyday a unique psychological and affective portrait of petromodernity at a turning point, one the culture was only barely
aware of, but one that would quickly emerge as a problem of global proportions.

Edward Burtynsky's Industrial Sublime

While the photographers of *New Topographics* reflect upon the conditions of petromodernity somewhat tacitly, perhaps no artist is as readily identified with depictions
of the oil industry in specific (and the global conditions of what I might call "petro-
postmodernity" more broadly) than Canadian photographer Edward Burtynsky.
Burtynsky's oeuvre frequently addresses the decimation of the landscape (both geographical and social) through the projects of "post-industrial" late capitalism. With
his signature large-format camera and attention to detail, he has produced large-scale
photographs that document tapped-out and abandoned rock quarries, the dehuman-
izing scale of sweatshop production in China, the deteriorating state of water as a
global natural resource, and the Third World industry surrounding the picking over,
breaking down, and repurposing of First World waste. But of all his projects, perhaps
the most famous is his *Oil* project. This project charts the life cycle of the petroleum
industry, from extraction to "the end of oil," and features depictions of the desolate
Alberta Oil Sands, where slick black pools blanket the landscape mirroring the clouds
above; the interiors of oil refineries, each an abstractly geometric tangle of gleaming
pipes and valves; a sea of identical white cars in an incomprehensibly vast Volkswagen
lot in Houston, Texas; and the rusted, rutted devastation of the SOCAR oil fields in
Azerbaijan, its profusion of crumpling derricks littering a landscape of oily puddles
and befouled earth. The images are both stunning (quite literally, for he chooses to
photograph those things that we rarely see, astonishing viewers with vast displays of
industry at its most gargantuan and systematic) and, for this author at least, deeply
disturbing.

Also distinct from the photographers of *New Topographics*, Burtynsky has made a name for himself both inside and outside the world of contemporary art. While he has exhibited in many conventional art venues (among them the National Gallery of Canada, the Art Gallery of Ontario, the Brooklyn Museum of Art, and the Corcoran Gallery of Art) and produced a number of fine art books, he is also the subject of the popular documentary film *Manufactured Landscapes* (2007, dir. Jennifer Baichwal), and has mounted exhibitions at nonart venues like the Boston Museum of Science and the Canadian Museum of Nature, where the institutional context would suggest less an aesthetic experience and more an informational one. Given this flexibility in his oeuvre and his advocacy for sustainable development, one might be tempted to think of Burtynsky's work as an exposé of sorts, revealing to his audience the dark side of the petroleum industry and, more broadly, the conditions of late industrial capitalism. Yet the artist's own statements about his work suggest a more complex reading. Time and again Burtynsky has emphasized that his aim is not to be didactic but to spark conversation. In his preface to *Oil*, for example, he speaks not of taking a particular political position on the petroleum industry, but rather of "contemplating the world made possible through [oil]."[14] And in a 2012 interview for the *Atlantic*, Burtynsky likened his photographs to a Rorschach test. He states: "If you see an oil field and you see industrial heroism, then perhaps you're some kind of entrepreneur in the oil business and you're thinking, 'That's great! That's money being made there!' But, if you're somebody from Greenpeace or whatever, you're going to see it very differently."[15] Burtynsky would, then, seem to simply present the visual information he has gathered (notably of sites and phenomena that many of his viewers have never seen or even imagined), and the work then pivots on the individual viewer's experiences and ideals as they form a response. But that response is prompted, in significant part, by the aestheticizing frame of Burtynsky's particular photographic perspective. And while that aestheticization certainly distances the viewer (as Mitchell's epigraph above suggests), it also opens these photographs to another kind of interpretation, one based again in the sensory terrain of the petro-postmodern world.

One of Burtynsky's most iconic and definitive works from the *Oil* project is a 2003 diptych *Oil Fields #19* (Figures 15.4a and 15.4b). The image is a landscape awash in the muted earth tones of dry grass, gravel, sand, and the golden reflections of late afternoon light. The surface of the ground is particularly present in the image because of Burtynsky's signature dramatically elevated point of view—he seems to point the camera down more than out, leaving in this work only the slightest strip of hazy sky and distant gray mountain running along the very top of the pair of images. The focus, of course, is the oil fields, signified by the pumpjacks, neatly distributed across that flat expanse of desert ground and stretching back almost as far as the eye can see. Using a

FIGURE 15.4A. Edward Burtynsky, *Oil Fields #19* (diptych), Belridge, California, 2003.
Photograph by Edward Burtynsky. Courtesy of the Nicholas Metivier Gallery, Toronto.

large-format field camera, Burtynsky achieves a striking amount of detail and deep
depth of field. Yet for all this detail, Burtynsky's elevated point of view tilts the ground
up toward the viewer, transforming the landscape into an abstract pattern, a stunning effect that both emphasizes the grand scale of this drilling operation and removes
the viewer from it by dissolving any sense of human scale. Because of the size of the
images (the prints in the *Oil* exhibition were 60 × 75 inches each) and their pairing as
a diptych, the composition fills the viewer's field of vision. Even in the book, which
reproduces the work both on its dust jacket and in the interior, the scale of the image
is grand and absorbing. The effect is dramatic and resolutely aesthetic, indeed, sublime. In this, as in most of Burtynsky's photographs, the viewer takes a god's-eye view
of petromodernity, from the massive oil drilling operations and refineries, to the auto
industry and the fast and cheap lifestyles and leisure pursuits it supports, to, finally, the
gorgeous and grotesque aftermath: mountains of discarded tires, drifts of bushling

FIGURE 15.4B. Edward Burtynsky, *Oil Fields #19* (diptych), Belridge, California, 2003. Photograph by Edward Burtynsky. Courtesy of the Nicholas Metivier Gallery, Toronto.

scrap metal, and the oily rusting hulks of broken-down container ships. The content of these images is startling and disturbing, evidence of the environmental degradation brought on by the petroleum industry on a scale most people have never conceived of, much less witnessed firsthand. And yet Burtynsky's artistry, his refined handling of detail, depth of field, color, and composition, takes a very real social and environmental problem and makes it lyrical, even beautiful. This aestheticizing gesture is, I think, no accident, but further evidence of Burtynsky's refusal to make his work overtly political. And yet this aestheticization is also critical to understanding the way that Burtynsky's work distills the affective conditions of late petromodernity.

To better understand how this works, it bears mention that Burtynsky's photographs are not only vast and sublime but also strikingly static. This may seem to state the obvious when talking about a photograph, but given the subject matter, Burtynsky's images of industry, like Baltz's images of commercial architecture, give us a working

landscape that fails to actually work, a space frozen or paralyzed. This effect is particularly striking in the documentary *Manufactured Landscapes*, where one sees Burtynsky
in the process of setting up and framing his photographs. In one scenario after another
we see an active landscape, a bustling factory floor or shipbreaking yard, for example,
that Burtynsky slowly and deliberately frames into one of his signature images. The
detail and composition of Burtynsky's image is then emphasized by the filmmaker's
slow pan across its surface, or careful pullback from a "punctal" detail.[16] This process
evokes Walter Benjamin's famous description of the daguerreotype and the way that,
"during the considerable period of the exposure, the subject, as it were, grew into the
picture in the sharpest contrast with appearances in a snapshot."[17] This effect is present as well in the *Oil Fields* diptych. While the image is, again, empty of any human
figure, thus making the space of the oil fields all the more static and abstract, close
examination reveals a slight blur on a couple of the pumpjacks in the foreground.
Though the detail disrupts the pristine crispness of the photo, it also draws attention
by contrast to the lack of motion anywhere else in the image, a landscape paradoxically littered with machines that, in the world of the image, at least, seem to be completely nonfunctioning. As with the shots of Burtynsky's process in *Manufactured
Landscapes*, the dramatic stillness of the *Oil Fields* diptych does seem to have about it
what Benjamin called an "aura" and an "air of permanence."[18] And this air of permanence, too, has a pronounced sensory effect. If *New Topographics* distilled its historical moment in its sense of unease and ambivalence, a tension between the domestic
comforts of petromodernity and a creeping realization of the finiteness of fossil fuels
on which the American dream is staked, then Burtynsky's photographs offer an unsettling sense of permanence in the postmillennial moment, a permanence that speaks
not to bottomless resources but to a failure to reimagine Western culture and prosperity any other way.

FEELING PETROAESTHETICS

In the epigraph to this essay, art historian W. J. T. Mitchell poses the "aestheticizing
distance" of the landscape in opposition to the more utilitarian formation of what
Raymond Williams calls "a working country." This would seem to suggest that the
landscape is at something of a remove from reality, a historical imagining imposed
upon the actual terrain. Yet within this distance, and particularly in the case of the
radically man-altered terrain of petromodernity, there lies a new site of possibility
for understanding landscape and its relation to history. Ultimately, neither *New Topographics* nor Burtynsky's *Oil* project presents the kind of overtly politicized "critical
realism" that one might locate in, for example, Allan Sekula's *Fish Story* (1989–95)
or Maria Whiteman and Imre Szeman's Sekula-inspired photo essay *Recovery: Life,*

Labour, Oil (2012).[19] Instead, what these two bodies of work do is to offer insight into the "structures of feeling" (to evoke another famous idea of Raymond Williams's) that define life in the West in the age of oil. With that in mind I would like to close with a discussion of one more image by Burtynsky. In this analysis I hope to take a step back, to consider not the artist's intentions per se, but his function as a kind of sounding board for his historical moment, and in so doing, to distill a bit of what LeMenager calls the "sensory and emotional values associated with North American oil cultures."[20]

In many ways, Burtynsky's *Breezewood* (Figure 15.5) is dramatically similar to Shore's aforementioned *Beverly Boulevard and La Brea Avenue*. The vivid color photograph depicts an increasingly common aspect of contemporary car culture—the freeway-adjacent complex of gas stations, hotels, and chain restaurants that serves primarily to accommodate truckers and road trippers en route to somewhere else. Despite its evocatively pastoral name, Breezewood is clearly a transitory space. Indeed, while it is now an exit on the Pennsylvania turnpike, it is a town whose economy has largely been based on stopover tourism going back to the early twentieth century. Like *Beverly Boulevard and La Brea Avenue*, *Breezewood* offers a selection of gas stations to motorists passing by, though Chevron and Texaco are here replaced by Exxon (two stations, one on each side of the cross street), Sunoco, Shell, and Valero. In addition to the iconic golden arches of McDonald's, *Breezewood* also features a Denny's (housed in a shiny replica of a midcentury diner), Quiznos, Pizza Hut, Perkins, Starbucks, Taco Bell, KFC, Subway, and a smattering of what seem to be locally owned dining establishments. Captured from Burtynsky's familiar elevated perspective (there is no horizon to speak of in this image, just a dense bank of forest on the far side of the highway overpass), this highway interchange becomes a maze of luminous signage jockeying for the highest register so as to be seen by passing motorists from the highway. And with so much crisp detail and depth of field, it is easy to get lost among all the potential consumption on display. Even the vehicles offer up corporate logos: Sam's Club and Walmart, Budget Truck Rental. In each corporate logo blazing brilliantly from a luminous sign, the viewer sees something familiar (even if one does not patronize these establishments, their ubiquity in the American landscape is undeniable). And yet in their sheer number and in the way Burtynsky's elevated point of view reduces these places only to logos (there is none of Shore's comfortable "lived in" aesthetic here), Burtynsky takes the local and makes it strange, foreign. Thus, while strikingly current to the time of this writing, the work maintains a majestically sublime aesthetic; at once beautiful and terrifying, it is ultimately incomprehensible. In so doing, *Breezewood* gestures toward an affective and imaginative dead end in our perception of oil culture. On the one hand, Burtynsky's work unequivocally documents

FIGURE 15.5. Edward Burtynsky, *Breezewood,* Pennsylvania, 2008. Photograph by Edward Burtynsky. Courtesy of the Nicholas Metivier Gallery, Toronto.

the gross excesses of petro-postmodernity, and his photographs capture something as familiar in North American culture as a highway truck stop in such a way as to make it grotesque, bloated with our desire for consumption and twenty-four-hour access. Yet, on the other hand, that excess seems to be the only way we can understand petro-culture, as limitless excess without grounds for comprehension or change. To return to Raymond Williams once more, Burtynsky's *Breezewood* is not simply an imagined landscape posed in opposition to a working country. Rather, *Breezewood* reveals that perhaps this terrain is not working in real time either. In its static beauty, the image envisions the working landscape of petro-postmodernity as something that simply cannot work: paralyzed, impossible.

NOTES

1. Peter D. Osbourne, "The Reverie of Power," *Travelling Light: Photography, Travel and Visual Culture* (Manchester: Manchester University Press, 2000), 16–51.

2. For a discussion of a variety of American landscape photographers working in this period, see Joel Snyder, "Territorial Photography," *Landscape and Power*, 2nd ed., ed. W. J. T. Mitchell (Chicago: University of Chicago Press, 2002), 175–201; and Alan Trachtenberg, "Naming the View," *Reading American Photographs* (New York: Hill and Wang, 1989), 119–63.

3. Stephanie LeMenager, "The Aesthetics of Petroleum, After *Oil!*," *American Literary History* 24, no. 1 (January 2012): 60.

4. Although the petroleum industry has made its footprint on the landscape on a global scale, it is my contention that the sensory impression constructed through the landscape photographs of petromodernity is one explicitly geared toward the Western (or more specifically, North American) experience. Thus I would suggest that even images like Edward Burtynsky's photographs of oil fields in Azerbaijan are in some way about the American perspective on oil and how the global nature of the industry figures into concepts of American nationalism and independence.

5. Raymond Williams as quoted in W. J. T. Mitchell, "Preface to the Second Edition of *Landscape and Power*: Space, Place, and Landscape," *Landscape and Power*, viii.

6. Mitchell, "Preface to the Second Edition of *Landscape and Power*," viii.

7. In a catalogue essay for the 2009 "revival" of the *New Topographics* exhibition, Britt Salveson quotes Jenkins as saying, "We were just doing what we thought a museum of photography ought to be doing." Britt Salveson, "New Topographics," *New Topographics* (Göttingen: Steidl, 2009), 13.

8. William Jenkins, "Introduction," *New Topographics: Photographs of a Man-Altered Landscape* (Rochester, N.Y.: International Museum of Photography at George Eastman House), 5.

9. Toby Jurovics, "Same as It Ever Was: Re-reading *New Topographics*," *Reframing the New Topographics*, ed. Greg Foster-Rice and John Rohrbach (Chicago: The Center for American Places at Columbia College Chicago, 2010), 1–12.

10. Finis Dunaway, "Beyond Wilderness: Robert Adams, *New Topographics*, and the Aesthetics of Ecological Citizenship," *Reframing the New Topographics*, 13–43.

11. The signs that do appear are for businesses with names like "Simcoa" and "Pertec." No particular product or service is ever identified, thus these names and the logos they bear signify simply "corporateness."

12. Fredric Jameson, *The Cultural Turn: Selected Writings on the Postmodern, 1983–1998* (New York: Verso, 1998), 13.

13. "Prologue," *New Topographics*, 9.

14. Edward Burtynsky, *Burtynsky: Oil* (Göttingen: Steidl, 2009), 9.

15. Edward Burtynsky with Geoff Manaugh and Nicola Twilley, "The Art of Industry: The Making and Meaning of Edward Burtynsky's New Exhibit, 'Oil,'" *Atlantic*, June 19, 2012, http://www.theatlantic.com/. In the same interview, Burtynsky tells an anecdote from his own experience that backs up this interpretation. While redoing his kitchen, Burtynsky recalls, he approached a particular quarry hoping to arrange a trade: a Burtynsky quarry photograph for a granite countertop. But the board of directors of the quarry refused, seeing in the photographs not majestic portraits of the tension between man and nature but a vision of their industry at "the end of the line."

16. Roland Barthes describes the "punctum" as the element of a photograph that grabs him, "pricks" and "wounds" him, with an affective quality that is profound, singular, and inexplicable. Roland Barthes, *Camera Lucida: Reflections on Photography* (New York: Hill and Wang, 1981), 26.

17. Walter Benjamin, "Little History of Photography," *Walter Benjamin: Selected Writings, Volume 2, Part 2,* ed. Michael W. Jennings, Howard Eiland, and Gary Smith (Cambridge: Belknap Press of Harvard University Press, 1999), 514.

18. Ibid.

19. The term "critical realism" is Sekula's. For a discussion of critical realism in Sekula's *Fish Story* as well as Whiteman and Szeman's photo essay on the Alberta Oil Sands, see Imre Szeman and Maria Whiteman, "Oil Imag(e)inaries: Critical Realism and the Oil Sands," *Imaginations* 3, no. 2 (2012): 46–67.

20. LeMenager, "The Aesthetics of Petroleum," 60.

16

Fossil, Fuel

Manifesto for the Post-Oil Museum

STEPHANIE LEMENAGER

"If the problem of death can be solved, the solutions to other problems cannot but follow."[1] These are the words of the Russian librarian and philosopher Nikolai Federov, as performed by David Wilson. Wilson is founder and curator of the *Museum of Jurassic Technology,* Los Angeles's ironic encomium to the Renaissance *Wunderkammern* that preceded Enlightenment-era museums of natural history. The "resurrection and resuscitation" in full body of all the dead who have gone before is a "task that will be realized by the museum," Wilson intoned in a voice-over at a recent screening of his film mash-up about Federov, *The Common Task* (2008). By this point in the screening it was unclear exactly who—Wilson or Federov—was commenting upon the museum's foundational desire. Natural history museums in particular have partaken in the desire to archive life itself, creating an exhibit space of "universal time," in the phrase of the cultural critic Tony Bennett, through the display of representative species from across the globe and across historical and geological periods.[2] Methods of natural history, systematic biology, and taxonomy made possible the very notion of biodiversity, according to historian Robert E. Kohler; Steven Conn argues that "ecological research is an updated version of nineteenth-century natural history" and that such research grows urgent in the shadow of mass extinction.[3]

Recent studies predict that as many as 30 percent to 50 percent of all species may be headed toward extinction by the middle of the twenty-first century.[4] In museum literature, the possibility of resurrecting these dead appears dependent upon both scientific methods like species inventories and their dissemination through new media. "Bio-informatics" signifies the digitization of museum collections such that global information sharing can keep scientists up to date about species loss.[5] "Interactives" or multimedia displays promise to generate creativity in museumgoers, fashioning a public primed for the flexible thinking demanded by ecological crisis.[6] The question of what endurance means for "nature," "history," and the natural history museum seems to break around what will continue to count as lifelike: information or object, media or bone.

Los Angeles offers a rich setting for a consideration of these questions, both because it is the world capital of entertainment media and because it hosts the world's largest archive of Pleistocene fossils, preserved in the asphaltum sinks that formed atop the Salt Lake Oil Field, itself the world's largest urban oil reserve. The La Brea Tar Pits serve up natural history in the raw, and they share kinship with the Renaissance curiosity cabinet and other "disreputable displays of objects," in Conn's words, such as those that might be found at fairs, freak shows, or circuses.[7] The sensory regime of these bubbling, odoriferous sinks exceeds the didactic visuality associated with nineteenth-century natural history, which Michel Foucault once described as "nothing more than the nomination of the visual."[8] In a way that may not be apparent to tourists, the tar pits represent a regional ecology. They are a petroleum archive within a city that exemplifies better than any other what oil could make of modernity. The effort to build a museum around the La Brea Tar Pits tells a story about Los Angeles's image of itself through the twentieth century, its desire to make history and the authenticity we associate with "nature" from within an immensely profitable pursuit of spectacle. One even can imagine the tar pits in the terms of the ecomuseum, the *ensembles ecologiques* developed most extensively in France during the 1970s as a means of honoring regional cultures.[9]

Asphaltum, the semisolid form of petroleum that bubbles in the misnamed tar pits, refers out to Los Angeles's freeway culture and to its "electric media" culture, to riff on an old McLuhanite term for film and television media powered by electricity—and indebted to fossil fuel. The apotheoses of automobility and electric media on display in Los Angeles spring from the peak of U.S. oil discovery in the 1930s. Oil played a foundational role in the making of two aspects of L.A.'s four ecologies, in Reyner Banham's terms: the ecology of *autopia* (the freeways) and the architecture of *the fantastic,* namely, the commercial storefronts designed to hail automobile drivers. Yet the La Brea Tar Pits indicate a petroleum reserve—again, the Salt Lake Oil Field—which has been removed from the "destructive circulation of commodities," in Andreas Huyssen's phrase, and from the transformative industrial work of the modern energy system.[10] In the midst of the world city whose infrastructure, architecture, and cultural production most bluntly express how fundamental oil has been to modern aspiration, the La Brea Tar Pits make oil visible as a material presence that cannot be easily reduced to the abstraction of "energy." Fences cordon off this dense petroleum that sits at the surface of the earth, displaying itself to the naked eye as an organic, irreducible treasure. In the shadow of the Los Angeles County Museum (LACMA), the tar pits can be contemplated as land art, with the potential to effect self-transformation that art implies to a middle-class public accustomed to refining its sensibility in art museums. The opportunity to think of oil as nature and/or art opens broad possibilities for thinking about how oil has expressed fundamental notions of humanness.

The tar pits share in the qualities of both carnival and museum. They are stinking, sticky objects indicative as much of the humiliations of embodiment as the transcendental notion of the human that has been articulated through technological prostheses, like computers or cars, which extend the human body—at an increasingly unsustainable expense. Unlike the Oil Sands Discovery Centre in Fort McMurray, Alberta, where bitumen, a substance quite similar to asphaltum, stars in a multimedia narrative whose first premise is the "transformation" and "refinement" of humanity through the heavy industrial labor of making a tarlike solid into oil, at Rancho La Brea oil has been decoupled from any clear idea of progress. The Alberta museum also displays fossils found in the bitumen—which the Canadian oil industry anxiously refuses to call "tar"—but there the fossils are incidental to the curatorial line, which reinforces the pleasures of disembodiment, of losing oneself in the big trucks and diggers that make tar sands extraction possible.[11] In Northern Alberta, tar ensures that the modern cultures and infrastructures that oil made possible in the twentieth century appear, improbably, as "the future." In contrast, Los Angeles's petroleum exhibit has been in the making since the early twentieth century, in a time and place where oil seeps were ordinary, frequent, and indicative of what looked like an endless supply of cheap energy. With this essential confidence in the fossil-fuel system, one could take petroleum as it presented itself, at earth's surface, and consider it as bothersome tar, as decaying microorganisms, or as media through which large mammals, including humans, might be preserved.

This manifesto considers the La Brea Tar Pits, which by the first decade of the twenty-first century have fallen far toward the disreputable status of the carnival, as the potential site of a post-oil museum. My gentle polemic draws upon Bennett's recognition of the natural history museum in particular as a "civic experiment" that reassembles objects in such a way as to create new entities "that can be mobilized . . . in social and civic programs."[12] In brief, I recognize museums as organizing material cultures that in turn generate new social narratives and civic behaviors. The La Brea Tar Pits offer a narrative about living with petroleum in petrotopia (e.g., Los Angeles) and the possibility of thinking beyond the explicit uses of oil that have defined modern energy.

Petroleum Media

At the George C. Page Museum of La Brea Discoveries, opened in 1977 among the tar pits in L.A.'s tony Hancock Park district, visitors learn about the problem of dying, if you are a mammal, and about what it means to be petroleum, which never dies. The mammal/petroleum difference generates a key aspect of the museum's drama, a locus of tension and anticipated resolution within the museum space, which relies upon petroleum-based media, including film and its famous asphaltum pits, to create a sim-ecology where mammals, too, might live forever, in the half-light of dioramas and continuously looping reels. Jonathan Spaulding, former associate curator of the Natural History Museum of Los Angeles County (of which the Page Museum is a satellite), argues that the curatorial voice of scientific authority that once organized such museums by taxon and chronology is "dead," citing "emotion," "sensation," and "rhythm" as key curatorial goals, to be accomplished through the spatial, acoustic, and kinaesthetic art of scenography.[13] Historian Alison Griffiths argues that such emphases on multisensory media in fact date back to the nineteenth century, citing museum features such as panoramas and the slightly later innovations of habitat groups and period rooms as examples of how "illusionistically re-created space can reinforce the sensory experience of space and time travel" that museums always promise.[14]

The Page Museum's natural advantage, its literally immersive scene over the fossil-rich Salt Lake field, may offer a rather low interactivity ratio in comparison to the new media pyrotechnics of younger and richer museums. But it is still the only museum famous for literally sticking to its visitors, forcing the issue of its oozy embodiment. In an era in which screen culture has become so ordinary as to make less of an impression on museum visitors than an apparently authentic experience, the tar pits present a discomfort that signals the "real."[15] On a recent tour, my docent, Gail, ended her introductory remarks with a promise to hand out recipes for tar removal at the end of our walk among the sumps.[16] As parents warned errant children to stay away from the museum's lawns and the distracted found their shoes sticking in place, it became clear that our clothing was at risk, and with it the social abstraction and status that distinguishes us from other mammals. At the Pit 91 excavation site, where some of the museum's most significant fossils have been discovered, the filthy T-shirts and jeans of volunteer excavators are on prominent display. Here nature presents itself quite viscerally, as stain.

If no single framing narrative can be derived from sticky shoes or blackened socks, such by-products of the La Brea experience at least suggest our vulnerability to and interrelationship with other life. The museum's displays reinforce the message of our vulnerability. Like modern "body genres" such as the horror film, exhibits invoke visceral mimicry in visitors by representing the distress of lone actors.[17] The first stop on the outdoor tour that is the highlight of the La Brea experience is "the Lake Pit," an asphaltum pit covered in water, the sort of mirage that the large mammals of the Pleistocene fatally ventured into for a drink. "The story of entrapment" that is the museum's leitmotif unfolds at the Lake Pit, according to a museum postcard: it is performed by "a fiberglass mammoth" (to scale) "struggling as she sinks into the large asphalt lake in front of . . . the museum. A father and baby mammoth watch from the shore."[18] Docents will add that the sinking Imperial Mammoth is "crying out," perhaps in "pain" or in "warning" to her appropriately nucleated family (Figure 16.1). At other points on the La Brea tour, and in the museum's story theater, visitors hear of the "screams" of entrapped mammals (the western horse, the Shasta ground sloth), most of which are now extinct. The museum's drama is discomfiting, all the more so because it is scaled to children. One of its few interactive displays consists of a child-height game wherein large metal pins might be pulled from asphaltum muck, although what we learn is that the muck relentlessly grips the pins and we cannot move them. In this low-tech game, entrapment—which can be conceived as little else than slow death—becomes corporeal knowledge. "Pull up on the handle, / Discover what it

FIGURE 16.1. Father and son enjoy family drama in the lake pit. Photograph by the author.

feels like to be trapped in tar," the exhibit beckons. While the single human skeleton at the Page, known as "La Brea Woman," has been retired to storage, there is little question that human death is on show. With a cloying ordinariness (it sticks to your shoe), personal extinction defines our "authentic experience," offering the downbeat to this museum's emotional rhythm.

Perhaps the museum's consistent reminder of the biological arc of singular lives (as Hannah Arendt tells us, the concept of mortality depends upon the elevation of the individual above species) makes the geological temporality represented in the tar pits more readable.[19] Imagining waves of extinctions that mark the edges of geological periods requires a shuttling back and forth between smaller- and larger-scale notions of loss. My docent at Rancho La Brea offered a strong argument for the difficulty of perceiving time at geological scales, as she explained when the Pleistocene occurred, some 2.6 million to 11,700 years ago, and how its now-foreign fauna reflect the fluctuations of Earth's climate resulting from the rotation of its axis closer or farther from the sun. "There are other theories of climate change," Gail noted, dismissively raising an eyebrow.[20] For her, this was not the place to throw the autonomy of geological process into question by discussing anthropogenic climate, although the petroleum patch could be an ideal site in which to raise such environmental concerns. Geological time, as a concept, has long been the natural history museum's ally, granting these museums what William Pannapacker celebrates as the aura of "things that are real."[21] Andrea Witcomb writes more cynically of the ways in which concepts such as aura and authenticity derive from "a museological practice which emphasized the importance of classification systems and taxonomies which attributed meaning to the object according to its physical characteristics." The nineteenth-century natural history museum's taxonomic systems claimed to express an immanent divine order, and they supported "a narrative of Western society as the pinnacle of civilization."[22] Through the 1960s, natural history museums offered an image of time's passage segregated from social turbulence.

But when, among nature's things, a natural history exhibit displays asphaltum, a form of crude oil that is a product of the Miocene seas that once covered the Los Angeles basin, a question arises for me: *Are some things that are real more real than other things that are real?* My point is auratic, rather than biological or ontic. By "aura" I mean the feeling and ambience of authenticity, which, as Benjamin taught us, cannot exist without the mechanical means of destroying "aura" by separating objects from their images in media such as photography.[23] At Rancho La Brea, the tar pits are in situ ecological features, therefore unlike the displaced natural objects typically associated with museums and laboratories, to which museums often have been compared. Although one can photograph the tar pits, their low visual readability makes for bad images, and they work best as multisensory media. They generate a physical ambience through the olfactory dimension of odor—specifically, the smell of hydrogen sulfide, which, along with the odorless methane gas that bubbles to the surface of the pits, betrays the formation of crude. Of course, too, they are viscous, and their stickiness offers a kind of bad patina, literally tracking their use over time by human excavators and unlucky animals.

Yet the tar pits extend into their geographic frame in a way that troubles their status as "real things." One only has to venture to the edge of Hancock Park to recognize their entanglement in a complex social history. The La Brea exhibit depends upon the high real-estate values of the surrounding neighborhood of Hancock Park, which made possible the preservation of this portion of the Salt Lake Oil Field from intensive drilling. Hancock Park, in turn, both derives value from and confers value upon Wilshire Boulevard, the historical shopping district that is known as L.A.'s Miracle Mile. It was on the Miracle Mile that the parking lot was essentially invented, as planners in the 1920s recognized the convenience of auto parking behind department stores in a city already poised to enable the individualized mobility of the car. Asphalt, which used to be made from asphaltum (and is now made directly from petroleum), paves "the mile" and the parking lots behind it. In the late 1960s, the parking lots of Los Angeles demonstrated the end of art and other *real things* in a series of aerial photographs that Ed Ruscha commissioned from a commercial photographer. Without a connection to the artist's own hand or to the promise of self-transformation evident in aesthetic categories such as the beautiful, Ruscha's *Thirty-Four Parking Lots* (1967) humorously enshrined the most ubiquitous indication of capital's homogenization of place. An art book of parking lots, photographed by someone other than the artist, empties the "art" concept, the "exhibit" or museum concept of the art book, and the idea of sacred space. Thinking about the parking lot as itself a kind of museum of modernity makes the conceit of the museum at La Brea, the asphaltum preserve, seem precious if not duplicitous. Does petroleum *ever* remain in situ, and what might be the ideological effect of showing it that way?

The original plan for "Pleistocene Park," an earlier incarnation of the La Brea outdoor museum, offered a clear alternative to car culture. This plan called for the screening of automotive traffic by landscape features and a single footbridge entrance to guard against auto touring.[24] Hailing from the 1920s, the "Pleistocene Park" concept marks the same decade that an *L.A. Times* editorial noted that all horses in Los Angeles "belonged in the tar pits" because this would be the automobile city. It was thought that climate itself dictated that cars should dominate Southern California, since horses were still recognized as more capable of negotiating the ice and snow of eastern winters.[25] In 1919, the first gas station in Los Angeles, run by the Gilmore family, opened near the current site of the George C. Page Museum. Now the Gilmores are known for "The Gilmore Collection," photographs of early fossil excavations on the La Brea ranch, which their descendants donated to the museum. At the intersection of modern media (automobiles, photographs) and the Pleistocene (via the Miocene), the La Brea petrol preserve manages to ally itself to both media time and geological time. Both are arguably forms of deep time, insofar as both involve a temporality irrelevant to the work of human memory. William L. Fox has written a fine essay about what the La Brea Tar Pits and a disaster movie featuring them, *The Volcano* (1997), tell us of the human sense of time, the way in which time has been "speeded up" by film narratives to the extent that longer-duration problems, such as oil scarcity and the increasingly devastating methods of oil mining, are not only imperceptible but indifferent to us.[26]

I admire Fox's essay, and I would add to its treatment of time the suggestion that geological and media time both offer styles of temporal imagining that enable the making of petroleum as a spectacle and, therefore, as a commodity. The Museum of La Brea Discoveries allows us to conceive of fossil fuel as fossil and fuel, and predominantly as fossil. We learn at the museum that what might have been an oil field became a paleontological site. Fossils and fuel are distinct forms of commodity whose value relies, in both cases, upon the occlusion of specific historical relationships— social and ecological histories—that confer value. The fossil is the unique object separated by museum display from other cultural meanings, while fuel offers an abstraction of matter, a dream of disembodied energy.[27] The dramatization of "fossil, fuel" at Rancho La Brea shows us how commodity values rely upon the segregation of the natural from history. In contrast, a natural history concept that places nature and history in dialogue has the potential to generate ecological action, as the methods of systematic biology now intervene in global habitat and species loss.

ECOLOGICAL DESIGNS

The making of the La Brea Tar Pits from a loosely defined exhibit into a museum coincides with the introduction of ecology into the design of the natural history museum in the 1960s. In New York City, the American Museum of Natural History's centennial exhibit "Can Man Survive?" opened in April 1969, marking a sea change in both the themes and presentation of natural history. The exhibit explored many of the ecological issues haunting North Americans at the time, including overpopulation and industrial pollution. "Can Man Survive?" employed cutting-edge mixed media, Alison Griffiths notes, offering an "immersive informational and experiential architecture" that turned visitors from mere viewers into active and uncomfortable participants.[28] Visitors walked through various stages of the show to be confronted by looping films of river pollution, bacteria, and other blight, without recourse other than to move deeper into the museum space. This premier exhibit on world ecological damage called for a total virtual environment in order to deliver its message, which, as the exhibit's title indicates, was that humanity was destroying itself.

Such doomsday ecological thinking, together with an amped-up commitment to making natural history into "hypermediated" space, introduces the moment when Pit 91 was reopened for excavation in 1969 at Rancho La Brea. It had been fifty-four years since the last major excavations at the site, and the events of 1969 signaled a civic commitment to reimaging the tar pits as a place of scientific inquiry. The reopening was coordinated by paleontologist George Miller, who shepherded a small army of volunteer excavators in what was billed by journalists as a quest for the origins of humankind in the Western hemisphere. Miller's team found evidence of human presence in California at approximately fifteen thousand years ago. Humanity showed up as trace, through decorative cuts made by tools on the leg bones of saber-toothed cats. Perhaps in celebration of this discovery, in 1973 Governor Ronald Reagan would name the saber-toothed cat as California's state fossil. Volunteers came together from all over Los Angeles to search for the origins of human culture in their own city. The "dig-in," as it was called, expressed civic pride and the flamboyant humanism of the late 1960s, mixed up with the hermeneutics of suspicion that also characterized the Vietnam War era.

In a letter to the *L.A. Times* of May 1970, a self-described "concerned citizen," Max Poschin, counts himself amongst the more than one hundred volunteers "privileged to have the opportunity of helping uncover paleontological treasures from the Ice Age" at Rancho La Brea. But Mr. Poschin's note of gratitude sounds a shrill concern: "What has happened to our priorities? . . . We have billions to prosecute an illegal war in Asia or to send men on perilous missions to the moon, but not even a tiny fraction of that amount to extend the knowledge of man here on earth."[29] Mr. Poschin explicitly condemns the lack of funding for public science, and he suggests more broadly that his participation in the search for human origins at the "dig-in" has not led him to imagine a long human future. Similar predictions that greater scientific knowledge of humanity would only reveal our near extinction appeared in U.S. popular culture in this period, blatantly in Paul Ehrlich's *The Population Bomb* (1968) and glancingly in popular fictions such as Franklin J. Schaffner's film *The Planet of the Apes* (1968). *The Planet of the Apes* series offers a wry commentary on the natural history endeavor. Like the AMNH exhibit "Can Man Survive?," the first *Planet of the Apes* film places human social history at the center of the problem of natural history, turning the tables on our curatorial ambitions by depicting the study and museum display of human subjects at a sophisticated research center overseen by orangutans. The fear that inspires the AMNH exhibit and the contemporary film, as well as Mr. Poschin's letter, is that humanity has written itself out of ecological relations, following the technologies of war into an inarticulate end-time. If we are lucky, evolved orangutans might someday tell the story of how we overstepped ourselves with weapons such as the hydrogen bomb. This moment of intense anxiety about human endurance on the planet generated some innovative attempts to remake the practice of natural history in such a way as to decenter Western cultures and human subjects.

More than 750,000 species, including many microfossils, were uncovered at Rancho La Brea on Dr. Miller's watch, testimony to his dedication to avoiding what he called the "stamp-collecting" of large mammal specimens that characterized the earlier twentieth-century excavations.[30] Following on the heels of the 1890s "dinomania," which was characterized by the investment of philanthropists like Andrew Carnegie in massive sauropod dinosaur skeletons for display in urban museums, early tar pits excavators focused on large Pleistocene specimens, namely, Columbian mammoths, ground sloths, American bison, saber-toothed cats, and American camels.[31] In contrast, Dr. Miller foregrounded the value of the asphaltum matrix that encased these trophy specimens, as it was the matrix that was the key to Pleistocene climate. The matrix offered plant microfossils. This was, again, the age of ecology, and Miller's volunteers were made aware that even tiny specimens might be relevant to the human story. Sadly, the Pit 91 project was facing bankruptcy by 1970, a problem later blamed on Dr. Miller's ambitious attempt to reconstruct a complete Pleistocene environment. When Miller was fired by the Natural History Museum of Los Angeles in 1972, the museum's director, Dr. Mead, stressed the importance of generating "more scientific information at less cost," an objective vaguely related to his corollary plan of making the tar pits into a "museum-like display."[32] The dual commitment of public science to education and research evident in this conflict is not unique to the Natural History Museum of Los Angeles. However, the question of how to make the tar pits into a "museum-like display" reflects the regional culture of Los Angeles as a place where spectacle, in Guy Debord's sense of screen images that obscure the origins and relations of things, has figured prominently.[33] The fine-tuned, relational thinking that Dr. Miller introduced to his field workers in the 1960s had little precedent in the history of designs intended for the La Brea site.

Let us consider, briefly, the various iterations of "museum" that were imagined at Rancho La Brea since the early twentieth century, in order to return to the current museum via ecological thought. For most of its history, the La Brea site was unmoored from ecological concerns, hitched instead to the theatrical problem of presenting the tar pits as a "wonder of the world." This was an effortful project, given that the pits essentially smelled and looked like mines to Angelenos accustomed to oil leaks and derricks in their neighborhoods. The famous "Lake Pit" where the museum's fiber-glass Imperial Mammoth now cries to her distraught baby became so deep and large because it had been a regular asphaltum mine, used for caulking, waterproofing, and fuel from at least the eighteenth century. The open pit mine was filled with water in the early twentieth century in order to offer a natural history feature to visitors of Hancock Park. This early act of outdoor scenography already implies a contest of meanings for what counts as natural history. The real ecological relations of locals to the asphaltum sumps, the use of these resources, could not be shown as "natural" in the sense that Pleistocene mammals could. The fiberglass animals now present in the park are products of a similar effort to make nature available to the senses not through its local uses—or even through the motile presence of the asphaltum itself—but through the representation of ancient mammals. The animal statuary suggests the conventional pleasures of natural history, referring to the more showy vertebrate fossils found in the pits. In contrast to the staining tar, the "animals" also create user-friendly interactive surfaces for parkgoers to touch, sit on, lunch beside, and photograph.

One of the most ambitious designs for the La Brea museum and park, and the design in which animal statuary figures most dramatically, is a WPA project referred to as the "Garden of the Giants." Howard Kegley wrote about it for the *L.A. Times* in 1940. "This stupendous undertaking, as a circus press agent might express it, will be super-colossal. In real fact it should out-Barnum old Barnum himself, for it will bring into permanent existence here, within a stone's throw of the world-famous Miracle Mile . . . a monumental memorial in masonry to a gosh-awful galaxy of prehistoric animals and birds."[34] These "permanent animals," crafted "in enduring rock," would ground an immersive, virtual landscape, through which "visitors, traversing the jungle-flanked paths, would unexpectedly come upon sculptured groups."[35] Fifteen years before the opening of Disneyland, the tar pits were conceived as a natural history theme park. Although WPA project planners stressed the "educational value such a monumental menagerie might possess," the project goal was sensational, making the effect of time and space travel that has been associated with museums into pure entertainment. Coming from the shops on Wilshire Boulevard into this spectacular park, visitors would experience a near-absolute departure from Los Angeles proper. They would wander into an "extinct" space-time. The petrol pits themselves have almost no meaning—social, sensory, or scientific—in this preserve of giant mammals. They are merely low-lying indices of an exotic *other* life, discontinuous with the Miracle Mile and modern humans.

The WPA ultimately did not approve the midcentury Garden of the Giants sculptural project because of the high cost of building over the active geology of Hancock Park. Gases arising from the pits were recognized as a perpetual fire hazard. Such perpetual fire suggests enduring life in a manner quite different from sculptural animals. The effort to express geological agency through sculptures of extinct mammals betrays a lack of imagination about where petroleum might fit within the natural history concept. It also suggests a human wish to make animals, and by extension ourselves, permanent—a task that mid-twentieth-century Americans liked to imagine might be feasible, given the right technology. Fiberglass, the material used to craft the latest set of Pleistocene mammals in Hancock Park, represents a chemical bid for endurance that, like many developed during World War II, utilizes petroleum-based plastics. The wish to be "forever animal" gets monumentalized, in oil.

For ordinary Angelenos, the La Brea Tar Pits have been close collaborators in human histories in a manner that park and museum designers tend to overlook. Vernacular use of the pits betrays long-standing appreciation for asphaltum as both a practical and an expressive material. While everyday use of the sumps for chores such as caulking faded with the dedication of Hancock Park as a scientific preserve in the early twentieth century, locals continued to interact with the tar pits in ways that exceed the directed goals of education or staged entertainment. The relationship between the people and their "tar" was intimate and sociable, if not ecological per se. For locals, the pits were matter that could convey messages, media in the most unmediated sense. Los Angeles newspapers brim with cases of individuals using the tar pits to make statements or to preserve strong feelings, presumably forever.

One celebrated case involved Mary Alice Bernard, daughter of the county livestock inspector and a "society" girl in the small-town Los Angeles of 1935. Newspapers report that Mary Alice suffered from "amnesia" and from "a nervous breakdown."[36] For months during her nineteenth year, she went missing. The Los Angeles County Sheriff's department searched the tar pits for Mary Alice's body, using a special apparatus to heat the contents of the pools so that heavy asphalt could be pumped out, and employing grappling hooks to drag the depths. The county was literally mining for Mary Alice. The cost of this was considerable, the news coverage sensational. The drama had begun when Mary Alice's mother remembered the young woman's parting hint: "Someday you will find me in the bottom of the La Brea pits." In fact, Mary Alice turned up several months later in Los Angeles, having gone to San Francisco to take a job as a domestic. The tar pits search closed, with newspaper photographs of sheriff's deputies bending over draglines. This young woman's flight from her family and her class did not grant her the permanence, as a local heroine, that she might have achieved had she really jumped into the pits. Countless others seized upon the tar pits as means of indelible self-expression. Three years after Mary Alice's vanishing, a man billed as the "Tar Pits Suicide Hoaxer" was jailed for making a false report of his own suicide in the tar pits in an embittered note to his ex-girlfriend.[37]

The tar pits present a handy means of preserving bad affects. They have served as local channels that broadcast sensory cues, such as smell, viscosity, and visual obscurity, that intimate human vulnerability. The story of the ten thousand-year-old "La Brea Woman" found in the pits mimics the noir-ish dramas of the twentieth-century tar pits insofar as paleontologists continue to speculate upon whether this woman was murdered due to her insanity—her skeleton suggests a sinus infection serious enough to affect brain function. Regardless of the validity of these speculations, the tar pits generate a good deal of thinking about what aspects of the mammalian body might be preserved, from its bones to its somatized feelings. In popular use, they suggest a rather dark ecology. They have served as sites for speculation about what might be done with human mortality and how the fragility of our own bodies might be used to make larger stories about humanity's relationship to itself, to geological time, and to other life.

When the Page Museum was built in the 1970s, its design expressed a brighter ecological mission. George C. Page, a former fruit-packing entrepreneur who recognized the La Brea Tar Pits as evidence of Los Angeles's singular heritage, commissioned an architectural design for the museum that preserved its historical status as an outdoor park. Rising like an indigenous burial mound from exhibit rooms built largely below ground level, the museum's external walls are covered with grass, making for sloping hills that visitors can sit and play on. Page responded to the "green" parks initiative that was popular in the Los Angeles of the 1970s. Too, his designers recognized that the asphaltum itself should be treated as a main event, signaling not only the original storehouse of the museum's fossils but also L.A.'s natural wealth as an oil town. The Page has been intended as an indoor-outdoor museum that combines some of the social ideals of the ecomuseum with the spectacular pleasures of earlier museum concepts, such as the erstwhile Garden of the Giants. The museum's mission speaks well to the twin ideals of ecology and entertainment that marked the moment of its completion. In the twilight of the fossil-fuel regime and the shadow of climate change, it simply needs to reanimate its vision.

Toward the Post-Oil Museum

A tour of the Page today makes clear that "life in asphalt" is no longer about ancient mammals, not really. The slick, black drums stacked up around Project 23 (a recent dig opened as a result of the neighboring LACMA's proposed expansion into fossiliferous ground) look like miniature oil drums, which essentially they are. "Is that the matrix?" one of my fellow visitors asked, a rising note of expectation in his voice. The blocks of shiny drums, and perhaps the popular film of the same name, grant this once-obscure product of the pits an almost glamorous persona (Figure 16.2).

FIGURE 16.2. The Matrix—life in asphalt. Photograph by the author.

Matrix was defended as a key to ecological knowledge of the L.A. basin by Dr. George Miller back in the 1960s, but forty years after Earth Day it has taken on new significance as a source of living microorganisms with potential "applications" for biotech, particularly biofuels, enhanced oil recovery, and bioremediation of oil spills. The U.S. Department of Agriculture and the EPA provided funding for a recent study of the living bacteria in La Brea matrix, conducted by environmental scientists at the University of California, Riverside. "We were surprised to find these bacteria because asphalt is an extreme and hostile environment for life to survive," reports Jong-Shik Kim, a postdoctoral researcher. "It's clear, however, that these living organisms can survive in heavy oil mixtures containing many highly toxic chemicals."[38] Entirely new branches on the so-called tree of life have been opened by the more than two hundred species of microorganism that Kim and his fellow scientists discovered. While the specific applications of this new life drive the research, the project produces a broad rhetoric of hope that life itself will persist through the most toxic scenarios of "hydrocarbon contamination and radioactive environments."[39] La Brea's living bacteria, progeny of soil microorganisms trapped in the petrol sumps tens of thousands of years ago, are *prepared*—for a worst-case anthropogenic climate, for the end of conventional oil, even for the bomb. The conceptual "application" for La Brea's new petroleum species is, simply, the future.

For museumgoers who witness this microscale bid for a future natural history—and it is possible to miss it, to walk right past Project 23, for instance—there will be questions, at least implicit ones, about how such a downsized future might accommodate the natural history museum. Microscale natural history unfolds in the privatized space of research laboratories and requires more explicit context than even the old-fashioned cabinet-and-text–based natural history museums provide. Then again, Steven Conn suggests that natural history museums survived a similar crisis of representation as early as the 1860s, when Darwinian theory shifted biological research "from science that could be conducted with the naked eye toward science that needed a microscope."[40] The Page Museum gestures toward an exhibit of the potentially human-life-saving bacteria in the tar pits, illustrating how the bacteria were discovered and what the discovery might mean for the long future. This display looks a bit like a drawing table and sits awkwardly against a wall, in the shadow of the fishbowl laboratory, the museum's public exhibit of paleontologists scraping and cleaning the recognizable bones (helpfully labeled "Zed's pelvis," etc.) of La Brea mammals. "What Does Bacteria Look Like?" asks a cartoonish laminated booklet. Potato-shaped figures of "bacteria" meant to enliven the laminated pages give us an easy answer: bacteria doesn't look like much, please imagine. While imagination and daydreaming can be more valuable modes of interactivity in museums than playing with high-tech displays, this exhibit fails to evoke dreams. BioArtists and BioArchitects have proven that microscale life can look like something beautiful, often through computer simulation and digital media.[41] It would be fabulous to see microorganismic life in oil, perhaps on screen, that rivals the beauty of the Pleistocene fossil. Perhaps the scarce funding that besets most natural history museums today, plus a traditional commitment to the sensual vertebrate skeleton, keeps a remarkably exciting aspect of the Page Museum—entirely new "branches of life" growing up in local petroleum—confined to marginal display.

The life forms that petroleum generates tell us something about how to live through oil scarcity and into an energy shift, about how to clean up our toxic spills, about living through the destruction that has complemented human creativity. The Page Museum could be a premier site of thinking through the limitations of petroleum culture with the North American public and the science community. Rancho La Brea has long been a "post-oil" museum insofar as it has considered its asphaltum as matter in excess of the transformative energy idea. Since its preservation as a public park, the La Brea site has challenged the meanings of oil and demonstrated how not to get stuck in any single logic of its value—consider, again, that the oil business that might have developed on the museum site was given over to "science" back in 1915, under the assumption that it was more profitable, here, to mine fossils than fuel. The museum might reconsider its framing narrative. Through the microorganismic life in the tar pits, the Page could rethink the definition of natural history or even life itself along the lines that the philosopher Bruno Latour has considered in his discussions of the agency of things, both "real" and fabricated, in the ongoing story of ecology.[42] Petroleum has bumped up against various forms of life at this site for millennia, generating death and treasure and a unique means of storing life's forms and the birth of new living species. The museum's most recent, Telly Award–winning film, "Ice Age Fossils of the La Brea Tar Pits" (2004), begins to consider the multiple vectors of ecological action and power by making a vial of petroleum its star.

Immediately outside the "Behind the Scenes" theater where the film "Ice Age Fossils" is shown, a weatherworn animatronic sloth creakily ducks up and down, attempting to free itself from an equally down-at-the-heels animatronic saber-toothed cat. These are not the new audio-animatronics of Disneyland, where a 2003 "Living Character Initiative" inspired engineers to create figures so lifelike that their speech appears spontaneous and they move autonomously through unregulated space.[43] At the Page, visitors press a red button to hear the ground sloth's simulated screams. Last summer, the fiberglass Imperial Mammoth sinking into the Lake Pit slipped off of her cement base, floating to the far end of the lake, where she appeared to be summoning her "husband" and "child" to a frolicsome swim. The Page Museum seems on its way to becoming a museum of obsolescent infrastructure, following its Pleistocene mammals into the tar. The asphaltum, for its part, entertains the saviors of a new geologic epoch, the Anthropocene, in which the Earth that humans damaged through our profligate use of fossil fuels just might be saved by microorganisms grown up in petroleum sludge. When petroleum becomes more important as habitat than as energy to simulate habitat, or preservative for our bones, then a new natural history concept is required. This one cannot be scaled to children or even humans, perhaps, but it can offer us the opportunity to reimagine life without ourselves at the center, which is a conceptual starting point for any sustainable design.

Notes

1. David Wilson, "Nikolai Federov, Konstantin Tsiolkovsky, and the Roots of the Russian Space Program," lecture at the University of California, Santa Barbara, April 13, 2011. For discussion of Wilson's ironic relationship to the natural history endeavor, see Lawrence Weschler, *Mr. Wilson's Cabinet of Wonder* (New York: Vintage, 1995).

2. Bennett is particularly concerned with how the relatively new time-scheme of the nation-state could be annexed to the universal natural history idea, in Tony Bennett, "The Exhibitionary Complex," *New Formations* 4 (Spring 1988): 88–89.

3. Robert E. Kohler, *All Creatures: Naturalists, Collectors, and Biodiversity, 1850–1950* (Princeton: Princeton University Press, 2006), 2, 3, 9; Steven Conn, *Museums and American Intellectual Life, 1876–1926* (Chicago: University of Chicago Press, 1998), 72.

4. Conn, *Museums and American Intellectual Life*, 72.

5. Some curators resent the emphasis upon information technology as the key to the continuing relevance of their collections, arguing, for instance, that "there is a tendency to divorce data from objects," "museums are the cornerstone of an object-oriented approach to the natural world," and "the useful life of . . . specimens" outlasts the analyses made of them. See, for example, Leonard Krishtalka and Philip S. Humphrey, "Can Natural History Museums Capture the Future?" *BioScience* 50 (July 2000): 612.

6. Alison Griffiths extensively historicizes interactive exhibits and the concept of "interactivity," suggesting alternatives to the explicitly digital or even technological application of the concept, in Griffiths, *Shivers Down Your Spine: Cinema, Museums, and the Immersive View* (New York: Columbia University Press, 2008). The journal *Curator* features frequent articles on interactivity and a 2004 special issue on the topic that includes useful reflections from museum practitioners, for example, John H. Falk, Carol Scott, Lynn Dierking, Leonie Rennie, and Mika Cohen Jones, "Interactives and Visitor Learning," *Curator* 47 (April 2004): 171–98.

7. Conn, *Museums and American Intellectual Life*, 40. For a key critical discussion of the popular exhibit versus the disciplinary role of the museum, see Bennett, "The Exhibitionary Complex," 86.

8. Michel Foucault, *The Order of Things: An Archeology of the Human Sciences* (London: Tavistock, 1970), 132; see also Tony Bennett, "Pedagogic Objects, Clean Eyes, and Popular Instruction: On Sensory Regimes and Museum Didactics," *Configurations* 6, no. 3 (1998): 345–71.

9. For a thoughtful general treatment of ecomuseums, see Peter Davis, *Ecomuseums: A Sense of Place* (London: Leicester University Press, 1999).

10. Andreas Huyssen, *Twilight Memories: Marking Time in a Culture of Amnesia* (London: Routledge, 1995), 33.

11. An overview of the rhetorical transformation of bitumen into "oil"—and brief reference to the Oil Sands Discovery Centre—can be found in Andrew Nikiforuk, *Tar Sands: Dirty Oil and the Future of a Continent* (Vancouver: Greystone, 2010). My general comments refer to a personal visit that I made to this museum, for research purposes, in August 2011.

12. Tony Bennett, "Civic Laboratories: Museums, Cultural Objecthood, and the Governance of the Social," *Cultural Studies* 19 (September 2005): 525.

13. Jonathan Spaulding, "*L.A.: Light / Motion / Dreams*: Developing an Exhibition on the Natural and Cultural History of Los Angeles," *Environmental History* 10 (April 2005): 295–313.

14. Griffiths, *Shivers Down Your Spine*, 168.

15. Andrea Witcomb offers a strong set of alternative definitions of "interactivity" and rebuttal to the screen-mania of multimedia exhibit promoters in *Re-Imagining the Museum: Beyond the Mausoleum* (London: Routledge, 2003). See especially chapter 6.

16. "Gail" (docent), "George C. Page Museum of La Brea Discoveries Outdoor Tour," April 17, 2011.

17. Linda Williams, "Film Bodies: Gender, Genre, and Excess," *Film Quarterly* 44 (Summer 1991): 2–13.

18. "Imperial Mammoth Struggling in Asphalt Pool," postcard, George C. Page Museum of La Brea Discoveries, n.p., n.d.

19. Hannah Arendt, *The Human Condition* (Chicago: University of Chicago Press, 1958), 201.

20. "Gail," "Discoveries Outdoor Tour."

21. Thomas H. Benton [William Pannapacker, pseud.], "Preserving the Future of the Natural History Museum," *Chronicle of Higher Education*, October 30, 2009; see also Thomas H. Benton, "Getting Real at Natural-History Museums," *Chronicle of Higher Education*, July 1, 2010; and "Getting Real at the Natural-History Museum, Part 2," *Chronicle of Higher Education*, August 1, 2010.

22. Witcomb, *Re-Imagining the Museum*, 102.

23. See Walter Benjamin, "The Work of Art in the Age of Its Technological Reproducibility: Second Version," in *The Work of Art in the Age of Its Technological Reproducibility and Other Writings on Media*, ed. Michael W. Jennings, Brigid Doherty, and Thomas Y. Levin; trans. Edmund Jephcott and Harry Zohn (Cambridge: Harvard University Press, 2008), 19–55.

24. Dana W. Bartlett, "Progress Made in Developing Tomb of Giants: Creation of New Park at La Brea Pits Gives City Unique Monument," *Los Angeles Times*, March 27, 1927. ProQuest Historical Newspapers.

25. Eva G. Taylor, "By the Way," *Los Angeles Times*, February 1, 1928. ProQuest Historical Newspapers.

26. William L. Fox, *Making Time: Essays on the Nature of Los Angeles* (Emeryville, Calif.: Shoemaker and Hoard, 2007), 34–37.

27. Although I see Adorno's discussion of the "museal" as ripe for critique, exhibited fossils speak rather literally to his sense that museum objects are dead, by virtue of separation from other cultural relations. See Theodor Adorno, "Valery Proust Museum in Memory of Hermann von Grab," *Prisms* (London: Garden City Press, 1967), 175–85.

28. Griffiths, *Shivers Down Your Spine*, 252.

29. Max Poschin, "What Has Happened to Our Priorities?," *Los Angeles Times*, May 2, 1970. ProQuest Historical Newspapers.

30. George Miller, quoted in John Kendall, "Fossil Dig to Be Phased Out: Firing Clouds La Brea Pit Study," *Los Angeles Times*, January 16, 1972. ProQuest Historical Newspapers.

31. See Paul D. Brinkman, *The Second Jurassic Dinosaur Rush: Museums and Paleontology in America at the Turn of the Twentieth Century* (Chicago: University of Chicago Press, 2010).

32. Quoted in Kendall, "Fossil Dig."

33. I refer to Guy Debord's now-classic *The Society of the Spectacle*, trans. Donald Nicholson-Smith (New York: Zone Books, 1995).

34. Howard Kegley, "Something Bigger than Barnum," *Los Angeles Times*, March 10, 1940. ProQuest Historical Newspapers.

35. Kegley, "Something Bigger than Barnum."

36. "Body hunted in Tar Pits: Missing Girl's Hint Recalled," *Los Angeles Times*, October 26, 1935. ProQuest Historical Newspapers.

37. "Tar Pits Suicide Hoaxer Jailed as Extortion Suspect," *Los Angeles Times*, January 23, 1939. ProQuest Historical Newspapers.

38. "Life in Asphalt: New Petroleum-Degrading Bacteria Found at Rancho La Brea Tar Pits in Los Angeles," *Astrobiology Magazine*, May 19, 2007. http://www.astrobio.net/Prelease/2336/life-in-asphalt.

39. Ibid.

40. Conn, *Museums and American Intellectual Life,* 44–45.

41. For example, see some of the BioArchitectural model cities in Lucas Cappelli and Vicente Guallart, eds., *Self-Sufficient City: Envisioning the Habitat of the Future* (Barcelona: Institute for Advanced Architecture of Catalonia, 2010).

42. Latour's discussion of the entangled agency of quasi-objects, nonhumans, and "matters of concern" runs throughout his work. One might begin with the explicitly political argument of a more recent book, for example, Bruno Latour, *The Politics of Nature: How to Bring the Sciences into Democracy,* trans. Catherine Porter (Cambridge: Harvard University Press, 2004). Here is Latour discussing the "tangled beings" that are the concern of political ecology: "Political ecology does not shift attention from the human pole to the pole of nature; it shifts from certainty about the production of risk-free objects (with their clear separation between things and people) to uncertainty about the relations whose unintended consequences threaten to disrupt all orderings, all plans, all impacts" (25).

43. Heather Wax, "Engineering the Magic: Disney Imagineers Give Robots Personality, Charm—and Even a Smile," *IEEE Women in Engineering Magazine* 2 (Summer 2008): 22.

PART V

The Future of and without Oil

17

Retrofutures and Petrofutures

Oil, Scarcity, Limit

GERRY CANAVAN

Fredric Jameson has written that "in our time . . . the world system . . . is a being of such enormous complexity that it can only be mapped indirectly, by way of a simpler object that stands as its allegorical interpretant."[1] In this chapter, I offer up oil, and oil capitalism, as one such interpretant for the historical world-system as a whole—and further offer up science fiction as a means to register the different meanings of "oil" that are available in different historical moments.

Oil's ubiquity and centrality within contemporary consumer capitalism suggests it as an especially useful allegorical interpretant. Oil is extremely local—as local as your corner gas station, as your car's gas tank—but at the same time it is the token of a vast spatiotemporal network of seemingly autonomous actors. In his story "The Petrol Pump," Italo Calvino evokes this immense, even sublime, interactivity: "As I fill my tank at the self-service station a bubble of gas swells up in a black lake buried beneath the Persian Gulf, an emir silently raises hands hidden in wide white sleeves, and folds them on his chest, in a skyscraper an Exxon computer is crunching numbers, far out to sea a cargo fleet gets the order to change course."[2] To Calvino's totalizing vision of oil's global interconnectedness we might add not only the million-year geological time scale necessary for oil's creation but also the immeasurably complex flows of money, power, and technology that make the current global economy, and U.S. hegemony within that order, possible. Indeed, both Timothy Mitchell and Dipesh Chakrabarty have recently noted that what we understand as "liberal democracy" is in many ways fundamentally indistinguishable from oil-based prosperity: "The mansion of modern freedoms," Chakrabarty puts it, "stands on an ever-expanding base of fossil-fuel use."[3] Oil has, in short, made twentieth-century technological modernity possible, in all senses, and the long-term continuation of that modernity appears utterly unthinkable in its absence. And yet, like it or not, we can now see quite clearly the end of this period of oil-fueled expansion. Oil got us here, but oil can't get us out. Capitalism's continuous need for ever-larger reserves of cheap energy—and the anticipation of the imminent end of our ability to fulfill that need through oil—is therefore a useful way to map the current situation and ultimate fate of contemporary technoculture more generally.

In *The Ecological Thought,* Timothy Morton suggests the name "hyperobject" for multitudinous entities like oil, plutonium, and the climate that exist on "almost unthinkable" spatial and temporal scales, and which seem in this way to exceed any kind of stable, bounded definition.[4] Steven Shaviro, in turn, has noted that science fiction's status as a consciousness-expanding "psycho-socio-technological cartography" makes it the perfect "focusing device" to attempt cognitive mappings of such hyperobjects.[5] This cognitive usefulness can be seen quite clearly, I think, in science-fictional treatments of oil and oil capitalism. Oil-as-hyperobject delimits our ability to both understand our historical past and imagine our possible futures, becoming the secret subtext of any number of futurological imaginings—and as the world-historical scarcity of oil has grown more and more obvious, the glittering techno-utopias of Golden Age science fiction become increasingly replaced by their psychic opposites: apocalyptic, post–peak oil horrors of deprivation and ruin.

Writing in *n+1,* Chad Harbach identifies the historical and ideological forces underpinning these kinds of science-fictional fantasies:

America and the fossil-fuel economy grew up together; our triumphant history is the triumphant history of these fuels. We entrusted to them (slowly at first, and with increasing enthusiasm) the work of growing our food, moving our bodies, and building our homes, tools, and furniture—they freed us for thought and entertainment, and created our ideas of freedom. These ideas of freedom, in turn, have created our existential framework, within which one fear dwarfs all others: the fear of economic slowdown (less growth), backed by deeper fears of stagnation (no growth) and, unthinkably, contraction (antigrowth). America does have a deeply ingrained, morally coercive politics based in a fear that must never be realized, and this is it. To fail to grow—to fail to grow ever faster—has become synonymous with utter collapse, both of our economy and our ideals.[6]

In a recent essay in *Harper's,* Wendell Berry makes much the same point, describing U.S. energy policy as a "Faustian economics" predicated on a "fantasy of limitlessness" that, when put under threat, produces intense claustrophobia and dread.[7]

Alongside all this we must of course note the *other* horn of our intractable futurological dilemma: the unintended and almost exclusively negative consequences of our oil-based economic system for the environment, particularly with respect to carbon release and climate change. Here our collective inability to think of a future beyond oil capitalism becomes all the more potentially catastrophic. I confess I have become perversely fond of the way Benjamin Kunkel frames the Janus-faced, intertwined crises of climate and energy we now face: "The nightmare, in good nightmare fashion, has something absurd and nearly inescapable about it: either we will begin running out of oil, or we won't."[8] That is to say: either we have peak oil and the entire world suffers a tumultuous transition to post–cheap-oil economics, or else there's plenty of oil left for us to permanently destroy the global climate through excess carbon emissions.

This chapter focuses on two orthogonal representations of oil in twentieth-century science-fiction narratives. In the first move, I discuss how the necessity of oil is put under erasure in much early and mid-twentieth-century science fiction (whereby oil is retrospectively imagined as a quickly discarded transitional technology). In these texts, I argue, we can see oil's inescapable centrality to twentieth-century liberal capitalism proven precisely through the fantasy of its painless transcendence. In the second move, I trace the subsequent breakdown of this fantasy, considering various recent manifestations of peak oil from the ecological science fiction of the 1970s through recent allegorical films like *Avatar, Moon,* and the little-seen *Daybreakers.* I argue that the one-time symbolic repression of oil has now been replaced in more recent science fiction with a doomed sense of its overriding, totalizing importance— which in turn generates for us a crushing sense of impending futurological limit, of resources and even time itself running out. Science fiction's happy promise of a post-oil, post-scarcity future, this is to say, has since the oil shocks of the 1970s been largely replaced instead with the creeping terror that technological modernity, and its consumer lifestyle, may in fact have no future at all.

Oil Ontology and the March of Progress

My approach here follows the lead of (among others) Imre Szeman, whose 2007 article "System Failure: Oil, Futurity, and the Anticipation of Disaster" in *South Atlantic Quarterly* suggests the possibility of a new understanding of the history of capitalism organized around what he calls *oil ontology.* "What," Szeman provocatively asks, "if we were to think about the history of capital not exclusively in geopolitical terms, but in terms of the forms of energy available to it at any given historical moment?"[9] Such a historiography of capital transitions us from steam capitalism (c. 1765) to oil capitalism (c. 1859), in which the massive reserves of solar energy stored in fossil fuels begin to be converted into mechanical power at staggeringly efficient rates of EROEI (energy returned on energy invested, a metric of the efficiency of a given energy source). For oil the numbers are truly staggering: they are initially 100:1, as late as the 1930s, meaning that for every calorie of energy expended in retrieving and extracting oil, a hundred calories were generated.

In the era of oil ontology, oil becomes synonymous with progress, even with the future itself. The discovery of oil in one's region means the quick infusion of cash and jobs; the discovery of oil on one's own property translates immediately and inevitably to indescribable riches. Oil, writes Ryszard Kapuściński, "creates the illusion of a completely transformed life, life without work, life for free, it expresses the eternal human dream of wealth achieved through a lucky accident . . . in this sense it is a fairy tale, and like all fairy tales a bit of a lie."[10] Oil is so valuable, of course, because it is a tremendous physical marvel; the cheap, essentially free energy stored in petroleum allows for a tremendous amplification of human powers. The invention of the internal combustion engine and its application in the "gas-guzzling" automobile creates the conditions for technological modernity and its fantastic potential to transform all aspects of life in the United States—a mode of production and consumption John Bellamy Foster has memorably named "the automobile-industrial complex,"[11] including not just the systems of production and delivery that make contemporary consumer capitalism possible but also the plastics commonly making up the very consumer goods themselves. Oil is the primary fuel for all the transformative technological wonders of the twentieth century; it is oil, of course, that powers the "racing car" that F. T. Marinetti famously described in his "Futurist Manifesto" as "more beautiful than the Winged Victory of Samothrace."[12] In her "Petroleum Song," Futurist poet Maria Goretti likewise celebrates oil as "the ardent blood of [our] conquests," describing petroleum drilling as a violent seduction of the Earth bordering on a rape: "the drill bores / sinks inseminates / bites crushes tears / the violated earth screams / but now now / the living blood / gushes / rises / rushes / toward the sky / elongated spurt / and sings." The explosion of the well is a sexual climax that is only the first of the future's joys: "all my body relaxes / quivers with pleasure / rejoices with splendor / but tomorrow tomorrow even more mine! . . . Tomorrow you will hear my new song / and it will be the song of azure airplanes!"[13]

In the ideology of ecstatic technological progress that oil ontology generates, it is only natural to assume that in due time oil itself will eventually be superseded by a new form of energy—something even more excessive and miraculous, allowing for even greater marvels and wonders. This assumption has the paradoxical result of relegating fossil fuels—so central to the workings of modern capitalism, so utterly transformative of every aspect of life in the twentieth century, and so generative of the fantastic new technologies science fiction lauds—to a short historical footnote in the *longue durée* of human progress: oil as mere transitional energy source. The "instructions" laid down for human civilization by Buckminster Fuller in his *Operating Manual for Spaceship Earth* (1969)—which borrows the common trope of interstellar travel from popular science fiction for its organizational logic—exemplifies this way of thinking. Energy that has been "deposited as a fossil-fuel savings account" serves the function of priming even greater energy production for human civilization:

> The fossil fuel deposits of our Spaceship Earth correspond to our automobile's storage battery which must be conserved to turn over our main engine's self-starter. Thereafter, our "main engine," the life regenerating processes, must operate exclusively on our vast daily energy income from the powers of wind, tide, water, and the direct Sun radiation energy. The fossil-fuel savings account has been put aboard Spaceship Earth for the exclusive function of getting the new machinery built with which to support life and humanity at ever more effective standards of vital physical energy and reinspiring metaphysical sustenance to be sustained exclusively on our Sun radiation's and Moon pull gravity's tidal, wind, and rainfall generated pulsating and therefore harnessable energies. The daily income energies are excessively adequate for the operation of our main industrial engines and their automated productions. The energy expended in one minute of a tropical hurricane equals the combined energy of all the U.S.A. and U.S.S.R. nuclear weapons. Only by understanding this scheme may we continue for all time ahead to enjoy and explore universe as we progressively harness evermore of the celestially generated tidal and storm generated wind, water, and electrical power concentrations. We cannot afford to expend our fossil fuels faster than we are "recharging our battery," which means precisely the rate at which the fossil fuels are being continually deposited within Earth's spherical crust.

All this is of course assumed to be perfectly achievable—"provided that we are not so foolish as to continue to exhaust in a split second of astronomical history the orderly energy savings of billions of years' energy conservation." "We cannot afford to expend our fossil fuels faster than we are 'recharging our battery,' which means precisely the rate at which the fossil fuels are being continually deposited within Earth's spherical crust."[14] Oil is a stopgap, a one-time hack. By definition it can be (one way or the other) only the briefest era in species history.

Here a teleological vision of history emerges that we should recognize as in many ways still the key futurological hope of our time: that "our" rapid usage of the world's entire reserve of fossil-fuel energy *today* is justified because we need it to develop the self-sustaining tidal/wind/solar/fission/fusion technology that will eventually replace it *tomorrow*—replacement technology that, we have been perpetually assured for decades, is "just a few years away." In his work, John Bellamy Foster is careful to note the class and geographical differences within that impulse toward a flattened, smoothed-out "our": "In 2008, Americans in the highest income quintile spent three to four times as much on both housing and clothing, and five times as much on transportation, as those in the poorest quintile. In Canada where consumption data is available in deciles, ecological footprint analysts have found that the top income decile has a transportation footprint nine times that of the bottom decile, and a consumer goods footprint four times that of the bottom decile."[15] Likewise, Naomi Klein's work on climate justice and activism has pointed out the "cruel geographical irony" that the countries most vulnerable to climate change having emitted the least amount of carbon into the atmosphere. She suggests 20 percent of the population of the planet has emitted 75 percent of total historical greenhouse gas emission, with the United States (5 percent global population) having emitted approximately 25 percent just on its own.[16] From this perspective, Fuller's technological optimism looks perhaps less like objective analysis and more like a rationalization for an incomprehensibly large theft on a geologic scale.

In the index for his exhaustive catalogue of the early publication history of science fiction, *Science Fiction: The Early Years,* Everett F. Bleiler captures the way this ideological assumption of energy progress is registered in the science fiction of the turn of the century. Noting that he has chosen not to record conventional energy and electricity systems at all—that is, fossil fuels—Bleiler goes on to list nearly three dozen futuristic energy sources, ranging from atomic energy and solar energy to "ether flow," Vril, and "Zodiacal force."[17] In the follow-up, *Science Fiction: The Gernsback Years,* which covers the decades following the initial publication of *Amazing Stories* in 1926, the list of "new and unusual sources" of energy adds cosmic energy, feline static electricity, the newly discovered so-lunar ray, repulsion vortices, and space mirrors to its list of possible replacements for oil and coal.[18] In his work, Szeman groups these kinds of fantasies—so influential that they remain the default futurological assumption of most political discourse surrounding energy policy, despite the fact that their grandiose promises of postscarcity existence have thus far entirely failed to deliver results—under the general category of "techno-utopianism." As he puts it:

> The utopia I have in mind here is the "bad utopia" of future dreamscapes and fanciful political confections—"utopia" not quite just as an insulting slur against one's enemies, but rather as a projection of an alternative future that is, in fact, anything but a "conception of systematic otherness." . . . All of our worst fears about the chaos that will ensue when oil runs out are resolved through scientific innovations that are in perfect

synchrony with the operations of the capitalist economy: problem solved, without the need for radical ruptures or alterations in political and social life.[19]

Oil's scarcity is thus recast as a mere technicality: we will simply invent something that can do everything oil can do without the drawback of being scarce! The perpetual-motion-machine logic of this fantasy recalls for me Kim Stanley Robinson's critique of the rhetoric of sustainability: "It doesn't even mean sustainable anymore. It means: let us continue to do what we're doing, but somehow get away with it. By some magic waving of the hands, or some techno silver bullet, suddenly we can make it all right to continue in all our current habits."[20]

The science-fictional fantasy of oil replacement goes further still, of course, by bracketing the negative externalities of oil as soon-to-be obsolete, and thus unimportant in the larger scheme of human history. All that is terrible about oil and other fossil fuels—including the brutal labor conditions required to extract them from the ground, the imperial violence needed to secure and stabilize their flows, the vast pollution they cause, *and* their inevitable, eventual exhaustion—is thus reimagined as but a temporary unpleasant blip in the long march of progress. H. G. Wells's 1914 *The World Set Free*—said to have inspired physicist Leó Szilárd in his discovery of real-world nuclear reaction in the mid-1930s—lays out this narrative of progress explicitly. Human history, reads the first sentence of the novel, "is the history of the attainment of external power," with oil imagined as an early turning point in that history.[21] But with the discovery of atomic power, oil and other fossil fuels are abandoned almost overnight:

It was in 1953 that the first Holsten-Roberts engine brought induced radio-activity into the sphere of industrial production, and its first general use was to replace the steam-engine in electrical generating stations. . . . The American Kemp engine, differing widely in principle but equally practicable, and the Krupp-Erlanger came hard upon the heels of this, and by the autumn of 1954 a gigantic replacement of industrial methods and machinery was in progress all about the habitable globe. Small wonder was this when the cost, even of these earliest and clumsiest of atomic engines, is compared with that of the power they superseded. Allowing for lubrication the Dass-Tata engine, once it was started cost a penny to run thirty-seven miles, and added only nine and quarter pounds to the weight of the carriage it drove. It made the heavy alcohol-driven automobile of the time ridiculous in appearance as well as preposterously costly. For many years the price of coal and every form of liquid fuel had been clambering to levels that made even the revival of the draft horse seem a practicable possibility, and now with the abrupt relaxation of this stringency, the change in appearance of the traffic upon the world's

roads was instantaneous. In three years the frightful armoured monsters that had hooted
and smoked and thundered about the world for four awful decades were swept away to
the dealers in old metal, and the highways thronged with light and clean and shimmer-
ing shapes of silvered steel.[22]

Indeed, nuclear energy in the novel is so obscenely cost effective that its sudden introduction (and the subsequent scrapping of the now-redundant sectors of the economy
linked to the earlier energy mode) causes a severe economic depression.[23]

The future-history novels of Isaac Asimov are similarly exemplary of this phenomenon. On the rare occasions that Asimov's characters mention oil and coal at all, it is
with the horror that a contemporary citizen of democracy might speak of cannibal-
ism or slavery. In Asimov's *Foundation* novels—first serialized in *Astounding Science
Fiction* beginning in 1942, and depicting the collapse of a Galactic Empire into a
millennia-long dark age—one of the earliest markers of this slide into barbarism is
the loss of the nuclear-powered economy and the return to fossil fuels: "After the disintegration of the First Empire, there came the fragmentation of organized science,
back, back—*past even* the fundamentals of nuclear power into the chemical power of
coal and oil" (emphasis mine).[24] Asimov's other novels echo this projective investment in a post-oil future. In *The End of Eternity,* a time-travel novel, one of the time
travelers is forced to make do with a "clumsy diesel engine" after being permanently
trapped in what he sees as the distant, barbaric past, our time;[25] at the end of *I, Robot,*
one of the markers of Atomic Age Utopia in Africa is precisely their possession of
"untouched" oil and coalfields, in contrast to every other major power in the world.[26]
Even in Asimov's sole attempt at writing for the comics medium, the unfinished
Star Empire, the notes for the artist describe the "panorama of a futuristic city. Tall,
graceful uncrowded buildings with lots of greenery between symmetrical developments" include a line indicating "there was no coal or oil smoke, no gas fumes or
ground traffic."[27] Asimov's sidelong references to oil and his repeated insistence on
oil's *absence* as a marker of human progress are unique in the period only insofar as he
bothers to mention the existence of oil at all—most science fiction of this period
simply takes for granted that the wonders of the future will be powered by clean
atomic energy "too cheap to meter," in the famously unfulfilled slogan of the nuclear
industry.

SCARCITY AND COLLAPSE

Of course not *all* midcentury science-fiction texts take such an uncomplicatedly rosy
view of the future—several key texts take early notice of the concerns over access to
oil that would come to dominate world politics in the late twentieth century. In Olaf

Stapledon's *Last and First Men* (1930), for instance—a hyperbolic view-from-ten-thousand-feet of the billion-year history of the human race and its subsequent evolution into the second, third, and fourth through eighteenth races of men—the basic scarcity of oil and coal becomes a long-term problem for the human race over millennia, driving both war and the formation of a world state to strictly regulate the use of oil only when absolutely necessary. Even this level of intervention is insufficient to prevent the final collapse of the civilization of the first men: the Second Men and all subsequent evolutions of *Homo sapiens* have to make do without any oil or coal at all.[28] In Wells's *The World Set Free,* discussed above, this is the fate facing mankind before the invention of atomic energy:

> It is part of my business to understand economics, and from that point of view the century before Holsten was just a hundred years' crescendo of waste. Only the extreme individualism of that period, only its utter want of any collective understanding or purpose can explain that waste. Mankind used up material—insanely. They had got through three-quarters of all the coal in the planet, they had used up most of the oil, they had swept away their forests, and they were running short of tin and copper.... The system was already staggering when Holsten began his researches. So far as the world in general went there was no sense of danger and no desire for inquiry. They had no belief that science could save them, nor any idea that there was a need to be saved. They could not, they would not, see the gulf beneath their feet.[29]

In *The Man Who Awoke* series of stories by Laurence Manning, beginning in 1933, later anthologized by Asimov for his collection *Before the Golden Age,* the same diagnosis is made, but the fantasized "solution" is on a far longer timetable. Here Norman Winters, a twentieth-century Rip Van Winkle pastiche, sleeps for millennia to visit all future eras of human development; when he awakes, the protagonist is confronted by a future that is still furious at the loss of these tremendously useful natural resources. The Chief Forester of a civilization three thousand years in the future denounces the ruinous cultural practices of our "age of waste":

> The height of the false civilization of Waste! Fossil plants were ruthlessly burned in furnaces to provide heat; petroleum was consumed by the billion barrels; cheap metal cars were built and thrown away to rust after a few years' use; men crowded into ill-ventilated villages of a million inhabitants—some historians say several million ... for what should we thank the humans of three thousand years ago? For exhausting the coal supplies of the world? For leaving us no petroleum for our chemical factories? For destroying the forests on whole mountain ranges and letting the soil erode into the valleys?[30]

Winters can only offer his apologies: "I understand you have a very poor opinion of my own times, due to our possibly unwise consumption of natural resources. Even then we had men who warned us against our course of action, but we acted in the belief that when oil and coal were gone mankind would produce some new fuel to take their place."[31]

Asimov later credited Manning for his own recognition of an impending oil crisis and the need for alternative energy to prevent it: "In the 1970s, everyone is aware of, and achingly involved in, the energy crisis. Manning was aware of it forty years ago, and because he was, I was, and so, I'm sure, were many thoughtful young science fiction readers."[32] Here science fiction's ability to diagnose, popularize, and educate becomes a point in its favor over and above more "mainstream" literary practices. In the same interview in which he denounces fashionable rhetorics of "sustainable development," Kim Stanley Robinson similarly links the sort of thinking under way in *The Man Who Awoke* to the ethical-political role science fiction might play in cultural discourse:

> It's almost as if a science fiction writer's job is to represent the unborn humanity that will inherit this place—you're speaking *from* the future and *for* the future. And you try to speak for them by envisioning scenarios that show them either doing things better or doing things worse—but you're also alerting the generations alive right now that these people have a voice in history.
>
> The future needs to be taken into account by the current system, which regularly steals from it in order to pad our ridiculous current lifestyle.[33]

Unhappily, however, such recognitions of the futurological consequences of oil's profound scarcity are comparatively rare in early and mid-twentieth-century science fiction; the index for Bleiler's *Science Fiction: The Early Years*, has no entries for either oil or fossil fuels, and *The Gernsback Years* notes only a handful.[34]

The crucial turning point for a more complete cognitive mapping of oil and oil capitalism in science fiction might well be Frank Herbert's *Dune*, published in 1965. *Dune* famously transmogrifies oil imperialism into a battle for control of the "spice" that makes interstellar navigation possible; spice, the necessary fuel for this futuristic multiglobal economy can only be found on the desert planet Arrakis, whose indigenous Fremen resist planetary occupation and, in the closing chapter of the first novel, having taken control of their planet's spice reserves under the control of a charismatic religious leader, threaten to unleash an intergalactic "jihad."[35] *Dune* not only anticipates later developments in Middle Eastern politics, but captures the nightmare of history Amitav Ghosh describes as resulting from the "oil encounter":

To the principal protagonists in the Oil Encounter (which means, in effect, America and Americans on the one hand and the peoples of the Arabian Peninsula and the Persian Gulf on the other), the history of oil is a matter of embarrassment verging on the unspeakable, the pornographic.[36]

We go to war in the Middle East, both Bush I and Bush II explained, in order to "protect our way of life," which is predicated on the low-cost availability of petroleum fuel (oil) and petroleum products (plastics); the spice, they may as well have said, must flow.

Dune registers not only the geopolitical fragility of the oil supply but also its general scarcity: spice is needed for galactic commerce but can *only* be found on Arrakis. In this sense *Dune* writes the history of the American 1970s before it starts. M. King Hubbert became nationally famous in 1970 when his 1956 prediction of peak U.S. oil production proved accurate. Hubbert defined a bell-shaped curve for annual oil extraction and production, which peaks when you have extracted roughly half the oil from your reserves.[37] On the upward part of the curve, oil appears plentiful—recall that at the start of the oil age in the United States oil was literally seeping up out of the ground, and could be scooped up in buckets—but as extraction continues oil becomes more difficult to find and retrieve. The peak marks the moment of maximum extraction; after this point, there is an inevitable decline. The EROEI of oil has trended sharply downward ever since the inauguration of the oil economy: from 100:1 to 30:1 in the 1970s to approximately 11:1 on average in the 2000s.[38] The historical narrative of oil capitalism is in this respect necessarily declensionist: as each year brings oil that is both harder to find and harder to extract than what had been available before, each year brings us closer to the end of capitalism as we have come to know it. Every American president dating back to Richard Nixon has used an "addiction" metaphor to describe the relationship of the U.S. economy to oil, especially foreign oil[39]—and this trend line too suggests the diminishing returns of the "addicts' high" (and perhaps that the referent of the addiction metaphor has slowly slipped from alcohol to heroin as the situation has progressed without material progress toward a solution).

The Club of Rome published its paradigm-shifting book, *The Limits to Growth*, in 1972. In 1973, the OPEC embargo on oil sales to nations that had supported Israel in the Yom Kippur War (also known as the Fourth Arab-Israeli War) led to national rationing and gas lines, as well as contributing to the worst domestic recession since the Great Depression (and the worst until the current crisis).[40] The year 1973 in general marks a striking moment of transition for postwar U.S. hegemony—it is the year of Watergate, the withdrawal from Vietnam, the collapse of the Bretton Woods

monetary system, and the passage of the Endangered Species Act, among other notable benchmarks of *limit*—but first and foremost, 1973 is the year of the oil shock, the
year the reality of capitalism's dependence on a finite, nonrenewable energy was made
inescapably clear and painfully immediate. Crucially for the study of science fiction,
1973 is also the first year since 1961 without a planned manned mission to the moon,
that "first step" toward the "High Frontier" of inevitable extraplanetary expansion
to which humanity has still never returned. Instead, our future has turned out to be
terrestrial, and quite literally running out of gas. "Nature guards her treasures jeal-
ously," proclaims the ad for Knox Oil and Gas that begins the 1983 film *Local Hero,*
directed by Bill Forsyth: "Just a decade ago these fields were beyond reach: we didn't
have the technology. Today a Knox engineer will tell you that he might need a little
time, but he'll get the oil. He knows that a little time is all we have left."[41]

Myriad science fictions of the 1970s, taking a far more pessimistic and political
tack than the earlier science fiction of Wells, Manning, and Asimov, attempt to register this crisis and imagine possible solutions to it, but frequently stumble both on the
scientific details and on the pessimistic intuition that there simply *is* no viable solution. In Asimov's own *The Gods Themselves* (1972), the energy crisis is solved by the
invention of a miraculous solar "pump" that would be the perfect green energy
source—if only it weren't stealing its free energy from the universe next door. In
Ursula K. Le Guin's "ambiguous utopia" *The Dispossessed,* the rich planet Urras is able
to survive its parallel energy crisis only by instituting onerous controls on all aspects
of consumer capitalism, especially oil. Earth in the novel made no such moves, and
a cataclysmic nuclear war was the result. In Ernest Callenbach's *Ecotopia,* which has
lent its name to an entire subgenre of science-fictional "happy endings," part of the
"utopia" involves the abolition of the automobile and a switch to renewable forms of
energy, including in the end a switch to a fantastic form of fully green energy derived
(somehow, miraculously) directly from chlorophyll.

Despite all sense of urgency, then, these fantasies ultimately endorse Szeman's
sense that "oil capital seems to represent a stage that neither capitalism nor its opponents can think beyond."[42] This postoptimistic sense of the future is registered in the
most undeniable terms in Sid Meier's *Civilization* franchise of computer games, from
the 1990s, which allow the player to run a millennial-spanning empire from the foundation of its capital city in 4000 BC to the launching of the first space colonists to
Alpha Centauri near the year 2000. But even in *Civilization,* it turns out, there is nothing after a petroleum economy. Once you hit our moment, the technological tree
simply stops progressing altogether; the only technologies you can research are called
"Future Tech 1," followed by "Future Tech 2," and so on. The progress that had once
seemed inevitable can now no longer be named at all.

In 1981, William Gibson likewise imagined that longed-for next stage of history—the sparkling but unrealized *Star Trek* cornucopia we once called "the future"—as a ghost haunting our dingier, dustier present:

> They were the children of Dialta Downes's '80-that-wasn't; they were Heirs to the Dream. They were white, blond, and they probably had blue eyes. They were American. Dialta had said that the Future had come to America first, but had finally passed it by. But not here, in the heart of the Dream. Here, we'd gone on and on, in a dream logic that knew nothing of pollution, the finite bounds of fossil fuel, or foreign wars it was possible to lose. They were smug, happy, and utterly content with themselves and their world.[42]

Our future, the implication is, will be anything but smug and happy. The first oil panic of 1973 is now forty years in the past, but we seem no further from the *Mad Max*–style ruins of a world without oil; indeed, we seem totally uninterested in any serious investment in energy despite both the inevitability of peak oil and the growing threat of climate change. Even knowing the oil is running out, even knowing the oil we have yet to burn will poison the climate, the only thing we can think to do is drill harder—and we need think only of the familiar "drill baby drill" refrain of the Tea Party movement to see the fantastic political appeal of this sort of denialism in action. "Nobody gets beyond a petroleum economy," one of the characters in Dan Simmon's extraplanetary *Hyperion* series remarks sadly. "Not while there's petroleum there."[43] The Larry Niven fantasy series "The Magic Goes Away" (1976) cleverly merges the melancholic anticipation of peak-oil collapse with the mournful nostalgia for the past that is characteristic of the fantasy genre: in that world it is not oil, but the manna that powers magic, that is running out. But of course this is our world exactly—the magic of the twentieth century is slipping through our fingers, the present already lost to history even as we live it.

In the absence of some sufficient substitute for oil's energy miracle—in the absence, that is, of a future that is both prosperous and *possible*—the only solution for the imagination seems to be to cast itself back into the past in search of the secret of what's to come. Isaac Asimov himself—having perhaps finally abandoned the teleological historical progressivism of his earlier work—takes up this direction in a 1977 *Time* essay titled "The Nightmare Life without Fuel," which finds the world of 1997 being slowly dismantled and demolished, automobiles vanished, and all energy substitutes stalled: "Anyone older than ten can remember automobiles. They dwindled." Meanwhile the country is filled with starving wretches felled by snowstorm and disease: "Where will it end? It must end in a return to the days before 1800, to the days before the fossil fuels powered a vast machine industry and technology. It must end in subsistence farming and in a world population reduced by starvation, disease and violence to less than a billion."[44] James Howard Kunstler (a leading Jeremiah of the coming post-oil "long emergency") explores this kind of "retrofuture" in his 2008 novel *World Made by Hand,* which sees only one conceivable alternative to the current global oil economy: a return to the hyperlocal artisan economy of the early nineteenth-century United States.

In Paolo Bacigalupi's 2009 science-fiction novel *The Wind-Up Girl,* set in the post-carbon, post–climate-change twenty-third century, we find another imagined endpoint of this downward trajectory. The end of oil is recognized in retrospect as the end to both globalization and U.S./Western hegemony, sparking a century-long period of breakdown and disaster known as "The Contraction." The novel's protagonist, Anderson Lake, a representative of food conglomerate AgriGen, is described early in the text as "Flotsam of the old Expansion. An ancient piece of driftwood left at high tide, from a time when petroleum was cheap and men and women crossed the globe in hours instead of weeks."[45] To the people of this time, the oil age is remembered as a distant "golden age"—but one that is permanently and hopelessly in the past, never to return.

Robert Charles Wilson's *Julian Comstock* (2009), too, similarly envisions the post-petroleum world as the return of obsolete historical social forms. The contemporary era has come to be remembered by the people of 2172 as the "Efflorescence of Oil," the word "efflorescence" describing the evaporating of water that leaves behind a thin layer of salty detritus.[46] Here that detritus is the ruined remains of our own twentieth- and twenty-first-century lives: the hardship and dislocation of global collapse, the inscrutable plastic junk that litters their countryside, their myths that man once walked on the moon, a generally ruined world. American life has become much more technologically constrained, along the lines of the pre-oil world; New York is considered the greatest city in the world in part because it still manages electrical illumination for four hours every day. The unholy combination of the end of oil with global warming has decimated the world's population through starvation, deprivation, and disease; the society that has ultimately emerged out of the disaster has abandoned science, reason, and democracy in favor of superstition, theocracy, and authoritarianism—science fiction recasting its future not as progress but as hopeless regression. Here, then, is what science fiction looks like without (or after) the future: the twentieth century is envisioned not as the launching pad for a glorious technofuture but as an anomalous moment of prosperity and historical possibility that quickly burns itself out, leaving in its place the worst combination of Manifest Destiny America, feudal Europe, and decadent Rome.

ROCK BOTTOM AND RECOVERY

If the end of oil is indeed, as Imre Szeman writes, "the biggest disaster that 'we' collectively face," it is only because its time scale seems to be the most immediate—the long century boil of climate change seems leisurely in comparison to a global petroleum peak that could come within a decade, or the next few years, or perhaps has already arrived. But this urgency carries with it an opportunity; the very immediacy of the crisis means that oil is simultaneously the Archimedean point best at hand from which we might intervene and start to move the world. The end of oil, as we have seen, fuels at once both Utopia and dystopia: it is the crisis that breaks the world into ruin but also the opportunity out of which the possibility of another world might emerge. Here again we can find science-fictional imaginings retreating from the cold, rational calculus of energy scarcity into something more like fantasy—only in our moment the fantasy is not of improbable technology and perpetual-motion hacking of the laws of physics, but rather that human beings might change the way they behave.

Moon, a film from 2009, announces in its opening minutes that capitalism has finally solved all the problems at last: a diegetic, in-universe commercial from Lunar Energy announces that helium mining on the moon allows enough essentially free energy to fuel the planet without drilling for oil, despoiling environments, and ruining the lives of smiling Third World children.[47] But it turns out that there is still a cost to all this. Literalizing the ethical rupture at the heart of Ursula K. Le Guin's short story "The Ones Who Walk Away from Omelas," we have now a single human being who is forced to live in misery to provide all this free energy to the world: Sam Bell, who has to live on the moon to oversee the mining operation, perhaps the worst job in Utopia. And it gets worse for poor Sam; while he believes he is only on the moon for a lucrative three-year contract, in fact he will never leave; he is clone #5 in a series of Sams, each of whom has been implanted with the memories of the original Sam and each of whom has approximately a three-year life span until he must be replaced. The horror of the movie is the shock of discovering the true costs of a world of plenty—and the ultimate utopian fantasy of the movie is that this situation *would* actually be rejected by the public at large, if only the truth were known. In the end the Sams are able to work together to get one of them off the moon and back to Earth to expose what is going on, and in the film's final shot it is revealed that Lunar Energy's stocks have tanked.

At the core of James Cameron's allegorical *Avatar* (2009), whatever else we might have to say about the film's lavish visual spectacle and its troubling politics of race, gender, disability, and indigeneity, there is a parallel fantasy that the smooth functioning of global (now *interplanetary*) capitalism and the familiar violence of its history might somehow be *interrupted.* What if, on Pandora, the normal circuit between oil

capitalism and imperial power could be broken? The desire for this radical transformation is so strong that it leads even the film's domestic audiences to root against a
"Space Marine" stand-in for the U.S. military as it invades the planet Pandora looking
to seize control of its valuable resources for the benefit of a desperate, dying Earth—
with our hero turning traitor, leading a successful guerrilla resistance, and ultimately
forcing the imperialists off the planet. At the end of the film his reward is to be permanently transferred into the body of the big-O Other—to, in essence, not have to be an
eco-imperialist any longer. Now living in another body, part of another global history,
he is free to lead another, better life.[48]

A similar miracle takes place at the end of little-seen box-office-flop *Daybreakers*
(also from 2009), which literalizes the metaphor famously employed by Karl Marx:
"Capital is dead labour, that, vampire-like, only lives by sucking living labour, and
lives the more, the more labour it sucks."[49] A decade after an outbreak that has turned
the national elite into vampires, in *Daybreakers'* 2019 there are no longer enough
unaltered humans left to feed the 1%'s unquenchable thirst for blood. Vampires who
go without blood for too long become, in a horror-movie figuration of drug addiction, monstrous "subsiders" who attack anything that moves; as the film opens, the
subsider crisis is just reaching the suburbs. Coffee shops advertise that they "still sell
20% blood"; "blood riots" rock the Third World. But the signs of the coming crash are
everywhere; with all hope for an energy substitute stalled, America has reached "peak
blood." The solution here is again personal transformation, but now with a twist. It
turns out that through controlled exposure to the sun—literal *enlightenment*—vampires
can be cured. And the "cured" vampires cannot be revampirized; in fact their blood
itself now contains the cure, turning any vampire who drinks from them into a cured
human as well. What is being imagined is a kind of viral social revolution, operating
through an epidemiological network—friend to friend, relative to relative, coworker
to coworker—with the power to slowly transform a society of vampire-consumers
back into human beings once again.[50]

The utopian, even quasi-religious, fantasy in all three narratives is the dream of
salvation: that the nightmare of exploitation, and our own complicity in these practices, might somehow be stopped, *despite* our inability to change. All three films
uncover this sense of hope in personal transformation, in epiphanies located in the
individual but potentially accessible to all. Since U.S. consumerism is so often framed
as an addiction, the ecological state of grace imagined by these films may well be
thought of as something like A.A.'s "Higher Power." As Kierkegaard put it in an epi-
gram sometimes invoked by Darko Suvin, "We literally do not want to be what we
are."[51] The task before us, then, would seem to be to transform that dream-wish into
waking act, to find ways to nourish and sustain the drive to change even in a world of

ordinary, nonmiraculous causation—which means, à la *Daybreakers,* somehow pushing those personal transformations past the level of the individual into collective experience and society-wide change. Naomi Klein has similarly captured the revolutionary political sensibility that is being called into existence by allegorical texts like *Avatar, Moon,* and *Daybreakers*:

> We all know, or at least sense, that the world is upside down: we act as if there is no end to what is actually finite—fossil fuels and the atmospheric space to absorb their emissions. And we act as if there are strict and immovable limits to what is actually bountiful—the financial resources to build the kind of society we need. The task of our time is to turn this around: to challenge this false scarcity. To insist that we can afford to build a decent, inclusive society—while at the same time, respect the real limits to what the earth can take.[52]

That, one supposes, or else begin practicing the apology to the future offered by Kurt Vonnegut in a late work: "Dear future generations, please accept our apologies. We were roaring drunk on petroleum."[53]

Notes

1. Fredric Jameson, *The Geopolitical Aesthetic* (Bloomington: Indiana University Press, 1995), 169.

2. Italo Calvino, "The Petrol Pump," *Numbers in the Dark and Other Stories,* trans. Tim Parks (London: Jonathan Cape, 1995), 174.

3. Dipesh Chakrabarty, "The Climate of History: Four Theses," *Critical Inquiry* 35 (Winter 2009): 208. See also Timothy Mitchell, *Carbon Democracy: Political Power in the Age of Oil* (New York: Verso, 2011).

4. Timothy Morton, *The Ecological Thought* (Cambridge: Harvard University Press, 2010), 19. The concept is further developed on pages 130–35 and developed further still in his book *Hyperobjects* (University of Minnesota Press, 2013).

5. Steven Shaviro, "Hyperbolic Futures: Speculative Finance and Speculative Fiction," *Cascadia Subduction Zone* 1, no. 2 (April 2011): 4.

6. Chad Harbach, "The Politics of Fear, Part III: Business as Usual," *n+1,* December 4, 2007, http://nplusonemag.com/politics-fear-part-iii-business-usual.

7. Wendell Berry, "Faustian Economics: Hell Hath No Limits," *Harper's,* May 2008, 36.

8. Benjamin Kunkel, "The Politics of Fear, Part II: How Many of Us?" *n+1,* March 18, 2008, http://nplusonemag.com/politics-fear-part-ii-how-many-us.

9. Imre Szeman, "System Failure: Oil, Futurity, and the Anticipation of Disaster," *South Atlantic Quarterly* 106, no. 4 (Fall 2007): 806.

10. Quoted in Retort (Iain Boal, T. J. Clark, Joseph Matthews, and Michael Watts), *Afflicted Powers: Capital and Spectacle in a New Age of War* (New York: Verso, 2005), 40.

11. John Bellamy Foster, *Ecology against Capitalism* (New York: Monthly Review Press, 2002).

12. F. T. Marinetti, "Futurist Manifesto," in *Critical Writings*, trans. Doug Thompson (New York: Macmillan, 2006).

13. Maria Goretti, "Petroleum Song," *Futurism: An Anthology*, ed. Lawrence Rainey, Christine Poggi, and Laura Whitman (New Haven: Yale University Press, 2009): 476–79.

14. Buckminster Fuller, *Operating Manual for Spaceship Earth* (Zurich: Lars Müller Publishers, 2008), 128–29.

15. Foster, *Ecology Against Capitalism,* 122.

16. Naomi Klein, "Climate Rage," NaomiKlein.org, November 11, 2009. http://www.naomi klein.org/articles/2009/11/climate-rage. This essay was originally published in *Rolling Stone*.

17. Everett F. Bleiler, *Science Fiction: The Early Years* (Kent, Ohio: Kent State University Press, 1990), 875–76.

18. Everett F. Bleiler with Richard Bleiler, *Science Fiction: The Gernsback Years* (Kent, Ohio: Kent State University Press, 1998), 638.

19. Szeman, "System Failure," 813.

20. Geoff Manaugh, "Comparative Planetology: An Interview with Kim Stanley Robinson," BLDGBLOG, December 19, 2007, http://bldgblog.blogspot.com/2007/12/comparative -planetology-interview-with.html.

21. H. G. Wells, *The World Set Free* (London: Macmillan and Co., 1914), 1.

22. Ibid., 41–42.

23. Ibid., 44–45. The rest of the novel details the creation of a utopian state on the back of the atomic energy revolution, beginning (as is characteristic of Wells in this period) with a devastating atomic war as the energy source is weaponized, and the subsequent creation of a World State to regulate the atom and demilitarize the planet.

24. Isaac Asimov, *Second Foundation* (New York: Spectra, 1991), 114. This event is mentioned in each of the first three *Foundation* novels, making clear the importance of transcending oil to Asimov's vision of the future.

25. Isaac Asimov, *The End of Eternity* (New York: Tor, 2010), 162.

26. Isaac Asimov, *I, Robot* (New York: Bantam Books, 1991), 211.

27. Isaac Asimov, notes for *Star Empire* (unpublished), Asimov Collection, Howard Gotlieb Archival Research Center, Boston University, box 7d.

28. Olaf Stapledon, *Last and First Men and Star Maker* (New York: Dover Publications, 1968).

29. Wells, *The World Set Free,* 263–64.

30. Laurence Manning, *The Man Who Awoke* (New York: Ballantine, 1975), 20–21.

31. Ibid., 25.

32. Isaac Asimov, ed., *Before the Golden Age* (Garden City: Doubleday, 1974), 344. Thanks to Michael Page for bringing this quote to my attention in his own excellent work on oil-age science fiction.

33. Manaugh, "Comparative Planetology."

34. Bleiler, *Gernsback Years,* 669.

35. Frank Herbert, *Dune: 40th Anniversary Edition* (New York: Ace Trade, 2005), 482. See also the well-known film adaptation: *Dune,* dir. David Lynch (1984; Orlando, Fla.: Universal Studios, 1998), DVD.

36. Amitav Ghosh, "Petrofiction—*The Trench* by Abdelrahman Munif," *New Republic*, March 2, 1992, 29.

37. M. King Hubbert, "Nuclear Energy and the Fossil Fuels," presentation before the American Petroleum Institute (March 7–9, 1956). Notes from this talk are available at http://www .hubbertpeak.com/hubbert/1956/1956.pdf.

38. See, for instance, Charles Hall, "Why EROEI Matters (Part 1 of 6)," TheOilDrum.com, April 1, 2008. http://www.theoildrum.com/node/3786.

39. A July 16, 2010, segment on *The Daily Show* offers an amusing supercut of each of these presidents making nearly identical statements about oil dependence. See "An Energy-Independent Future" at thedailyshow.com.

40. This moment of "oil shock" is of course repeated in 1979, after the Iranian Revolution, and echoed again both in the 2000s oil price spikes and in the gas lines found in New York and New Jersey after 2012's Hurricane Sandy.

41. *Local Hero,* directed by Bill Forsyth (1983; Burbank, Calif.: Warner Home Video, 1999), DVD.

42. Imre Szeman, "System Failure: Oil, Futurity, and the Anticipation of Disaster," *South Atlantic Quarterly* 106, no. 4: 806–7.

43. William Gibson, "The Gernsback Continuum," *Burning Chrome* (New York: Harper Voyager, 2003), 24–37 (34).

44. Dan Simmons, *Hyperion* (New York: Spectra Books, 1990), 449.

45. Issac Asimov, "The Nightmare Life without Fuel." *Time,* April 25, 1977. Available at time.com. We are now, alas, almost as distant from Asimov's "Nightmare" 1997 as he was when he made this call for urgent change:

And what can we do to prevent all this now?
Now? Almost nothing.
If we had started 20 years ago, that might have been another matter. If we had only started 50 years ago, it would have been easy.

46. Paolo Bacigalupi, *The Wind-Up Girl* (San Francisco: Night Shade Books, 2009), 16.

47. Robert Charles Wilson, *Julian Comstock: A Novel of 22nd-Century America* (Tor Books, 2009), 30.

48. *Moon,* directed by Duncan Jones (2009; Los Angeles: Sony Pictures Classics, 2010), DVD.

49. *Avatar,* directed by James Cameron (2009; Los Angeles: Twentieth Century Fox, 2010), DVD.

50. Karl Marx, *Capital, Vol. 1* (New York: Penguin Books, 1976), 342.

51. *Daybreakers,* directed by Michael Spierig and Peter Spierig (2009; Santa Monica, Calif.: Lionsgate, 2010), DVD.

52. Quoted in Darko Suvin, *Defined by a Hollow: Essays on Utopia, Science Fiction, and Political Epistemology* (London: Peter Lang, 2010), 218.

53. Klein, "Climate Rage."

54. Quoted in Jacqueline Blais, "Vonnegut 'still had hope in his heart,'" USAToday.com, April 13, 2007.

18

Crude Aesthetics

The Politics of Oil Documentaries

IMRE SZEMAN

How does the problem of oil appear in documentary film? In what follows, I examine the manner in which oil is represented in three "feature" documentaries released over the past five years: Basil Gelpke and Ray McCormack's *A Crude Awakening* (2006), Joe Berlinger's *Crude: The Real Price of Oil* (2009), and Shannon Walsh's *H2Oil* (2009).[1] As might be expected, while each has oil at its core, these documentaries differ substantially both in subject matter and form. Berlinger's *Crude* deals with a protracted legal case against the activities of Chevron in Ecuador; Walsh's film examines the ecological and social impact of the Alberta oil sands, specifically its effects on the communities that rely on the water used in conjunction with bitumen processing; and Gelpke and McCormack offer an overview of the politics and economics of oil, together with the environmental damage it causes and the potential crisis of the end of oil. By examining them together, I want to consider the range of ways in which these documentaries frame oil as a problem for their audiences, and what resources they offer as possible solutions to this (historically unprecedented) social and ecological problem. These documentaries both reflect and are a source of the social narratives through which we describe oil to ourselves; it is revealing to see both the limits and possibilities of the narratives they proffer, which are pieced together out of the fragments of concepts and discourses dating back to the Enlightenment concerning nature, the social, and human collectivity.

As is the case with documentaries on a wide range of social issues, these films about oil understand themselves as important forms of political pedagogy that not only shape audience understanding of the issues in question but also hope to generate political and ecological responses that otherwise would not occur. This production of an outcome or change in societal imperatives is a long-standing desire of the kind of politically and ethically committed documentary filmmaking that for publics has to a large degree become identified with the function of documentary as such—even if there may be relatively scant evidence of the hoped-for translation of audience awareness of film themes into political action outside the theater.[2] While it nonetheless remains productive to critically assess the political efficacy of such documentaries— whether by considering the formal or stylistic approaches each makes to its subject matter,[3] examining their capacity to effectively expose "the gap between self-professed norms and behavior,"[4] or by probing the generic politics of such "commodity biographies"[5]—my aim here will be to consider what these documentaries tell us about the social life of oil today. In what follows, I will treat these films as providing examples of narrative and aesthetic choices through which the problem of oil is framed—or *can* be framed—not only within the films but within the social more generally. The site of politics I will focus on is not the success or failure of any given documentary to constrain or mobilize a political response, but rather what the discursive, narrative, and aesthetic strategies employed suggest about the dominant ways in which the problem of oil is named and solutions to it proposed. Fredric Jameson famously describes cultural texts or artifacts as "symbolic acts" in which "real social contradictions, insurmountable in their own terms, find a purely formal resolution in the aesthetic realm."[6] It is in this sense that I will offer readings of these three documentaries as aesthetic acts that, in their own specific manner, have "the function of inventing imaginary or formal 'solutions' to unresolvable social contradiction"[7]—unresolvable in perhaps a stronger and more determinate way than the social contradictions to which Jameson referred.

My essay proceeds in three parts. First, by offering readings of these documentaries, I draw out the ways in which each narrates the social life of oil. In her recent discussion of human rights films, Meg McLagan argues that these films are developed around the axiom that to expose hidden forces and problems to the light of film is to generate the capacity in publics to address the situations the films uncover.[8] One of the reasons for focusing on these three films in particular is that while they, too, might have this axiom at their core, they proceed with the awareness that the importance of oil to social life is already well known, that publics have yet to adequately respond to its demands and looming crises, and indeed, that they may be entirely unable to respond even if they adequately understand the issues. As my analysis of these three films will show, the "solutions" these films offer to the social contradictions generated by oil are made difficult by the fact that the place of this resource in our lives seems to defer the politics one hopes to generate from the production of a documentary about it—and not just the politics directly connected to documentary practice, but to broader ideas that persist about the relationship between belief and action in the operations of social life more generally. In the second part, I draw out some key discursive and conceptual claims made within these documentaries about the unprecedented social problem of oil. Finally, I conclude with an exploration of exactly what kind of "unresolvable social contradiction" oil might be. In "Two Faces of the Apocalypse," Michael Hardt productively explores the antinomies that define and separate the anticapitalist and environmental movements.[9] The insights offered by these films suggest that the problem of oil has the potential to destabilize the aims of both movements. As surprising as it may sound, it is the socially taken-for-granted physical substance of oil—and, of course, the practices that it supports and enables—that has to be placed conceptually and discursively at the heart of both movements if either is to realize its ambitions.

OIL ON FILM: *A CRUDE AWAKENING*, *CRUDE*, AND *H2OIL*

There are an increasing number of documentary films that address the role, function, and impact of oil in the world today. The three films that I will discuss here attempt to map the social ontology of oil—the how, why, and wherefore of oil in our social, cultural, and political life. *A Crude Awakening* alerts publics about the degree to which contemporary global society is dependent on a natural resource necessarily in short supply. *Crude* and *H2Oil* each examine the environmental consequences of oil exploration, with a focus on its effects on those indigenous communities who live in proximity to the resource and who thus have to endure both the ecological traumas of ongoing drilling and the sludge and slurry left at past drill sites. What distinguishes *H2Oil* from *Crude* is that the former includes brief lessons on peak oil as part of its overall narrative and makes this an element of its case against the Alberta oil sands; *Crude*'s focus, in contrast, is on the dynamics of law and corporate power as these play out in relation to a commodity at the heart of capitalism's profit logic. In what follows, I probe the "lessons" each provides for thinking about oil by drawing out the (implicit or explicit) ways, both thematically and formally, in which they address the problems this substance generates.

A Crude Awakening: The Oil Crash is divided into ten sections (introduced by intertitles) that provide a narrative of the significance of oil for contemporary global society. It takes the form of a social documentary intended to identify and explain a contemporary problem hidden from view. The secret exposed here is the depth of dependence of contemporary social and economic systems on oil—a nonrenewable resource whose era of abundance and easy access is now past, even if this fact seems little acknowledged by the manner in which it continues to be used and exploited.

The film conveys the gravity of our historical moment with respect to oil through three techniques. First, it showcases testimonials about oil from a large number of experts. The range of expertise on which the filmmakers draw is impressive, as is the attention to the politics of each of these talking heads. Two of the most prominent speakers are Matthew Simmons, an energy investment banker, author, and adviser to President George W. Bush, and Roscoe Bartlett, a Republican U.S. congressman from Maryland. The film is careful to include voices from the oil industry, as well as from academics and scientists who deal with the issues the film raises from the vantage point of their own specialties. Notable for their absence—with the sole exception of attorney Matthew David Savinar, who until recently ran a website on the politics of peak oil—are those activists or environmentalists (or even Democrats!) whom one might expect to find in a film awakening us to the challenges of peak oil.

A second technique is the communication of information about oil through the use of facts and statistics. These come directly from the mouths of the experts themselves, and they are invariably alarming (e.g., each calorie one eats requires 10 calories of fossil fuel to produce; by 2030 the planet will have to bring 200 million new barrels of oil on stream *per day* in order to deal with the depletion of existing wells as well as growth in demand; and so on). Finally, the context of peak oil is framed through the formal decisions made with respect to the images and sounds that fill up the space between the talking heads. There are numerous points one could make with respect to the particular use of montage and fast-cutting in many of the sequences in the film. The speed of much of the visual evidence, especially against the backdrop of Philip Glass's minimalist soundtrack, suggests "a life out of balance," as do the many images meant to evoke oil culture: sheiks walking through fancy shopping malls, sludge-filled rivers and oceans, battlefields on which wars have been fought over oil, and the mess of drill sites all over the world. At times, *A Crude Awakening* interlaces these images with older footage of car ads, instructional videos, and clips from celebratory corporate documentaries, all of which appear in hindsight as not just shortsighted but as obscene testaments to humanity's waste and (in the case of the clips from the instructional videos) the very different relationship of supposedly objective knowledge systems (i.e., science and documentary film) to oil in the not-too-distant past.

The ten sections of the film build an effective case against oil. They link oil to geopolitical conflict (section 4: "A Magnet for War"), identify its centrality to daily social life (section 2: "We Use It for Everything!"), and explore the reasons for concern about the end of oil (section 6: "Peaking Out"). What it does not do is offer a solution or resolution to the coming oil crash. The third section of the film looks at three spaces of oil production that have experienced the traumatic passage from oil boom to bust (McCamey, Texas; Maracaibo, Venezuela; and Baku, Azerbaijan). These are microcase studies intended to provide examples of what might soon happen on a macroscale. What we see are images of formerly flourishing towns and cities, now semiabandoned and ugly. The images of Baku's oil fields (which have been captured iconically in the photographs of Edward Burtynsky)[10] are especially haunting: the screen is filled with the remnants of old wooden derricks running up and spilling into the Caspian Sea, fresh oil still staining the ground. If these cases are meant as object lessons, one might expect them to be followed by information as to how it might be possible to manage the down cycle of oil that will soon be experienced on a planetary scale. *A Crude Awakening*, however, seems intent on informing its viewers that there is no way of offsetting a planetary crisis. The penultimate section of the film (section 9: Technology to the Rescue?") presents possible options—electricity, hydrogen, biomass, nuclear, wind, and so on—only to have technology experts rule each of them out on the basis of inefficiency (at present it takes 3 to 6 gallons of gas to create enough hydrogen to enable us to drive the same distance as 1 gallon of gas), scale (10,000 nuclear plants would be needed to replace oil), or lack of resources (with that many nuclear plants in existence, uranium reserves would be exhausted in one to two decades). The film lays open the consequences of a civilization based on oil in order to present audiences with some insight into the why and how of the conflicts and pressures of the near future—a future about which there is little of substance that can be done because of the weight of existing infrastructure and the realpolitik of power in contemporary political and economic systems.

There appears to be a deliberate decision in *A Crude Awakening* to avoid directly linking the narrative of peak oil to the impact of petrochemicals on the environment. The question of whether or not continued oil use—at either current or higher levels—will damage the environment is suspended, one suspects, in order to focus on the necessity of oil to current ways of living and being, and to preclude challenges to the film that might emerge from the growing contingent of climate skeptics. By contrast, *H2Oil* and *Crude* each explores specific examples of the impact of oil exploration and production on the environment and human communities. What we learn from these cases are not only the manner in which oil damages both ecological and human health but also the degree to which the interests of elected governments, national legal systems, and multinational corporations are intertwined in ways that make difficult the possibility of addressing some of the specific (as opposed to systemic) impacts of oil.

Crude examines a landmark legal case against the consequences of the oil exploration and extraction conducted by Texaco (purchased by Chevron in 2001) in Ecuador from 1964 to 1993. There are two main anchoring narratives in the film. The first follows the actions of Ecuadorean lawyer Pablo Fajardo and his American counterpart,

Steven Donziger, over a two-year period (2006–7) as they pursue a suit against Chevron on behalf of thousands of members of the Cofán indigenous community. The second is a single moment in the trial in which plaintiffs, defendants, and the presiding judge in the trial visit the Lago Agrio oil field as part of the evidentiary process. In the first narrative, we witness the political and cultural struggles in which Fajardo and Donziger engage in an effort to generate awareness and legitimacy for their case. In addition to on-the-ground fights within the Ecuadorean legal system, this includes actions at Chevron shareholders meetings, talks with the New York legal firm that is funding the suit, and engagement with the (then) new left-wing government of Rafael Correa. In January 2007, the public relations battle they conduct in conjunction with the legal proceedings is accelerated as a result of the commission of a *Vanity Fair* article on Fajardo's fight against Chevron on behalf of the Cofán, which leads to the involvement of pop singer Sting and his wife, Trudi Styler, and results in Fajardo being given a CNN Hero's Award in 2007. Even though the plaintiffs build legitimacy and support for their case in the media, legal maneuvering by Chevron means that a case that had at the time of the film's release (2009) already been in process for fourteen years would continue for another ten: the documents collected in the trial's evidence room are so numerous that it is difficult to imagine any judge being able to work through them in a meaningful way even in the decade estimated by the film at its conclusion.

The perspective of the film is clear: Chevron is at fault and is using its immense power as a multinational corporation ($204 billion in revenue in 2010) to make a conclusion to the trial impossible. The dirty soil and water, and the numerous health problems of the Cofán (infant deaths, cancers, skin lesions, and more), contrast starkly with the talking-head segments with Chevron scientists and lawyers, whose mobilization of scientific data attesting to the safety of their drill sites cannot but seem little more than corporate lies (indeed, the film points out that Ricardo Reis Veiga, the Chevron lawyer interviewed in the film, was indicted for fraud by the Ecuadorean government). Despite the fact that Donziger is shown to work the system in sometimes ethically questionable ways (he whispers to Trudi Styler to mention Chevron as frequently as possible in her comments on the situation of the Cofán, and the New York firm for which he works stands to make a fortune if the case is successful), his relentless indictment of Chevron's corporate malfeasance mirrors the film's own perspective on both the situation in Ecuador and in the world at large.

However, the second guiding narrative of the film complicates this easy indictment of Chevron's actions. In this section, Fajardo and Chevron's attorney, Adolfo Callejas, move around the Lago Agrio oil field, each making points as to what might constitute physical evidence for use in the trial (contaminated water, oil-soaked soil, etc.). While

Callejas uses numerous tactics to shield Chevron from responsibility for the site, all return ultimately to the question of ownership. Callejas argues that while Fajardo and the plaintiffs make numerous claims, they provide no substantial evidence. Chevron disputes the claim that the water is contaminated by oil, or argues that such contamination as does exist introduces no health risk; they insist that it is impossible to link water contamination to oil that they own (as opposed to oil that might have seeped into river or groundwater from other drill sites or through natural means); and they make numerous legal points in relation to property rights. Property begets responsibility; and so Chevron argues that Petroecuador assumed responsibility for the site when they took it over, that the site was always a Texaco-Petroecuador consortium (such that the latter shares whatever responsibility is assumed for the former), that Texaco no longer exists as a company and so cannot be held responsible, and that the area in which the Cofán live was designated as an oil-exploration site by the government in the 1960s, and so no people should be living there to begin with. Taken together, these points (and there are others in a similar vein) offer a confusing defense. Rather than building a coherent case, it is as if they are being thrown out in the hope that one or another will stick. After all, if there is no pollution, then does it matter who owns the oil? If it were truly the case that Petroecuador has had responsibility for the site since 1992, why would Chevron be anxious about the level of pollutants in the area? And doesn't the government bear ultimate responsibility for the Cofán if it has allowed this indigenous group to live in an area not intended for people? So why mount a defense about "safe" levels of oil in water *or* who owns what, when? From the perspective of the film, such confusing and overlapping arguments constitute further evidence against Chevron. But from another perspective, the claims made by Callejas and other Chevron employees draw attention to the metanarrative of the film, which is less about oil than about the constitutive, systemic gap between, on the one hand, social responsibility, equality, and justice, and, on the other, the legal and political mechanisms that are in place to address the very real crisis faced by a community that now lives on in the barely concealed sludge of former drill sites.

As its title indicates, *H2Oil* is also about what happens when oil finds its way into water as a result of industrial oil extraction. In the main, this film looks at the effects of the Alberta oil sands on the First Nations (Athabasca Dene) community in Fort Chipewyan, a hamlet situated on Lake Athabasca near the terminus of the Athabasca River. The Athabasca runs through the primary site of bitumen extraction and constitutes an important element of the process by which oil is recovered from the near-solid "tar" that makes up the oil sands. Based on the recorded levels of polycyclic aromatic hydrocarbons (PAH) and arsenic in the water of both the river and the lake, the Dene and environmental scientists argue that the Athabasca River is absorbing the chemicals left behind by the extraction process, whether through deliberate action or errors and accidents in the retention of tailings. The main body of the film moves back and forth between claims and counterclaims about the level of toxins in the Athabasca by the Dene and the Alberta government, and in so doing explores the larger dynamics of corporate and political power in the province as it follows attempts by members of the Fort Chipewyan community to draw attention to the serious environmental and health problems they face.

While it is committed to the exploration of the problems of Fort Chipewyan, *H2Oil* makes use of this case to outline the larger political, economic, and ecological entanglements generated by the oil sands. Well-known critics and commentators on Canada's oil policy (and its connection with climate change), such as Tony Clarke, Dr. David Schindler, and Dr. Gordon Laxer, are given an opportunity to weigh in on the implications of current government decisions (or lack thereof) on greenhouse gas emissions, water and soil contamination, and national resource independence. There are also short cartoon segments included that provide quick instructional overviews of the mechanics of oil sands, the implication of the Security and Prosperity Partnership of North America for Canadian water and oil, and the places to which the end product of the oil sands are pumped (*all* of the oil is currently exported to the United States). If *Crude* emphasized the role of corporations in the narrative of oil and water, the antagonists in *H2Oil* are in the main government agencies and ministries, whose representatives argue that they are behaving in a responsible and efficacious manner to address health and ecological concerns. The Ministry of the Environment disputes every one of the facts and figures on cancer rates, oil seepage, and carcinogen levels in the water proffered by scientists critical of their practices. A secondary narrative concerning the problems generated by a drill site for a spring-water company based in Hinton, Alberta, amplifies this criticism of government, highlighting how difficult it is even for businesspeople outside the oil industry to bring attention to the overuse and contamination of groundwater as a result of oil exploration and extraction.

Notable for its absence in *H2Oil* is the oil industry itself. With few exceptions, its presence is signaled only by the frequent images inserted into (what have become) a form of generic montage about the oil sands: enormous, glowing refineries, made up of systems of pipes, exchangers, and condensers of almost unimaginable complexity; slow aerial pans of the vast extraction sites, framed against the edges of boreal forest now fast vanishing in their wake; and the slow-motion movement of grasshoppers (oil-pumping units) conjoined with (in a fashion similar to *A Crude Awakening*) sped-up images of consumer modernity—driving, building, shopping. The film is careful to highlight the close connection between industry and government in Alberta.

The Office of the Environment is located in the Petroleum Tower in downtown Edmonton, and Assistant Deputy Minister of the Oil Sands Sustainable Development Secretariat, Heather Kennedy, is identified as a former employee of oil giant Suncor. Nevertheless, in contrast to the intimacy with which *H2Oil* engages with the Dene and others (e.g., Fort Chipewyan's medical doctor, John O'Connor), oil corporations are filmed at a distance, figured as inhuman Goliaths in comparison to the all too-human Davids living in Northern Alberta who are dependent on water that makes them sick.

Taken together, these three films and the critical discourses that they mobilize— multiple in each case, and neither dogmatic nor simplistic—provide insight into how the problem of oil is framed and negotiated, both within documentary but also beyond it. These investigations of oil on film generate three insights into the discourses and narratives of the politics and problems of oil. In an earlier essay, I argued that there are three broad social narratives through which the futures of oil (and so approaches to its present) have been articulated: strategic realism, techno-utopianism, and eco-apocalypse.[11] These three documentaries interest me in particular because they do not fall easily into any of these categories; nor are they examples of the kinds of formally inventive, reflexive documentaries on the problem of oil to which I have devoted attention elsewhere.[12] While they share some of the conclusions of these latter documentaries, their commitment to a more expository or observational documentary form places them to one side of my earlier taxonomic scheme—neither abandoned to the realpolitik of struggles over diminishing resources, nor advocating a miraculous technological solution, nor accepting the disastrous fate of the end of oil even while critiquing the manner and extent to which we late moderns use it. Even while they are cautious not to promote "solutions" to oil (even in the case of *Crude* and *H2Oil*, films for which redress for the affected indigenous communities might constitute at least a small step forward), they avoid the (sometimes too easy) discourse of eco-apocalypse. In all three films, conclusions are suspended in order to better map the nervous system of oil capitalism.

System Failure, Antinomy, Scale

What insights, then, emerge from these films about how we narrate and respond to the problem of oil?

1. System Failure. All three films make clear that our existing social systems are inoperative. Though it might seem obvious to say it, oil is only a problem because of the larger systems through which it flows. The injustices faced by the Dene and the Cofán cannot and will not be resolved through existing mechanisms of law, property, electoral politics, or knowledge (i.e., science). The struggles waged by both indigenous groups regarding the scientific establishment of levels of pollutants in their water highlight the malleability of knowledge when it bumps into the imperatives of government and business. Systems of property and ownership overwrite questions of corporate or ethical responsibility: one rejoinder by Chevron lawyer Callejas is that it is impossible to identify the oil in the Cofán rivers as belonging to Texaco because "it doesn't have a trademark on it." There is no suggestion in *Crude* that a different legal outcome might come about if the U.S. government legislated oil companies differently: the jump of the case to Ecuador is an attempt to see if corporate laws might be stronger elsewhere, but the film is careful not to suggest that even in Correa's government property laws might be jettisoned. In *H2Oil*, government hypocrisies are not linked to this or that party in power—such that an electoral shift would open up new possibilities—but to the operations of power around a commodity that will be excavated no matter what the health or environmental outcomes. *A Crude Awakening* is most directly about system failure: whether or not large social systems develop a greater awareness and more concerted direction about their energy futures, there is little sense that they can in fact meaningfully address the impacts of oil *or* manage to offset the looming civilizational crisis of oil ontologies. Existing systems have failed precisely by working all too well.[13]

On the evidence of this film, two axioms drive the social toward this "successful" system failure. The first is accumulation. Even at levels that have recently (March 2014) caused U.S. drivers to pull back on filling up their tanks, oil remains cheaper than drinking water or a Starbucks latte (which *A Crude Awakening* estimates at $50/gallon).[14] It remains a primary commodity in global production and consumption systems that depend on an ever-increasing expansion of GDP as a measure of social wealth and of progress—the reason economic growth trumps action on the climate in Alberta (and almost everywhere else). A second axiom operates at the level of the subject. One might ask: Why do people work for Chevron? Or Petroecuador? Why do workers and technical experts flock to the spaces of oil production? It is unlikely that it is because there is a strong support for the imperatives and initiatives of oil extraction and the economies it supports, but rather the need for work and fiscal security in an era in which the first axiom no longer encounters the impediments and strategies of a good (i.e., Keynesian) state. Lianne Lefsrud and Renate Meyer have studied the mechanism by which scientists involved in the Alberta oil fields explain to themselves their involvement in a process that they understand to have a climate impact: the availability of work enables a denial of scientific evidence even by scientists themselves.[15] There is a telling moment in *H2Oil* in which spokespersons for an oil sands company are sent to address the concerns of the Dene First Nations. Their response to the criticisms by the Dene: they are only doing their jobs and not intending to hurt anyone. Their refusal to drink the local water suggests that they, too, suspect that the companies they represent are in fact causing damage to the environment and its human inhabitants. But they work for them anyway.

2. *Antinomy?* The identification of a failure in the capacity of a broad range of social systems to address anything as serious as the crises generated by oil can lead only to one conclusion. If existing systems cannot address the problems these films bring up, *everything* has to change—new systems have to come into existence guided by new axioms. But how to move from here to there? There is an expected suggestion in each of these films that it is through education and the transmission of information to publics, which in turn will generate change through official and unofficial social and political networks, that politics "happens"—in other words, the gesture that politically committed documentaries tend to make toward the pedagogic effects of "seeing is believing." The opening segments of *A Crude Awakening* address bluntly the limits of knowledge about peak oil (Congressman Bartlett: "Not one in fifty, not one in one hundred people in our country have an inkling of the potential problem we're facing"); part of the intent of the film is to transform this small minority into a majority. The additions of the didactic segments to the narrative of *H2Oil* confirm director Shannon Walsh's hopes for the film to play a role in "educating a public who hadn't yet heard of the tar sands, and creating a context for further activism," and the ominous subtitle of *Crude* speaks to a similar desire to explain the "real price of oil."[16]

But even if the films never disavow this fundamental political aim, they recognize the complexity of the situations they encounter and represent, and they are cautious about the degree to which they are willing to figure their politics solely in relation to this pedagogic mode of knowledge transfer. These films frame two antinomies—first, that of the constitutive gap between knowledge and action, and second, between aesthetics and politics. While neither of these may be an antinomy in the strong sense of the term (they are not the same as the Kantian puzzle of the divide between natural causality and human free will, for instance) the suggestion of a blockage that seemingly no amount of conceptual thought *or* political activity looks likely to undo generates a genuine problem for knowledge and aesthetic practice. Antinomy here is meant to describe a stark social contradiction that emerges out of the messy activity of innumerable social systems. Generating an awareness of the structuring role of oil in civilizational processes, and so, too, its obscene primacy over both human needs and ecological ones, produces on its own no resolution, even as it indicts the poverty of the present. As a genre, political documentaries like these three films might be seen as the invention of an imaginary solution to a social contradiction—the "imaginary" being the phantasmic liberal public sphere it imagines into existence, that supposed space in which debate and discussion leads to a resolution that maximizes (say) individual freedoms within the demands of social necessity. These films gesture in this direction, but the substance they each address—oil—does not allow them to imagine that they do more than give evidence of the social contradiction produced by this sticky substance. The politico-aesthetic at work in these films gestures toward the possibility of audiences "doing something" because of the conditions in their world, while at the same time being unable to commit themselves fully to a belief that they can produce either an increase in knowledge or political action—less as the result of failure of political will than because of a recognition of the constitutive nature of the world they produce and represent on film.

The productivity of antinomy is that it gestures to an overcoming that is present in the terms of the structuring division—one that requires only the right insight into the dynamic that produces the division to begin with. A crude, reductive (which is not to say unproductive) way in which to think about oil is to understand it as foundational to contemporary social form. The social contradiction is that the founding premise of society *as such* is draining away and cannot be replaced. Where exactly can one find synthesis in such a system, even if one were to undo the "enlightened false consciousness" that generates the gap between knowing and doing, evidence and action?[17] As a result of the demands of its subject matter, social contradiction in these films remains on their surface, whether they try to generate an imaginary resolution to it or not.

3. *The Failed Sublime, or, Scalar Aesthetics.* This final point emerges out of the previous one. A dominant aesthetic strategy in reference to oil is to emphasize scale. This is perhaps an obvious approach to a site like the Alberta oil sands, which are estimated to be the size of Florida and which include numerous surface mining sites and vast tailings ponds that permit a direct visualization of environmental destruction.[18] But there are other ways to visualize and narrativize the scale of oil, too, including images of old drill sites on which derricks are clustered as tightly as bees in a hive, or the flow of traffic along freeways and through cities all over the planet. These images of cities and traffic are prominent in *H2Oil* and *A Crude Awakening*, and identify the civilizational dependence on oil that will lead to crisis as its last dregs are used up; the former images of environmental impact, present in all three films, point to the astonishing degree to which human beings have remade a space as big as a planet, and continue to do so in ever more visible ways.

The use of scale in these documentaries is intended to add to knowledge and to generate an affective response. Is this not an appeal to the Kantian sublime in both of its aspects, the mathematical and the dynamical? Again, as with antinomy, the correlates are inexact: the palette of cities, however many different images of sped-up traffic we are shown, is not without limits, and the images of the oil sands are not of Nature but of its antithesis: "nature" after its encounter with humanity. Nevertheless, the gesture these films make toward representing oil through the visualization of scale do seem to have as their endpoint the same gesture as Kant's analysis of the sublime: to bring into cognition even that which seems to supersede and fall outside of it. We are placed in awe of scale not so that we give up in the face of the vast existing infrastructure that depends on oil, or that we concede to an ever-expanding tear in the face of the earth (one now said to be visible from space), but that it provoke a closure of that gap between knowing and acting described above.

And yet this gap persists. Has the possibility of a politics through such scalar aesthetics collapsed? Throughout the history of film theory (starting with writers such as Jean Epstein and André Bazin), there is an insistence on the capacity of film to record what is otherwise inaccessible to vision, opening up reality to that quotidian experience that cannot help but miss reality's full ontological presence and depth. One should not disavow the capacity of documentary to bear witness to reality in just this way, both at the level of form and content; the sublime of oil culture that these films visualize does not readily appear to everyday experience, which is one of the reasons the consequences of the end of oil are neither feared nor acted upon. One can see Kant's sublime as a domesticating process that renders what might well be alien to thought amenable to existing schema. The fact that the sublime fails is then not an issue, since its capacity to control and contain filmic images of traumatic scale in fact drains the latter of its effects, which is the exact opposite of what one might want. At the same time, however, abandoning oil to mathematical incomprehension or the terror of destroyed nature on a vast scale, the way in which a scalar aesthetics might be thought to do its work, seems to abandon thought to the inaction of what Slavoj Žižek has termed "cynical reason": awareness without action, even in the face of disaster, since we cannot possibly act on something that exceeds our comprehension. In the end, what is incomprehensible is not the scale of our action on the world, but that our social world has as its foundation a substance demanded by our quotidian infrastructures, an input whose time has come, and soon will be gone. It's unclear what action one could take, even if one wanted to.

THE POLITICS OF DOCUMENTARY IN AN ERA OF SCARCITY

Michael Hardt's "Two Faces of Apocalypse: A Letter from Copenhagen" draws attention to the similarities in and differences of the politics of the anticapitalist and environmental movements. Superficially, one might expect these two movements to be more similar than different, or even as occupying the exact same ground: a visual representation of this relationship would be less a Venn diagram in which there is a zone of overlapping concern (and so zones of exclusion, too), but of two perfectly congruent sets that appear to be distinct only because each group spends more time in one part of the field than the other, thus misrecognizing the extent of their shared interests. Reflecting on his experience at COP 15 (the 2009 United Nations Climate Change Conference), however, Hardt recognizes that there are significant differences that would have to be addressed before each movement can operate fully in conjunction with the other.

Hardt identifies three antinomies that define and separate the anticapitalist and environmental movements (about the points of intersection—an opposition to property relations and their joint challenge to traditional measures of economic value—I'll say no more). The first and defining one has to do with "a tendency . . . for discussions in the one domain [environmental movements] to be dominated by calls for preservation and limits, while the other is characterized by celebrations of limitless creative potential."[19] A second has to do with the question of knowledge. While "projects of autonomy and self-governance, as well as most struggles against social hierarchies, act on the assumption that everyone has access to the knowledge necessary for political action," Hardt writes, "the basic facts of climate change—for example, the increasing proportion of CO_2 in the atmosphere and its effects—are highly scientific and abstract from our daily experiences. Projects of public pedagogy can help spread such scientific knowledge, but in contrast to the knowledge based in the experience of subordination, this is fundamentally an expert knowledge."[20] The final antinomy grows out of a different relationship of each to time. For anticapitalist movements, radical change that would bring about the end of days is the opening to a new (and better) world. By contrast, for environmentalists, "the end of days is just the end," as the radical change that is likely on the horizon is one of "final catastrophe."[21]

The second antinomy is the one on which documentaries of the kind that I have been exploring here hope to do their work, either by translating expert knowledges into lay language or by producing accounts of damage to the environment that can be narrated and made visible, moving audiences from the specific (a film or a specific case) to the general (a confrontation with the issues facing the globe as a whole). When the subject matter is oil, it is impossible not to reflect on the terms of the first antinomy—that is, on limits, not only that of Earth's environment but also of one specific element of it whose use has resulted in an assault on the environment even as it has contributed to or amplified the (apparent) limitlessness of human productive and imaginative capacity.

But it is the third antinomy that haunts documentaries on oil. The division Hardt points to in this third moment is, at least from one perspective, the least convincing. Does the end of days always already signal the effective destruction of the earth's environment? Or can it not also speak to the possibility of a new world in which the antinomy between limitlessness and limit has been resolved (which is to say: What kind of revolution today could imagine that it has passed the end of days if it has not conceptualized what it means to live within limits?). As long as it is figured in terms of climate change, the apocalyptical imagination of environmental movements continues to operate with an understanding of final catastrophe as temporally distant. When one thinks of catastrophe as the end of oil, however, the time horizon is pulled much closer, even as its politics are more difficult to cognitively map. For where should we place oil within this opposition of limit and limitlessness, the environment and the common? Oil is limited, and its use pulls closer that larger limit of the earth's environment, of which it is simultaneously a part (limit) and an other (catastrophe) that the future would be better off without.

And what of the common and its limitlessness? A radical change to the present may well be precipitated by the evaporation of a commodity on which the common depends more than it might want to believe. "Fossil fuels helped create both the possibility of modern democracy and its limits";[22] given the problems of modern democracy, its evaporation alongside that of the energy inputs that helped fuel it might be welcome. But there is no guarantee that the new world on the other side of the end of oil will be one made in the image of revolutionary groups and their labors. The oil documentaries that I've explored here struggle with Hardt's antinomies and the political antinomies of crude aesthetics that I describe above, leaving open the question of how to resolve them (or even *if* they can be resolved), and refusing to offer solutions that would do little more than affirm that which they would seek to deny. Does this constitute a form of political success or failure? Or, perhaps their politics lie in the evidence they provide of the limit of what can be said about a socially ubiquitous substance that remains hidden from view—even today, and even in the process of bringing it to light.

NOTES

1. *Crude: The Real Price of Oil*, dir. Joe Berlinger (DVD; Red Envelope Entertainment, 2009); *A Crude Awakening: The Oil Crash*, dir. Basil Gelpke and Ray McCormack (DVD; Lava Productions AG, 2006); and *H2Oil*, dir. Shannon Walsh (DVD; Loaded Pictures, 2009).

2. See Jane M. Gaines, "Political Mimesis," *Collecting Visible Evidence*, ed. Jane M. Gaines and Michael Renov (Minneapolis: University of Minnesota Press, 1999), 84–102.

3. Salma Monani, "Energizing Environmental Activism? Environmental Justice in Extreme Oil: The Wilderness and Oil on Ice," *Environmental Communication: A Journal of Nature and Culture* 2, no. 1 (2008): 119–27.

4. Meg McLagan, "Introduction: Making Human Rights Claims Public," *American Anthropologist* 108, no. 1 (2006): 192.

5. Jennifer Wenzel, "Consumption for the Common Good? Commodity Biography Film in an Age of Postconsumerism," *Public Culture* 23, no. 3 (Fall 2011): 573–602.

6. Fredric Jameson, *The Political Unconscious: Narrative as a Socially Symbolic Act* (Ithaca: Cornell University Press, 1982), 64.

7. Ibid.

8. McLagan, "Introduction," 191–95.

9. Michael Hardt, "Two Faces of Apocalypse: Letter from Copenhagen," *Polygraph* 22 (2010): 265–74.

10. See Edward Burtynsky, *Oil* (London: Steidl, 2011).

11. Imre Szeman, "System Failure: Oil, Futurity, and the Anticipation of Disaster," *South Atlantic Quarterly* 106, no. 4 (2007): 805–23.

12. Imre Szeman, "The Cultural Politics of Oil: On *Lessons of Darkness* and *Black Sea Files*," *Polygraph* 22 (2010): 3–15.

13. See Eric Cazdyn and Imre Szeman, *After Globalization* (London: Wiley-Blackwell, 2011), 134–52.

14. Mark Shenk, "Oil Drops Below $100, Gasoline Tumbles, on U.S. Supply Surge," *Bloomberg Businessweek*, May 11, 2011. http://www.bloomberg.com/news/2011-05-11/crude-oil -falls-for-first-day-in-three-on-projected-gain-in-u-s-supplies.html.

15. Lianne Lefsrud and Renate Meyer, "Science or Science Fiction? Experts' Discursive Construction of Climate Change," *Organizational Studies* 33, no. 11 (November 2012): 1477– 1506.

16. Claire Ward, "The Future Is Inside Your Sock: How People, through Documentaries, Can Make a Difference," NFB.ca (blog), May 10, 2011. http://blog.nfb.ca/2011/05/10/the -future-is-inside-your-sock-how-people-through-documentaries-can-make-a-difference/.

17. See Peter Sloterdijk, *Critique of Cynical Reason*, trans. Michael Eldred (Minneapolis: University of Minnesota Press, 1987).

18. One notable instance of the documentary use of scale is Peter Mettler's *Petropolis*, which consists entirely of aerial shots emphasizing the size and scope of Northern Alberta oil extraction. *Petropolis: Aerial Perspectives on the Alberta Tar Sands*, dir. Peter Mettler (DVD, Greenpeace Canada, 2009).

19. Hardt, "Two Faces of Apocalypse," 271.

20. Ibid., 272–73.

21. Ibid., 273.

22. Timothy Mitchell, *Carbon Democracy: Political Power in the Age of Oil* (New York: Verso, 2011), 1.

Oil and Dust

Theorizing Reza Negarestani's Cyclonopedia

MELANIE DOHERTY

Reza Negarestani's 2008 work *Cyclonopedia: Complicity with Anonymous Materials* has been described as a work of "theory-fiction" and "Geotrauma" by those who have worked to theorize the text.[1] An enigmatic work by an enigmatic Iranian philosopher and writer, *Cyclonopedia* employs a mix of fictional prose, academic critique, and observations culled from technology studies, global political theory, and contemporary philosophical debates to assemble a Deleuzian critique of global capitalism and petroleum politics. Negarestani's previous work has been featured on *CTheory.net*, an online journal edited by technology theorists Arthur and Marilouise Kroker,[2] as well as in the independent journal *Collapse,* edited by Robin Mackay, which features essays that "bring together a mix of philosophers and theorists, artists and scientists to explore fundamental themes and ideas which academic philosophy, in its tendency towards specialisation and partisanship, increasingly fails to address."[3] *Cyclonopedia* was published in 2008 by re.press, an Australian open-access publishing house devoted to contemporary philosophy and fiction. Accessible in print and digital format, the hybrid text has gained a wide and heterogeneous following of technologically savvy scholars, independent philosophers, and artists. As Zach Blas notes in a review of *Cyclonopedia,* "The text has become an inspiration to academics and intellectuals, spiraling out into the disciplines of media theory, continental philosophy, queer theory, and political thought; the creative industries of architecture and the arts have also embraced the text . . . either making work as a direct response to *Cyclonopedia* or collaborating with Negarestani. A 2011 *Cyclonopedia* symposium at The New School in New York City brought these splintering engagements together, demonstrating the profound and varied impact the book is continuing to make."[4]

In what follows, I will address *Cyclonopedia* as an experimental attempt to outline an alternative theoretical perspective on oil, petrocapitalism, and oil-based American foreign policy, a perspective that playfully engages the tenets of Speculative Realist philosophy and Middle Eastern mythology and contests the imperial discourses and dominant media narratives underpinning American oil imperialism. As we will see, the various components of *Cyclonopedia* (fictive narrative, assembled documents, communiqués, republished web content, and so on) advance a nonanthropocentric imagining of oil, the extractive landscapes of the Middle East, and the oil wars pursued by the United States. Urging us to take a step back from semiotics to readdress the material conditions of our world, the text uses the trope of the "blobjective" to reimagine oil as a nonhuman entity invested with its own agential power. At the same time, *Cyclonopedia* uses the trope of "dustism" to refigure the Middle Eastern landscape as a radically autonomous being. The text employs the theoretically charged theme of the cyclone, finally, to conjure the specter of a massive war effort spinning free of its national origins. Considered together, these three textual impulses yield a new outlook on oil that diverges from, and attacks, the ideologies of technological mastery and imperial militarism that drive contemporary petrocapitalism. Pursuing these impulses through a range of tenuously connected textual forms, *Cyclonopedia* assembles a cacophonous account of the oil economy and petropolitics that works simultaneously to counter the linear narratives and one-way soliloquies that comprise dominant media discourses on oil.

In following out these various imperatives, *Cyclonopedia* draws on the theoretical insights of Gilles Deleuze and Félix Guattari. In part, *Cyclonopedia* theorizes itself as a work of Deleuzo-Guattarian minor literature that attempts to transform the geopolitical narratives of the contemporary oil-driven global economy by burrowing into its plot holes and exploding them from within. Deleuze and Guattari define the three characteristics of minor literature as "the deterritorialization of language, the connection of the individual to a political immediacy, and the collective assemblage of enunciation."[5] A minor literature "doesn't come from a minor language; it is rather that which a minority constructs within a major language."[6] A major literature has smooth and fixed identities, often in line with the leading hegemonic ideologies of its moment, while a minor literature actively scrambles the very concept of any fixed identities. In a similar way, *Cyclonopedia* functions less as a traditional novel with predictable psychological characters or linear plot lines slouching toward resolution, and more as an act of schizoanalysis undermining and reorganizing the reader, particularly Western readers, using language that is weaponized, in a very literal sense, as a minority-literature counterattack on the state language of the U.S.-occupied Middle East.

At the same time, the text functions as a complex assemblage that problematizes its own status as an "encyclopedia" and works instead as a "cyclonopedia," emphasizing the cyclonic process that not only undermines any singular readings of itself as "theory" or "fiction," but also underscores the cyclonic power of Deleuze and Guattari's concept of the War Machine, which argues that war becomes a self-propelled machinic force that eventually exceeds State identities to take on a life of its own: "We have watched the war machine grow stronger and stronger, as in a science fiction story; we have seen it assign as its objective a peace still more terrifying than fascist death . . . we have seen it set its sights on a new type of enemy, no longer another State . . . but the 'unspecified enemy.'"[7] In the glossary of Negarestani's text, this concept of War-As-Machine is questioned as politically dangerous: "The model of war-as-a-machine evades discourses which institutionalize war as a social, economical and political object within the anthropocentric judicial system. War has an economy, a politics, a socius and a population of its own."[8] Continuously invoking theorists whom the text will also inherently critique, *Cyclonopedia* unfolds as a shifting, paradoxical, and heteroglossic work, and this may be one of its primary tactics. As Deleuze and Guattari note in their discussion of minor literature: "How many styles or genres or literary movements, even very small ones, have only one single dream: to assume a major function in language, to offer themselves as a sort of state language, an official language."[9] They suggest instead that an idealized minor literature does not allow itself to be co-opted by the languages of power: "Create the opposite dream: know how to create a becoming-minor."[10] *Cyclonopedia* does this by simultaneously positing itself as a narrative of oil and also critiquing the forms of culture, power, and politics made possible by oil.

After meeting the metafictional Reza Negarestani online and following a series of mysterious messages, the American character Kristen Alvanson travels to Istanbul to meet the elusive Negarestani, only to discover instead a manuscript in her hotel room: the eponymous *Cyclonopedia*. This text turns out to be an oddly cobbled-together collection of notes from a recently disgraced professor at Tehran University, "archaeologist and researcher of Mesopotamian occultural meltdowns, Middle East and ancient mathematics, Dr. Hamid Parsani,"[11] as well as war communiques from a Delta Force deserter, Colonel Jackson West, who descends into madness as he sides with nomadic insurgents and depicts the Middle East as a "sentient entity" made of oil and dust. Along with these texts and figures at the interstices of fiction and theory, the reader also encounters the writings of an online research team, the Hyperstition Collective, which is described in the first lines of the manuscript:

11 March 2004. Somewhere amid the fog of the Net, behind a seemingly forgotten website, in Hyperstition's password-protected laboratory—a location for exploring a diverse

range of subjects from the occult to fictional quantities, from warmachines to bacterial archeology, heresy-engineering and decimal sorceries (Qabalah, Schizomath, Decimal Labyrinth and Tic-xenotation), and swarming with renegade academics, pyromaniac philosophers and cryptogenic autodidacts—there is a tumultuous discussion.[12]

The "Hyperstition laboratory" and its heated online debates weave in and out of the interstices of the text, and the URLs provided by Alvanson in the prologue allow the reader to view some of the archived discussions online. This intersection of book and Internet becomes a unique experience for the reader, as there are moments throughout the text and the actual websites online that blur reality and fiction. Furthermore, Alvanson's counterpart in the real world is an American artist and academic who happens to be married to Reza Negarestani.[13] As we shadow Alvanson reading the *Cyclonopedia* manuscript, we learn that the Hyperstition online collective may in fact be behind her trip to Istanbul, rather than the metafictional Negarestani. Lulled by waves of a vague dread and sentient darkness, the metafictional Alvanson becomes the reader's reference point via footnotes throughout the text, and potentially one of the contributors to the text itself, as she finds herself compulsively writing ominous phrases in the margins of the manuscript: "Reading about misauthorship and inauthenticity, I am compelled to write on the left margin of a page of the manuscript: 'Am I not the most generous parent who feeds her children with her own meat?'"[14] At later moments in the text, Alvanson reemerges in various permutations as an inherent critique of phallogocentrism. She appears both through footnotes that undercut the text, as a love interest, and as a rather more threatening demon, Aisha Qandisha, a type of female Jinn who "possess men, yet they do not occupy or colonize their hosts. Instead they lay open male hosts to the Outside, an openness in the sense of laid, cracked, butchered open."[15] We learn that Aisha Qandisha is the "most fearsome" of the female Jinn: "When she possesses a man, she does not take over the new host, but opens the man to a storm of incoming Jnun and Jinns, demons and sorcerous particles of all kinds; making the man a traffic zone of sweeping cosmodromic data. This is why she is feared. And she never leaves—she always resides in the man to guarantee his total openness, which is not always pleasant."[16] As we follow Alvanson's narrative, we discover with her, in a *mise en abyme* moment, the manuscript of *Cyclonopedia* itself, which constitutes the remainder of the text. As she teeters on the edge of consciousness in her Istanbul hotel room, Alvanson wonders whether the metafictional Negarestani, whom she has only communicated with online, actually exists, or if he is perhaps the voice of a collective. As the intrepid Alvanson attempts to find Negarestani, she searches the *Hyperstition* website, the URL of which is included in the text for the reader.[17] As a collection of emails and blog posts, the site defines the term "hyperstition" as "a myth that makes itself real." Quoting Alvanson's monologue in the text:

> Research Reza Negarestani online:
>   Find an article entitled "John Carpenter's *The Thing*: White War and Hypercamou-
> flage," but the text has been replaced with this message: "This page is not available";
> a piece with a similar title is in RN's manuscript. Even though the post is not available,
> the comments are, and the first comment from RN reads: "Identities are the plot holes
> of someone else's curriculum vitae."[18]

Alvanson's reality crisis becomes one of many moments in the text when identities are questioned and challenged. This happens on the level of the individual, as well as at the level of institutional identities, as we see in the monologues of the various academic and military leaders scattered through the text. A reversal of superstition, using Middle Eastern myths to interfere with the regularly scheduled broadcasts of Western culture, the definition of hyperstition suggests that fictions enact a material reality all their own, deployed as memes that swarm and reorganize culture. Alvanson's character, as the one who discovers the eponymous *Cyclopedia* in text and attempts to read the manuscript, informs us that although she was unable to find Reza Negarestani, she has decided to "take the manuscript back to the US with the intention of publishing it."[19]

Cyclonopedia functions in part as a playful literary critique of various contemporary philosophical movements that have been organized under the term "Speculative Realism." Following a conference at Goldsmiths, University of London, in 2007 involving discussions between Quentin Meillassoux, Iain Hamilton Grant, Graham Harman, and Ray Brassier, the debates have hinged most notably on contemporary philosopher Quentin Meillassoux's critique of what he has termed "correlationism." The critique of correlationism argues that contemporary philosophy needs to break out of the human-world binary that has organized humanistic inquiry. As Meillassoux defines the term in his book-length essay *After Finitude*, he critiques what he calls "the correlationist two-step," which consists of the "belief in the primacy of the relation over the related terms; a belief in the power of reciprocal relation."[20] As Meillassoux posits: "For it could be that contemporary philosophers have lost the *great outdoors*, the *absolute outside* of pre-critical thinkers: that outside which was not relative to us, and which was given as indifferent to its own givenness to be what it is, existing in itself regardless of whether we are thinking it or not; that outside which thought could explore with the legitimate feeling of being on foreign territory—of being entirely elsewhere."[21] Throughout *Cyclonopedia*, the text suggests that access to Meillassoux's nonhuman "absolute outside" and communication with pure exteriority happens through the "blobjective" perspective of oil itself. Meillassoux's critique of correlationism asks us to think a world without humans, to speculate on the absence of ourselves from the picture, to think a radical objectivity. Exteriority in the text also exceeds cultural differences of the West and the Middle East to think beyond the human, to anticipate "xeno-communications" from outside the realm of the human. Meillassoux's logic of radical objectivity resonates in *Cyclonopedia*, as the term "blobjectivity" and the blurring of lines between subject and object become part of complex debates about oil and dust.

Furthermore, one of the more radical perspectives lumped under the signifier "Speculative Realism" belongs to the work of Ray Brassier. Posited in his book *Nihil Unbound*, Brassier's "eliminative materialism" rests on the argument that philosophy always already works from the truth of human extinction because of the impending death of the sun and the expansion of the universe. Negarestani's solar capitalism and apocalyptic language takes on new resonance in light of Brassier, although at points *Cyclonopedia* vacillates between articulating a political critique with an eye toward revolutionary practice and a ludic critique of Brassier through Bataille-inflected apocalyptic hyperbole, including references to the aesthetic nihilism of Nick Land's *Thirst for Annihilation*.

Cyclonopedia appears to pose a question to the reader: If we abandon the correlationist human-world view, what political opportunities would the resulting philosophy actually provide? Certainly, as correlationists, we can only talk *about* matter, but does that mean we cannot posit how matter is constitutive of the very conversation we are having? According to Speculative Realism, we have only a limited correlationist model for discussing how the specific properties of sunlight or oil organize human lives; instead, matter becomes something inert upon which humans project semiotic meaning. What if, instead, we take, as the text suggests, a "blobjective" perspective? How do we think the nonhuman role of oil? The text playfully examines this question of thinking the nonhuman, and also uses references to pulp science fiction, including the work of Dean Koontz, H. P. Lovecraft, and John Carpenter, to describe a closed-down Western xenophobic paranoia about the Middle East as radical "other."

As a response to global petrocapitalism, the text has been categorized in the growing field of "weird fiction," which take cues from a mixture of pulp fiction and cultural and critical theory, such as China Miéville's *The City and the City*; however, its very opacity and Bakhtinian heteroglossia also encourages a variety of politically engaged readings. As a work of theory-fiction, *Cyclonopedia* uniquely problematizes its own references to other texts, including works by Deleuze and Guattari, Nick Land, Freud, and notably H. P. Lovecraft, in a way that echoes the text's articulations of radical openness as a critique of paranoid identity formation: "H.P. Lovecraft is frequently accused of propounding a 'heavily fetishized archaic terror mixed with extreme racial paranoia.' This compulsively consistent racism oozes into his works and thoroughly pervades them. In his Cthulu Mythos, this droning racism is promulgated on a cosmic plane as a prokaryotic horror-population, the Old Ones, lurking as the avatars of absolute exteriority. . . . The more closed the subject, the more brutally it is opened."[22] In other words, Negarestani's references to Lovecraft are not exactly celebratory. Instead, they posit Lovecraft's work as an example of an imaginary informed by profound racism, masculinist castration anxiety, and paranoid xenophobia.

As an amalgam of historically nomadic and disparate cultures, the Middle East unfurls in *Cyclonopedia* as a sentient chronic and cthonic disturbance to the identities supported by petroleum-based global capitalism. The text repeatedly suggests that *Cyclonopedia* is a collective work, and Negarestani's real-world performance art also playfully pokes holes at literary, academic, and nationalistic identities. Discussions of "hidden writing" and the voices of the Hyperstition collective, as well as Alvanson's failure to meet the metafictional Negarestani in the text, bolster the reader's suspicions and recall Deleuze's third characteristic of minor literature in that "in it, everything takes on a collective value. Indeed, precisely because talent isn't abundant in a minor literature, there are no possibilities for an individuated enunciation that would belong to this or that 'master' and that could be separated from a collective enunciation."[23] In fact, for Deleuze and Guattari, the collective (and its debatable lack of talent) is precisely what makes a work of minor literature inherently political: "Indeed, scarcity of talent is in fact beneficial and allows the conception of something other than a literature of masters; what each author says individually already constitutes a common action, and what he or she says or does is necessarily political, even if others aren't in agreement."[24] At many points throughout *Cyclonopedia*, the text is opaque, messy, and punctuated with so much stylistically divergent commentary that it leads the reader to question whether it *is* truly the work of one author, or if it is indeed culled from a collection of online debates. Occasionally, one also wonders at what time of night or under what chemical influences "they" might have been communicating. The text's definition and discussion of Hidden Writing repeatedly suggests the presence of this lurking cyber-multitude: "The propagation of plot holes in hidden writing is not merely the evidence of actual independent plots beneath the text, beneath and through the so-called main story (book within a book). More importantly, it is the indication of the active inauthenticity and anti-book distortions that Hidden Writings carry. . . . Shifting voices, veering authorial perspectives, inconsistent punctuations and rhetorical divergences bespeak a crowd at work, one author multiplied to many."[25] The text suggests the history of its own iterative creation online through a series of bloglike discussions, such as the "X/Z Dialogues," which appear throughout the book.

In these dialogues, the anonymous "X" and "Z" debate the scope and definition of petropolitics. They debate both the theoretical as well as the practical history of oil as power in the United States and Middle East. Oil functions as a lubricant that "eases narration and the whole dynamism toward the desert."[26] Furthermore, "oil is the undercurrent of all narrations, not only the political but also that of the ethics of life on earth."[27] X observes, "Bush and Bin Laden are obviously petropolitical puppets convulsing along the cthonic stirrings of the blob," as Z replies:

> In the case of the Islamic front, oil has been mutated into a kind of constructive parasite through which economical, military and political brotherhood emerges. For middle-eastern countries there is a strategic symbiosis between oil as a parasite and mono-theism's burning core, because oil wells up on an "Islamic Continent," not a mere geopolitical boundary. In other words, Islam has made for a petropolitical network fueled and meshed by Jihad and its monotheistic protocols. Jihad positively participates with oil both in feeding blob-parasites (i.e. western and eastern oil-mongering countries) and fueling its body to propel forward. At this point, the Islamic Apocalypticism of Jihad as a religio-political event and the role of oil as harbinger of planetary singularity overlap. . . . Islam does not perceive oil merely as motor-grease—in the way Capitalism identifies it—but predominantly as a lubricant current or a tellurian flux upon which everything is mobilized in the direction of submission to a desert where no idol can be erected and all elevations must be burned down—that is, the Kingdom of God.[28]

The "X/Z" debates resonate with Deleuze and Guattari's discussion of "War Machines" in *A Thousand Plateaus*. The network of war machines eventually exceeds the control of the nation-state. In both the material and semiotic planes, *Cyclonopedia* questions the autonomy of these machines of power, communication, oil, and culture.

The Hyperstition team depicted in the text suggests the work of the controversial Cybernetic Culture Research Unit (CCRU) out of Warwick University in the 1990s, with theorists like Nick Land, Sadie Plant, and Mark Fisher arguing about "accelera-tionism" as a response to capitalism by accelerating its uglier logics from within. Accelerationism takes its title from a term borrowed from Deleuze and Guatarri who argued in the 1970s that, for the sake of political efficacy, capitalism should not be resisted but rather accelerated. According to Fisher, the 1990s variant of accelera-tionism used the work of Manuel De Landa and others to argue that we should "draw a distinction between markets (as bottom-up self organising networks) and capital (as oligargic and predatory control)."[29] In other words, instead of *withdrawing* from capitalism, especially in light of an ecological crisis that is fully under way, the best response is to *accelerate* capitalism to will on its own repressed Freudian death-drive desire for collapse. Fisher questions this approach: "Was accelerationism merely a new cybernetic mask for neoliberalism? Or does the call to 'accelerate the process' mark out a political position that has never been properly developed, and which still has a potential to reinvigorate the left?"[30] While there are echoes of nihilism in *Cyclonopedia*, the playfulness of Negarestani's text poses similar questions as it problematizes such an approach and subtly critiques it as simply another variant of Western neoliberalism.

Indeed, Manuel De Landa's *A Thousand Years of Nonlinear History* (1997) theorizes the role of material networks of power established by the state and the military through an analysis of Foucault's biopower and Deleuze and Guattari's concept of the "Body without Organs." In the conclusion of his text, he points to the role of the nascent Internet. As a "meshwork," De Landa notes: "Although the Internet (or rather its precursor, the Arpanet) was of military origin (and its decentralized design a way to make it resistant to nuclear attack), the growth of its many-to-many structure was not something commanded into existence from above but an appropriation of an idea whose momentum sprang from a decentralized, largely grassroots movement."[31] For De Landa, the tension between "meshworks" and "hierarchies" spans human history, and he quotes from Deleuze and Guattari's *A Thousand Plateaus* as they describe the Body without Organs: "The organism is not at all the body, the BwO; rather it is the stratum on the BwO, in other words, a phenomenon of accumulation, coagulation, and sedimentation that, in order to extract useful labor from the BwO, imposes upon it forms, functions, bonds, dominant and hierarchized organizations, organized transcendences. . . . The BwO is that glacial reality where the alluvions, sedementations, coagulations, foldings, and recoilings that compose an organism—and also a signification and a subject—occur."[32]

These references to both Deleuze and De Landa's sedimentary stratification and flows of cultural and nationalistic identities are evoked in *Cyclonopedia* both in terms of the Deleuzian War Machine, which creates its own "cyclonic" self-sustaining vortices of power that propel themselves beyond the control of the nation-state, and also the very activity of drilling for oil. *Cyclonopedia* contrasts increasingly bizarre communiques from Colonel West, who has gone AWOL with local insurgents in the desert, with the obsessional research of Dr. Hamid Parsani, who engages in archaeological digs into Deleuzian "holey-spaces" and raises questions about the role of media communications within the hegemony of petrocapitalism. In a section of *Cyclonopedia* entitled "Machines Are Digging," the reader is introduced to Dr. Parsani's work through various nested narrative levels via a reference to an article entitled "Another Academic in Exile?" from the "Swedish Multilingual Journal of Middle Eastern Studies." The article's critique of Dr. Parsani reads:

If John Nash was disassembling everything he found in Time Magazine into diagrams and equations in a schizophrenic search for alien intelligence, Parsani has opted for the reverse process; whatever he encounters is immediately traced back to only one thing, Petroleum. Books, foods, religions, numbers, specks of dust—all are linguistically, geologically, politically and mathematically combined into petroleum. For him, everything is suspiciously oily. Therefore, his approach is fittingly paranoid rather than schizoid.[33]

The stance of paranoia becomes a refrain throughout the text, and here we see Parsani's obsession with oil as a form of paranoia. In his own research, he closes down other potential readings of culture: "Whatever he encounters is immediately traced back to only one thing, Petroleum." Here, through a purely materialist and anticorrelationist reading that ties all power back to oil, the Speculative Realist philosophers appear to be critiqued through the character of Dr. Parsani. The Hyperstition team further unpacks Parsani's academic work for the reader:

> Making an effort to disentangle Parsani's oil-thickened texts and to explain his obsession with petroleum-saturated subjects, the article elaborates how Parsani develops a political pragmatism of the Earth. The article argues that according to Parsani, only through this simultaneously political and pragmatic model of the Earth is the investigation of the Middle East as a sentient entity possible.[34]

What does it mean to think the Middle East as a "sentient entity"? The Hyperstition team further discusses the strange politics of oil and expands on Dr. Parsani's research:

> Parsani's breakthrough was coincidental with the ongoing discussion at Hyperstition's laboratory crisscrossing between the Deleuze-Guattarian model of the "war-machine" and desert-nomadism.... The discussion at Hyperstition ultimately developed into what would later be defined as "blobjectivity," or the logics of petropolitical undercurrents. According to a blobjective point of view, petropolitical undercurrents function as narrative lube; they interconnect inconsistencies, anomalies, or what we might simply call the "plot holes" in the narratives of planetary formations and activities.... To grasp war as machine ... we must first realize which components allow Technocapitalism and Abrahamic monotheism to reciprocate at all, even on a synergystically hostile level. The answer is oil: War on Terror cannot be radically and technically grasped as a machine without consideration of the oil that greases its parts and recomposes its flows; such a consideration must begin with the twilight of hydrocarbon and the very dawn of the Earth.[35]

As Deleuze and Guattari note in the "Treatise on Nomadology—The War Machine" chapter of *A Thousand Plateaus*, monotheistic religions become swirled into and complicit with the vortices of wartime ideologies:

> Let us take the limited example and compare the war machine and the State apparatus in the context of the theory of games. Let us take chess and Go, from the standpoint of the game pieces.... Chess is a game of State.... Chess pieces are coded; they have an internal nature and intrinsic properties from which their movements, situation, and confrontations derive.... Go pieces, in contrast, are pellets, disks, simple arithmetic units, and have only an anonymous, collective, or third-person function: "It" makes a move. "It" could be a man, a woman, a louse, an elephant. Go pieces are elements of a nonsubjectified machine assemblage with no intrinsic properties, only situational ones.[36]

The theory of the Go pieces is echoed in theories of "dustism" postulated in *Cyclonopedia*. Throughout the text, there is a stringent critique of the organized subject, which may become a place to critique the text about its own political efficacy. Notably, the Hypersition lab determines: "This cartography of oil as an omnipresent entity narrates the dynamics of planetary events. Oil is the undercurrent of all narrations, not only the political but also that of the ethics of life on earth.... To grasp oil as a lube is to grasp earth as a body of different narrations being moved forward by oil."[37] This logic of the multitude and Deleuze and Guattari's analysis of ideological constructions of power as contrasting tensions between chess pieces sanctioned by the State and Go pieces that do not follow a predetermined system of hierarchical stratification is echoed in the sections of the text on "dust" and "dustism." As Dr. Hamid Parsani descends into madness, he articulates a profound philosophy of both dust and military communications:

> Dustism is the middle-eastern way of renewing and becoming new for the earth, a course of action which is not taken in favor of... solar capitalism and the Sun's hegemony. Dustism favors the Earth's clandestine autonomy and its rebellion against the domination—whether vitalist or annihilationist—of the sun. The cult of dust celebrates the Insider. Dust inspires a radical and concrete approach to the Outside without becoming a solar slave or a Sun cultist, a blind disciple lacking autonomy or a native inhabitant of the terrestial sphere. And... well, this is exactly where capitalism, particularly the US, eats the Middle East's dust.[38]

The discussion of dust and the logic of sandstorms and swarms in *Cyclonopedia* open onto a strange radical materialism. Robin Mackay's discussion of "Geotrauma" in the introduction to *Leper Creativity*, the published anthology of essays from the *Cyclonopedia* Symposium held at Parsons/The New School in 2011, suggests a possible approach to Negarestani's work:

> The "Speculative Realist" racket provided a perfect opportunity; capitalizing on the vogue for imagining one can subtract theoretical thought from the human imaginary, from narrative and from sense. Through Negarestani we are able to inject it, precisely, with the narrative element that is, as paradoxical as it may seem, an integral part of the procedure. Signification cannot be crushed without following plots that tell ever-new stories of the earth. It's not a matter of using science or a new metaphysics to eradicate such tales, but of constructing a science of real plots, which is what Geotrauma—in Negarestani's hands—becomes.[39]

Mackay himself may well be an unreliable narrator, but his introduction gives the reader a lens through which to read the text. He reiterates the perceived but perhaps contrived tension between the Speculative Realists and Negarestani's work; contrived especially given that Mackay also wrote the first blurb on the back cover of Brassier's *Nihil Unbound*, calling it "[A] powerfully original work which determinedly sets in motion profound and searching questions about philosophy in its relation to the universe described by scientific thought . . . forcibly disabusing use [*sic?*] of the assumption that we have somehow dealt with the problem of nihilism."[40] As an aside, it's also worth noting that the roster of speakers who attended the New School *Cyclonopedia* Symposium appears suspiciously similar to the description of the Hyperstition research team described in the first lines of the manuscript, "swarming with renegade academics, pyromaniac philosophers and cryptogenic autodidacts."[41] But perhaps that's simply paranoia speaking.

As a counterpoint to the previous Speculative Realist theorists, Jane Bennett's work grapples with the concerns of new materialist theorists. Although she also foregrounds the political importance of addressing the material conditions of our contemporary world, her tone is not one of humanistic nihilism. For example, she offers the following description of her encounter with the material in her book *Vibrant Matter*:

> On a sunny Tuesday morning on June 4th in the grate over the storm drain to the Chesapeake Bay in front of Sam's Bagels on Cold Spring Lane in Baltimore, there was: Glove, pollen, rat, cap, stick. As I encountered these items, they shimmied back and forth between debris and thing—between, on the one hand, stuff to ignore . . . and, on the other hand, stuff that commanded attention in its own right, as existents in excess of their association with human meanings, habits, or projects. . . . I realized that the

capacity of these bodies was not restricted to a passive "intractability" but also included the ability to make things happen, to produce effects.... For had the sun not glinted on the black glove, I might not have seen the rat; had the rat not been there, I might not have noted the bottle cap, and so on. But they were all there just as they were, and so I caught a glimpse of an energetic vitality inside each of these things, things that I generally conceived of as inert. In this assemblage, objects appeared as things, that is, as vivid entities not entirely reducible to the contexts in which (human) subjects set them, never entirely exhausted by their semiotics.[42]

Her seemingly simple observation represents some of the most compelling aspects of new materialist arguments for both literary and philosophical analysis. The contingency of the objects encountered here, as Bennett so nicely puts it, are "*never entirely exhausted by their semiotics.*" They are, in other words, not simply the *background* of Bennett's subjective landscape, but rather they are in fact *constitutive* of it. It's important to note that Bennett is not rejecting the semiotic here, but she does point to the way in which it remains the privileged way the humanities encounter and engage with the world. She continues: "The items on the ground that day were vibratory—at one moment disclosing themselves as dead stuff and at the next as live presence: junk, then claimant; inert matter, then live wire. It hit me then, in a visceral way how American materialism, which requires buying ever-increasing numbers of products purchased in ever-shorter cycles, is *anti*materiality."[43]

In much of *Cyclonopedia*, oil and the geopolitical landscape of the Middle East itself become the full expression of apocalypse wrought by the machinations of global consumer capitalism. In Bennett's observed "antimateriality" of rampant consumer capitalism, she highlights the denial that (specifically, American) cycles of consumerism create. She points to the detritus that is blithely overlooked in pursuit of the latest laptop or cell phone upgrade. The thinginess of things, particularly of trash, confronts her in the above quote and shocks her out of the semiotic loop. Bennett addresses the importance, especially in this historical moment of peak oil and global warming, of fully addressing the material finiteness of our world by "semiotically" including scientific and sociological discourses outside the traditional humanities. Bennett's materialist approach contrasts with Ray Brassier's nihilistic and (ironically) human-centric tone at the conclusion of *Nihil Unbound*, in which he laments: "But to acknowledge this truth [of human extinction], the subject of philosophy must also recognize that he or she is already dead, and that philosophy is neither a medium of affirmation nor a source of justification, but rather an organon of extinction."[44] On the one hand, we see Bennett arguing that we need to address the finiteness of the human as species and global resources at this moment in history. On the other hand, if one fully embraced Brassier's philosophy on a political level, why bother recycling or reducing petroleum dependence if we are all already dead? For all its dark hyperbole, Negarestani's text suggests an inherent acceptance of human and planetary finiteness, but pokes holes in Brassier's nihilism as it offers a Dadaesque critique of the human-all-too-human self-absorbed subject who laments the end of humanity. The "blobjective" perspective of oil and the tentacles of global capitalism ooze into the human subject and deny it the "authenticity" that Alvanson searches for in the opening of the text.

With its cacophony of contradictory voices, *Cyclonopedia* also playfully enacts the cybernetic logic of noise. Aural noise pervades the text with "solar rattles," sandstorms, and "vowelless nomad glossolalia," but the text is also rife with debates on the function of noise in electronic communications. In the extensive glossary of terms at the end of the text, the definition of the Solar Rattle reads:

> On a global scale, the Solar Rattle is the ultimate musicality: It registers any message-oriented culture or signaling data-stream as a parasitic sub-noise ambient within itself. The Solar Rattle rewrites every datastream as an Unsign, even beyond any pattern of disinformation. . . . It is not an accident or an invention of contemporary pulp-horror fictions that the sonic cartography of Near and middle-eastern occult rituals (i.e. summoning, conjuration and xeno-communication) is essentially constituted of incomprehensible audio-traumatic murmurs and machinic ambience. The Solar Rattle and its Cthonic auditory agitations were already embedded within the immense capacity of vowelless alphabets of Middle and Near Eastern languages (Aramaic, Hebrew . . . etc.) to artificialize a diverse range of molecular sounds and sonic compositions. Sorcerers and summoners know very well that to communicate radically with the Outside, they must first strip their communication networks (cults?) from informatic signaling systems, grasping communication at the end of the sign and informatic reality. This is where the Solar Rattle installs communicative channels along with the Sun-Cthell axis of electro-magnetized hell.[45]

Noise in *Cyclonopedia* functions in the cybernetic sense, or as a result of its viral deployment in the world. As Gregory Bateson argues: "All that is not information, not redundancy, not form and not restraints—is noise, the only possible source of new patterns."[46] In other words, at least in cybernetic terms, noise is positive. Rather than getting rid of noise in the channels of communication, noise in information is a *productive* aspect of communication. It gives rise to new ideas, new lines of flight, new objects, and new critiques. Without noise, there is only signal repetition, causing the repetition of the same institutional structures, the same ideas, the same essentialized nationalistic subjectivities. In the above passage, the Solar Rattle is one of many weapons that underscores the miscommunication and misrepresentation of global mass-media systems in the employ of petrocapitalism. How do we introduce noise into these systems? How do we introduce the Solar Rattle? How do we disrupt the smooth media narratives so that they can no longer produce the same reductive identities? State power and sedentary distribution require us to reduce all noise as much as possible and produce massive redundancy. As Michel Serres argues in his analysis of noise in *The Parasite*:

> Noise destroys and horrifies. But order and flat repetition are in the vicinity of death. Noise nourishes a new order. Organization, life, and intelligent thought live between order and noise, between disorder and perfect harmony. If there were only order, if we only heard perfect harmonies, our stupidity would soon fall down toward a dreamless sleep. . . . There are two ways to die, two ways to sleep, two ways to be stupid—a head-first dive into chaos or stabilized installation in order and chitin. We are provided with enough senses and instinct to protect us against the danger of explosion, but we do not have enough when faced with death from order or with falling asleep from rules and harmony.[47]

Similarly, in *The Reality of the Mass Media*, Niklas Luhmann argues that recursivity and noise can be used to test the limits of the reality being represented by mass-media systems. Paranoia circulates around nationalistic, economic, religious, and political systems that engender a form of compulsive identity formation. In other words, the more closed down a communication system becomes, the more it is likely to implode. If the system of mass media is closely tied to oil (recall the embedded CNN reporters during the Gulf wars), then there is very little room for the perturbation of that closed system to actually transmit new or reliable information. As Niklas Luhmann notes:

> This was demonstrated very well by the successful military censorship of reports about the Gulf War. All the censorship had to do was operate according to the ways of the media; it had to contribute to achieving the desired construction and exclude independent information. . . . Since the war was staged as a media event from the start and since the parallel action of filming or interpreting data simultaneously served military and news production purposes, de-coupling would have brought about an almost total loss of information in any case. So in order to exercise censorship, not much more was required than to take the media's chronic need for information into account and provide them with new information for the necessary continuation of programmes. Thus, what was mainly shown was the military machinery in operation. The fact that the victims' side of the war was almost completely erased in the process aroused considerable

criticism; but most likely only because this completely contradicted the picture built up by the media themselves of what a war should look like.[48]

In Luhmann's terms, introducing noise into a system of communication helps to keep useful information flowing by perturbing systems that have become too redundant. While he is ultimately pessimistic about the possibility of reliable communication happening between separate spheres (such as the military and academia, or the United States and the Middle East), at the very least, the injection of noise makes more complex communication possible. One imagines the Solar Rattle disrupting the broadcast models of Fox News or CNN.

Through this form of what we might call "noise fiction," *Cyclonopedia* lays out an *n*-dimensional blueprint for a politics of noise, especially timely against the backdrop of protests like the Arab Spring and Occupy Wall Street, both of which use grassroots social-media tactics, the logic of swarms, and the many-to-many broadcast model of the Internet to counteract the one-to-many broadcast model of corporately owned mass-media outlets. As Douglas Rushkoff wrote in a 2011 editorial on the Occupy Wall Street protests: "This is not a movement with a traditional narrative arc. As the product of the decentralized networked-era culture, it is less about victory than sustainability. It is not about one-pointedness, but inclusion and groping toward consensus. It is not like a book; it is like the Internet."[49] In much the same way, *Cyclonopedia* is not like a book; it is like the Internet, offering a decentralized network model of literary production and rejecting traditional literary as well as political identities as "plot holes in someone else's curriculum vitae." However, Negarestani's nonhuman oil-entity lurks in very material ways beneath the production of culture, even, ironically, underneath its own production and distribution as a text. In the case of #occupywallstreet, even if the revolution is not televised but in fact is live-streamed or tweeted on the Internet, it is always already floating on oil. The fleeting act of viewing a website releases anywhere from 20 to 300 mg of CO_2 each second, and research has shown that running the Internet itself releases enough CO_2 each year to send the *entire population* of the UK to the United States and back annually. Twice.[50]

Cyclonopedia introduces a literary and political strategy of noise. It echoes the intertexual logic of Kathy Acker's parasitic literary remixes of white patriarchal narratives and William Burroughs's cut-up method. As Burroughs famously observed in an essay on writing: "The Word is literally a virus, and that it has not been recognized as such is because it has achieved a state of relatively stable symbiosis with its human host. . . . The Word clearly bears the single identifying feature of virus: it is an organism with no internal function other than to replicate itself."[51] As a machinic text, in the Deleuzian sense of the word, and much like Deleuze and Guattari's own writing,

Cyclonopedia uses text as a factory rather than a theater of representation, text as a material object, opaque and enigmatic, irreducible to one message and producing a series of diverse effects in the world. *Cyclonopedia* is a form of textuality designed not to *represent* the world, but to act virally *in the world*, to circulate thinglike throughout the world, producing effects by simultaneously scrambling existing codes, disrupting expectations, and casting the reader outside the covers of the book to gather even more experiences online, thus opening up spaces where new forms of practice and critique can take flight. The paradox is that Negarestani's texts, transmitted both online and through the distribution systems of print publishing, are also tied to oil and the ideologies of oil. Echoing Dr. Hamid Parsani's paranoia, *Cyclonopedia* warns the reader that on a variety of levels, "Oil is the undercurrent of all narrations."

Notes

1. Nicola Masciandaro, "Gourmandized in the Abattoir of Openness," 182; and Robin Mackay, "A Brief History of Geotrauma," 32; both appear in *Leper Creativity: Cyclonopedia Symposium*, ed. Ed Keller, Nicola Masciandaro, and Eugene Thacker (Brooklyn, N.Y.: Punctum Books, 2012).

2. See http://www.ctheory.net/home.aspx.

3. *Collapse* is available online at http://www.urbanomic.com/publications.php.

4. Zach Blas, "Hidden Writing," *American Book Review* 33, no. 3 (March–April 2012): 10.

5. Gilles Deleuze and Félix Guattari, *Kafka: Toward a Minor Literature*, trans. Dana Polan (Minneapolis: University of Minnesota Press, 2006), 18.

6. Ibid., 16.

7. Gilles Deleuze and Félix Guattari, *A Thousand Plateaus: Capitalism and Schizophrenia*, trans. Brian Massumi (Minneapolis: University of Minnesota Press, 2003), 422.

8. Negarestani, *Cyclonopedia*, 240.

9. Deleuze and Guattari, *Kafka: Toward a Minor Literature*, 27.

10. Ibid., 27.

11. Negarestani, *Cyclonopedia*, 9.

12. Ibid.

13. http://www.kristenalvanson.com/.

14. Negarestani, *Cyclonopedia*, xix.

15. Ibid., 120.

16. Ibid., 121.

17. See http://hyperstition.abstractdynamics.org/.

18. Negarestani, *Cyclonopedia*, xiii.

19. Ibid., xx.

20. Quentin Meillassoux, *After Finitude: An Essay on the Necessity of Contingency*, trans. Ray Brassier (New York: Continuum, 2008), 5.

21. Ibid., 7.

22. *Cyclonopedia*, 201–2.

23. Deleuze and Guattari, *Kafka: Toward a Minor Literature*, 17.

24. Ibid.

25. Negarestani, *Cyclonopedia*, 61.

26. Ibid., 19.

27. Ibid.

28. *Cyclonopedia*, 20–21.

29. http://k-punk.abstractdynamics.org/archives/011658.html.

30. Ibid.

31. Manuel De Landa, *A Thousand Years of Nonlinear History* (Cambridge: MIT Press, 2005), 252.

32. Ibid., 260.

33. Negarestani, *Cyclonopedia*, 41.

34. Ibid., 42.

35. Ibid., 16.

36. Deleuze and Guattari, *A Thousand Plateaus*, 352–53.

37. Negarestani, *Cyclonopedia*, 19.

38. Ibid., 92.

39. Mackay, "A Brief History of Geotrauma," 32.

40. Ray Brassier, *Nihil Unbound: Enlightenment and Extinction* (New York: Palgrave Macmillan, 2010), back cover.

41. Negarestani, *Cyclonopedia*, 9.

42. Jane Bennett, *Vibrant Matter: A Political Ecology of Things* (Durham: Duke University Press, 2010), 4–5.

43. Ibid., 5.

44. Brassier, *Nihil Unbound*, 239.

45. Negarestani, *Cyclonopedia*, 148.

46. Gregory Bateson, *Steps to an Ecology of Mind* (Chicago: University of Chicago Press, 2000), 416.

47. Michel Serres, *The Parasite*, trans. Lawrence Schehr (Minneapolis: University of Minnesota Press, 2007), 127.

48. Niklas Luhmann, *The Reality of the Mass Media*, trans. Kathleen Cross (Stanford: Stanford University Press, 2000), 8–9.

49. See "Think Occupy Wall St. Is a Phase? You Don't Get It," *CNN Opinion* (October 5, 2011). http://www.cnn.com/2011/10/05/opinion/rushkoff-occupy-wall-street.

50. See "What's the Carbon Footprint of . . . the Internet?" *The Guardian*, August 10, 2010. http://www.guardian.co.uk/environment/2010/aug/12/carbon-footprint-internet.

51. William Burroughs, *The Adding Machine* (New York: Arcade Publishing, 1993), 47.

20

Imagining Angels on the Gulf

RUTH SALVAGGIO

The story of how oil culture has wracked ruin all across the Gulf of Mexico begs to be told by angels—at least two of them. The first of these winged creatures steps out from the pages of a famous twentieth-century essay, "The Angel of History." Witness to the havoc of the past, this angel looms large in modern consciousness. Its wings are helplessly extended by storm winds propelling it into the future, while it stands immobilized in a gaze riveted on the wreckage of the past. The second winged creature is much older than this Angel of History and has resided on the planet for millions of years—the Pelican. It, too, looms large in human consciousness, emerging at an auspicious historical moment in the catastrophic years of the twenty-first century, when a deepwater oil rig exploded in the Gulf. Pelican Angel also displays extended wings, but they are weighed down with a thick glaze of oil. Yet unlike our Angel of History, Pelican Angel still tries hard to lift. Propelled by instinct instead of a consciousness of past plunder, the pelican needs to get on with its work—fishing, nesting, flying low over the Gulf waters, or as the poet might say, keeping the cloudy winds fresh for the opening of the morning's eye.

The Angel of History and Pelican Angel now pivot on the Gulf of Mexico where some fifty thousand oil wells pock its wetlands, coastal shelf, and deep waters. This sheer number alone—*fifty thousand*—is enough to give pause, and eerily matches the number of brown pelicans thought to be thriving along the Louisiana coast when Standard Oil set up shop there in the early years of the twentieth century. That time frame also eerily coincides with the time when Walter Benjamin was writing about the Angel of History and reenvisioning history itself not as a chain of events but a catastrophe that keeps piling up wreckage. The storm that sucks this Angel of History into the future, Benjamin says, is what we call "progress."[1] Pelican Angel, rising from the Gulf waters and drenched in oil, seems at once the most recent incarnation of the Angel of History and the bearer of the history piled up from twentieth-century progress. Pelican Angel also stands in shocking contrast to the image of the pelican stamped on Louisiana's state flag, where a mother pelican is pictured feeding her young. Now, in the beginning years of a scary new millennium, she changes roles with our Angel of History, who seems incapable of nourishing either the young or the imagination. Here they both appear before us, perched somewhere along the wetlands of the Gulf where millions of winged creatures migrate each year—both angels compelling us to ponder the productive limits of history and the imagination. In an oil culture that has created piles of wreckage in the span of a single century and that remains consumed with high-finance investments and stock futures, we might turn to these two angels to take stock of the past and the future by investing in winged creatures who have long put the human imagination to work. Angels may be precisely what we need now to work our way out of this mess. The BP blowout that filled the Gulf with oil marks a rupture in big oil's most sacred territory, but it is also a rupture in the territory of petro-disaster response. Each angel charts a path here, and we humans are left to find our way along both of these historical and imaginary paths—where humans and angels have often feared to tread.

THE ANGEL OF HISTORY

The history of exploitive and extractive economies all along the Gulf of Mexico and its river corridors is more than anyone can easily take in. No surprise that in the midst of such wreckage, we would invent angels who can hardly bear the sight of the past. It is not a pretty picture. In her poem "Midnight Oil," Sheryl St. Germain, a native of the Louisiana oil lands, explains that she is tired of "way too many pictures / of oiled birds and the oiled waters of this dear place," of all the "pundits and politicians" who analyze and blame, and of all those who simply joke: "let's call the Gulf *the Black Sea.*"[2] Much of the ruin in fact remains darkly visible: objects concealed underwater and in the thicket of wetlands, labyrinthine pipes buried in the alluvium, an endless series of canals cut through for pipes and ships and providing a ready path for saltwater intrusion. Our Angel of History here on the Gulf might take a quick flight that reveals only the tips of those fifty thousand oil wells that have been drilled in the Gulf waters during the past century—including some four thousand in its deep waters, and now more than sixty in its ultradeep expanse. Not a single one of these wells existed before 1900. They all stand as distinctly twentieth-century specters, many of which have long run dry and been abandoned, thousands still pumping for all they are worth—which is estimated at $2 trillion,[3] a figure hard to imagine in the imaginative dead zone of money and high-finance capital. But since much of this wealthy empire is submerged, we do need to conjure up its material structures, its body parts, as our Angel of History might see them: thirty-three thousand miles of pipeline, an endless vista of refineries that you can smell during the day and see clearly at night as they light up the sky like some modern vision of Dante's Inferno, thousands of terminals and huge oil tanks, some on land and some in the form of oil carriers roaming the Gulf, ships and shipyards, oil platforms, oil rigs, oil barrels, almost everything except the oil itself, which remains curiously concealed. Even as we can smell the gas that we put into our cars, or sometimes see the black substance that gets poured into car engines or heating furnaces, oil itself remains a spectral substance—until something ruptures and it all comes pouring out.

What is also difficult to register, even if you are an Angel of History, is the actual historical development of this empire, year by year, decade by decade: the arrival of Standard Oil in the early 1900s, the production of hydrocarbons and construction of petrochemical plants as early as 1915, the emerging massive network of oil infrastructure throughout the 1920s and 1930s, the first offshore well built by Brown and Root in 1938, the first "out of sight" well drilled by Kerr-McGee in 1947, the astonishing accumulation of nearly *twenty-seven thousand wells* by 1977, the first deepwater well drilled by Shell Oil in 1975, and by 2000, the rise of BP—the initial name for British Petroleum that would soon don the name Beyond Petroleum—as it seized the reins of empire and overtook Shell Oil's dominance in the Gulf. As deepwater and ultra-deepwater drilling outpaced shallow-water extraction, and as government regulation of the Gulf oil empire grew weaker and weaker (as early as 1978, the Gulf region was declared exempt from most federal oversight and regulation), the picture here becomes even more difficult to see.[4] The very disappearance of the wetlands themselves, sliced through with canals for transport and oil pipelines and eaten away by encroaching saltwater, remains obscure. We are left with odd and disturbing signs strewn throughout the swamp—sick marsh grasses, dead stumps that were once giant oak and cypress trees, the roofs of houses abandoned to encroaching waters, even old cemeteries, as Mike Tidwell describes in his documentary book *Bayou Farewell*, "tumbling brick by brick into the bayou water . . . like slow-motion lemmings dropping over the fateful edge, one after the other."[5]

One need not travel far north of the Gulf waters and wetlands to see and smell big oil's chemical partners that have taken up residence all along the Mississippi River corridor. Our Angel of History should fly low here, low enough to take in the stench of it, and to view just a few of the dumping sites where chemical waste finds an easy path into streams and rivers—as if its entry into the delta subsoil were not sufficient. In fact, the very existence of petrochemical plants that sit perched along this river corridor, stretching for some fifty miles between New Orleans and Baton Rouge, remains unknown except to those who have any reason to travel either side of River Road. Tour buses that take people to the remnants of sugar plantations that once thrived here tend to follow the quickest and easiest route from nearby I-10 rather than drag their customers through too much of the waste of oil refineries. But River Road is precisely the route to follow if one is trying to come to terms with the past. Some 130 chemical plants dot the edges of the snaky Mississippi River, as if they were blisters all along the edge of its skin. Studies in recent decades reveal much of what cannot be seen here: millions and millions of pounds of hazardous wastes either burned in incinerators or deposited in vast chemical landfills, or simply "released" into the air as if they, like the slaves who worked this land a century before, were only in search of freedom. Historically, this petrochemical empire is known for the production of polyvinyl chloride, or PVC as it is commonly known, although chemical plants that produce everything from nitrogen fertilizer to synthetic rubber now share space on this tragic delta land once razed in the quest for sugar and now poisoned in the pursuit of plastics.

The catchall term *plastics* seems almost too simple to describe this vast empire built on refined oil and all its progeny. After World War I, Standard Oil began to isolate hydrocarbon chains that made possible the production of myriad synthetic materials that were derived from alchemical substances with the names of di-isobutylene, acetylene, butadiene, toluene, trinitrotoluene—all tracking their lineage back to oil. As one writer succinctly puts it, "Ethyl alcohol, isopropyl, methyl ethyl ketone, naphthenic acid: all of these chemicals—used for the plastic coatings of bombers' noses, the manufacture of penicillin, the coating of fabric, napalm, anti-mildew agents for tents— were produced at the Standard Oil (Louisiana) plants on the Mississippi River."[6] When Standard Oil had arrived in Louisiana in 1909, even its elite echelon of managers could hardly have suspected that their investment in the extraction of oil would itself become extracted and funneled into such a profitable and vast chain of chemical by-products and their synthetic formations. Louisiana's Ascension Parish alone, where our Angel of History might find a suitably named home, is ranked first in the state for toxic emissions, including those from Vulcan Chemicals, which recorded twenty accidental chemical releases within a single six-month period. Some of these releases in Ascension Parish hardly pave the way for an ascent to heaven, or for ascending even barely beyond the toxic fumes that permeate the region. One such release at Borden Chemicals in 1996 let loose "over 8,000 pounds of ethylene dichloride, hydrogen chloride (HCl), and vinyl chloride monomer, a known cancer-causing agent."[7] The connection between vinyl chloride and cancer, not to mention a host of other pulmonary disorders and related life-threatening illnesses, is supported by substantial scientific documentation. Yet PVC products permeate our culture, and any association linking them to cancer and respiratory illness here along the Gulf, where all the distillation occurs, continues to be aggressively challenged by scientists employed by oil and chemical industries.[8] In the midst of all the studies and litigation, what remains as evidence is the very name ascribed to this petrochemical corridor—now known to most people simply as Cancer Alley.

Yet other connections beg for recognition by our Angel of History. The insidious linkage between oil, petrochemicals, and sugar also remains largely unseen—until it emerges in plain view of tourists who visit the refurbished mansions of former sugar plantations that once thrived all along this river corridor. Oil tanks and chemical pipelines now surround the remaining homes of the "sugar kings,"[9] and petrochemical plants dispose of their toxic waste in the waters and air of old farming communities, where many of the descendants of slaves who worked the sugar fields continue to live. Farther down River Road, south of New Orleans, our Angel of History could take a glance at the Murphy Oil Refinery, where hurricane Katrina's surge spread Murphy's toxic sludge for miles all around this once-abundant delta land still dotted with orange and satsuma orchards. A few miles away, in Chalmette, one catches a glimpse of the old Domino Sugar Refinery, a huge and hulking rusty reminder of past economies that are still kicking here in the midst of petroculture. All around, canals slice through these deltas, making way for ships and pipelines that transport big oil's extractions, and for yet more hurricane surges that will wash the refuse far and wide. Historians should pay more attention to sites such as these, where one rapacious economy sits atop the ruins of another. Angels might help us see more clearly what remains unseen—the profits "released" from the labor of bodies, the sweetened grounds on which petrochemical plants took root, environments ravaged for sugar cane and fossil fuels, the connections that bind bodies to lucrative landscapes and that bind landscapes to sugar kings and oil magnates.

Among the residents of Cancer Alley is NORCO, a handy acronym for the New Orleans Refinery Company, which sits just north of the Louis Armstrong Airport, where thousands of people arrive every week to visit this historic city. Descending here by plane, one rarely conjures up the sugar plantations and chemical plants that play their part in this history. But driving north on Airline Highway as it extends into River Road, you can count them sprouting up amid the remaining blotches of sugar cane—Globalplex, ADM, Cargill, Marathon, and then suddenly, as if somehow out of place, the refurbished San Francisco Plantation welcoming tourists and surrounded by oil tanks in the background. The bad smell is now everywhere. As a child growing up in the Ninth Ward of New Orleans, I could sometimes sense the chemical odor that was notoriously associated with NORCO. The actual odor could have come from any nearby refinery, but NORCO remains a catchall term for the petrochemical aura pervading the region. When we would rarely drive alongside this veritable chemical kingdom, we knew simply to hold our noses because NORCO was coming up.

Now I know that in 1997 alone, more than two million pounds of toxic emissions were released there, many of them recognized carcinogens.[10]

In the annals of the environmental justice movement, one encounters endless narratives that give some face to this disaster zone—stories of communities torn apart and then poisoned by chemical plants and their waste products. Our Angel of History might volunteer for work with these environmental justice groups that enable a clearer vision of this wreckage. Here we meet up with people who regularly flee their homes and run out into any nearby field in response to the alarms that sound whenever chemical plants release their toxic refuse into the air. But for many who live in these poor communities, there really is no escape. One witness testified, "we have run and run and run," but only to the point of exhaustion. Nor is there any escaping the billions of pounds of hazardous waste injected underground in Louisiana, some 15.5 billion pounds alone in 1988, one-third of the nation's total injection. Our Angel of History might spend special time in Geismar, a small town in Ascension Parish, specifically targeted for development by the petrochemical industry. As Barbara Allen explains in documenting all these matters in her book *Uneasy Alchemy*, the citizens of these largely African American towns are "descendants of freed slaves living on Freedmen's Bureau land grants," but whose farming and fishing communities were poisoned to the root when Ascension Parish became "one of the largest concentrations of chemical plants in the world."[11]

I recently took the I-10 exit for Geismar leading to River Road and began counting and listing the petrochemical plants here—Momentive, Baker and Hughes, Rubicon, Praxan, BASF, Shell Chemical Company of Geismar—all appearing as a tangled conglomeration of oil tanks and smokestacks and pipelines, with special pipes extending across River Road to their ports along the Mississippi River, where ships will arrive to drop off and pick up their booty. North of here, in Baton Rouge, sits the massive Exxon refinery that anchored the petrochemical empire in this region. Beyond that is the campus of historically black Southern University, where, in 1998, students "protested the burning of leftover Vietnam-era Napalm at the Rhodia plant near the school."[12] The sad legacy of such burn-offs extend back in history and across the globe. They also afford a scary premonition of BP's massive burn-off of oil on the surface Gulf waters after the Deepwater Horizon blowout, where dark clouds floated up and away in the Gulf stream to who knows where—a grand release if ever there was one.

But a different image emerges not far from the Exxon refinery in Baton Rouge. It is the image stamped on the state flag, flying high at the capitol—the image of a mother pelican feeding her young. Some have suggested that a more appropriate image for the state flag would be the "Exxon tiger or the Texaco star,"[13] or, we must now add, the series of loops forming a delicate green star marked with the almost inconspicuous initials—*bp*. What a refreshingly green way to imagine British Petroleum, not to mention the corporate promise of a world Beyond Petroleum, right smack here in the ruins of the Gulf. Dig deep into the history of these green initials alone and one might discover all sorts of wreckage once covered in endless fields of sugar cane and now covered up by crafty management feeding America's endless thirst for oil. The mother pelican on the state flag seems almost lost among these images of petro-dominance. But she looms large in the history of the Gulf.

PELICAN ANGEL

The October 2010 issue of *National Geographic*, entitled "Special Report: The Spill" and featuring an oiled pelican on the cover, also included a haunting centerfold image—a picture of a huge pelican being lifted from a thick glob of orange-brown oil sludge. The parish official who seized the bird, we are told, "impulsively rescued this severely oiled brown pelican" by clasping its closed beak in one hand and grasping the top of its extended wing with his other hand. Its full body stretched out before us, the bird appears twice caught—ensnared in oil, and seized, or rescued, by human hands. The caption accompanying this image quotes the official's statement: "'You could see the life draining out of it,'" followed by the reporter's comment: "The bird lived."[14]

FIGURE 20.1. Photograph from *National Geographic*'s "Special Report: The Spill." Courtesy of Joel Sartore/National Geographic Stock.

Looking at this image, we ourselves become caught, and caught up, in the pull of history and imagination. History calls us back in time, locating and describing sites of damage, accessing the wreckage. We can learn from history or not. But imagination pulls us into the wreckage. We are called upon to enter the photograph, to imagine immersion in oil, to feel the sheer draining away of life. This same kind of aesthetic pull calls us to the dark heartbeat of St. Germain's poem "Midnight Oil," where a factual description of oiled birds taken from Wikipedia undergoes poetic transformation, offering us words now capable of igniting both sensation and the possibilities of sensory response:

> Penetrates plumage, reduces insulating ability, makes birds vulnerable to temperature fluctuations, less buoyant in water. Impairs bird's abilities to forage and escape predators. When preening, bird ingests oil that covers feathers, causing kidney damage, altered liver function, digestive tract irritation. Foraging ability is limited. Dehydration. Metabolic imbalances. Bird will probably die unless there is human intervention.

If history asks that we witness the past with an eye on the future, a photograph and a poem demand of us the distinctly aesthetic rescue work propelled by imagination. They call up the tangible images of history, and they call out to humans capable of processing the sensory imprint of these images. *"Bird will probably die unless there is human intervention."* Human intervention may save the day for this particular rescued bird, but human intervention also caused this awful scenario in the first place. The limits and possibilities of human intervention unfold before us as images in need of processing. Seeing these images and sensing the bird's predicament, we are urged to conjure up and feel its breathing, its body striving to lift. And when we read that the bird lives, we ourselves almost feel a breath of relief, a breath that links us to this damaged but still-breathing creature in a shared sensorium.

That images take root in the senses is a notion embedded in the very term "aesthetic," deriving from the Greek term *aesthesis,* to perceive through the senses, to provide a path for sensory impulses from external organs to the brain and nerve centers. Its opposite effect would be *anesthesia,* the widely used medical term for the process of deadening the senses to stimulation. The "severely oiled pelican" gets "impulsively" rescued because of this animated sensory impulse, an impulse ignited in a shared and potentially wide-ranging sensorium—what Jacques Rancière calls the "distribution of the sensible."[15] Seeing the image, we can sense the life draining out of the bird. We sense this suffocation in oil that *"penetrates plumage, reduces insulating ability, makes birds vulnerable"*—and by extension, suffocates us, makes us vulnerable. If human intervention is required, in this case, to lift the bird from the very oily glob caused by humans, that is because we have access to this shared sensory habitation that incites our mutual impulses.

Yet Rancière's notion of "distribution" also accounts for the selective channeling of access to the sensed world, just as oil itself is channeled in invisible underground pipelines and throughout the machinery and vehicles it fuels. Most people never see or smell oil and all its chemical distillates, while others live under conditions where they smell and breathe it every day, are drenched in its refined vapors, taste its residue in the water supply. This distribution of the sensible also explains how oil and its toxic effects can remain shut off to sensory response most of the time and for most of the people, and yet become shockingly visible and acutely sensed in the midst of eco-disasters. In her volume of poems entitled *Breach*, Nicole Cooley describes the "afterlife" of objects distorted by disaster, in her case by the floodwaters that breached the levees in Katrina's surge. For her, each disfigured object becomes an image that demands sensory response. In one poem, she asks: "Don't you want to touch it?"[16] The desire to touch damaged objects from a great modern flood, or to want to touch, take into hand, lift the damaged pelican from the oiled waters—this desire breaks through the normal sensory distribution of oil, which maintains its invisibility, and suddenly, shockingly, opens a space where we can sense its presence and effects, where we can see and touch and smell all the plunder. Even as chemical dispersants are used specifically to redistribute thick masses of oil so that they become minuscule and virtually invisible, the pelican drenched in oil reminds us that all is not well. It emerges as an awful specter, bears the wreckage of twentieth-century oil history, and demands that we enter into a sensory domain that puts us in touch with that wreckage. Without human intervention, the bird may die—and many, many in fact did. But without sensory access to oil and its omnipresence in the Gulf, the cleansing of birds will remain a temporary fix to a persistent lethal problem that gives the appearance of continually disappearing.

One wonders if the pelican's consciousness, fueled by its own acute visual sensory abilities, has enabled this grand bird to survive in ways that we humans have yet to understand. And we might wonder, too, if the Gulf oil disaster has called upon us to extend and "distribute" our own human sensory capabilities throughout our mutual habitations—in the Gulf and elsewhere. Pelicans have thrived for some thirty million years on this planet, and have probably made their home in the Gulf marshes for as long as this alluvial landscape has existed in its current form, at least over the past two millennia. A large and commanding sea bird that feeds and nests throughout these marshes and waters, it shares space here with northern gannets and laughing gulls—comprising with them the three bird species that, according to the official report of the National Commission on the BP Deepwater Horizon Oil Spill, were the three most affected by the Gulf disaster.[17] In describing their rich wetland habitat, Mike Tidwell gives us some glimpse of the abundance of birds who extend their wings in the Gulf environment—including "353 species of birds residing here at some point during the calendar year. Wading birds, waterfowl, songbirds, raptors, shorebirds, colonial nesting birds," birds lifting their wings—"from the awkward takeoff of double-crested cormorants in the lakes and bays to the six-foot wingspan of brown pelicans nesting on the barrier islands."[18] But the fifty thousand brown pelicans once thriving along the Louisiana coast at the beginning of the twentieth century, when Standard Oil arrived, were brought to near extinction by the 1960s as a result of the use of the chemical pesticide dichlorodiphenyltrichloroethane, commonly known as DDT. Thanks to the scientific and image-making work of Rachel Carson, who published her book *Silent Spring* in 1962, DDT was banned a mere decade later, when the nearly decimated brown pelican population was brought back to Louisiana by transferring birds from a Florida colony. Since the time of that particular human intervention, pelicans have not only recovered but thrived along the Gulf coast wetlands, and were finally removed from the endangered species list barely a year before the BP blowout. "The bird lived."

Carson's image of a silent spring, devoid of birds and birdsong, incited the cultural imaginary in ways that few images have ever managed to do. Now, fifty years after the publication of her book, it is astonishing to recall the impact of this single image—a spring day without the sounds of birds. The image seized the social imaginary, incited political and legislative action, and ushered in nothing less than the modern environmental movement. It reclaimed the Gulf as habitat for some of the oldest winged creatures on earth. We might pause to consider here the power of a distinctly aesthetic response, one fueled neither purely by cognition and scientific analysis nor by some easy appeal to sentimentality—but by the very sensation of silence. In her poem "Midnight Oil," St. Germain describes "times we need silence as much as we need news / or a poem that creates a silence / in us where we can feel again." Carson gave us that poem and that space, her own redistribution of what is sensible, astonishingly and quietly audible—where we can feel an encroaching silence and almost hear its awful emptiness. She gave us access to this sensorium not through some naturalized image of a postcard landscape, but in the very midst of gathering chemical clouds.

Still, something else tugs at the human imagination when we see an oiled bird rise from the Gulf. The specter of wounded birds has long haunted the imagination— birds shot midair who fall to the ground, ensnared birds, sick birds, caged birds, birds incapable of flight. So powerful is the traditional linkage of birds and human imagination that we might say the flight of imagination itself seems propelled by birds. Literature alone is infused with their beating wings—the soaring eagle, the falcon who cannot see the falconer, the wild swan gliding along a smooth lake, the nightingale and its dark song. From Homer's winged word to Maya Angelou's caged bird who sings and Emily Dickinson's robin who continues to await crumbs after her death, birds seem an image of imagination itself. When a famous ancient mariner shoots an albatross for no good reason, he is doomed to tell his story forever—as if passing on some secret message about life. But the more recent image of birds soaked in oil gets mired even in imaginative flight, stuck in both the oil itself and in our imaginative resources. The image seems almost to have taken up residence in the human imagination, and gets called up ritually nowadays whenever another accident that is never supposed to happen again does, in fact, happen yet again. From the Santa Barbara oil spill to the Exxon Valdez to Deepwater Horizon, the image of an oiled bird takes center stage. The image, in fact, seems never to disappear entirely, as if ready to rise up again whenever another pipeline breaks open or another rig explodes. In turn, media portrayals of petro-disasters inevitably circle around this haunting image of the oiled bird. DDT was banned by legislation, but oiled birds keep appearing—almost as if they remain on call for certain command performances, before they will be cleaned up in the popular imaginary when all the cameras simply direct our attention elsewhere.

This ritual and repetitive imaging of oiled birds, especially in its most recent incarnation in the commanding presence of the Gulf pelican, disturbs even the traditional human imaginary of the wounded bird. It presents an image that no one wants to see yet that needs to be registered and processed. It forces into focus not only a sensory repetition of the bird's plight but also the sticky sensory substance of oil itself, which otherwise remains invisible in pipelines and tanks and in the bellies of ships. DDT managed to elude the imagination. But the image of oil sticks. It lurks around, keeps surfacing on the otherwise undisturbed surfaces of water and consciousness. We rarely see it, but we benefit from its effects everywhere. We consume it without any sense of taste or touch, and yet we consume it voraciously and repetitively. The oiled bird gets stuck in this repetitive imagistic cycle that we keep consuming. Cameras seize the bird's image again and again, as if we cannot get enough of watching this great damaged winged creature. The image gets plastered on glossy magazine covers with such portentous titles as *National Geographic* and *Time.* Time in fact has become frozen for this scene of our trashed national geography at the edge of North America. CNN keeps delivering the image of the sea bird, again and again, so that the image impresses itself on the mind in some loop of memory, like a recurring nightmare.

In the midst of being stuck in this nightmare of recurring bad images, we humans still want, almost impulsively, to rescue the bird. This same impulse provides us with a rare if not tragic opportunity to rescue and extend the human sensorium in the midst of a great petro-disaster. The oiled bird at the center of *National Geographic* seems doomed, but whether it will live or die, it still animates our imaginations. *Anima* is a fascinating term that gives rise to the words we use to name both animal and spirit—from the Latin root designating air, breath, life. It is also a word that we use to describe breath-infused actions—*to animate, to inspire or be inspired, to stir, to incite*. To animate, a creature would seem to be possessed of both animal and spiritual qualities—like angels, or like winged animals. Drenched in oil, caught and caught up in a recurring bad image that gets played again and again in filmic media, winged animals occupy a liminal place in the relentless unfolding of disaster images, functioning at once as imaginative objects of our vision and embodied creatures who maintain their status as birds. Poised here, they inhabit the borderlands of the two realms we associate with the human and the animal and also with the imaged and the real. Unlike human actors, who at least give the impression of transporting themselves fully into what they represent, animals still somehow maintain their status as actual animals, never fully escaping the "weight of the real."[19] We worry that they may be abused or exploited in filmic representations, that they may be subjected to pain. We worry about bodies that seem real despite being cast in performative roles. Yet this weight of the real, especially as it bears down on oiled pelicans playing out their tragic roles in the midst of petro-disasters, may incite the very *anima* that stirs the spirit and that animates a capacious sensorium inhabited by both winged animals and the animals we know as humans. The human-animal divide gets breached as we sense and become caught up in the bird's plight. We humans come to life as both animated and animal creatures.

But there remains work to be done here. Linking the Latin term *anima* to the ancient concept of a soul, Julia Kristeva offers a diagnosis of a distinctly modern predicament that she calls "new maladies of the soul"—the inability to form images, the "inability to represent," which in turn blocks access to the bodies of ourselves and others, and to the fulsome work of sensation, namely, interpretation. In her essay "The Soul and the Image," she explains this modern malady as taking root in our stress-ridden and money-driven lives, if not in actual experience of trauma, leaving us depressed. Enter here our Angel of History, incapable of animation. Those suffering this malady, she suggests, seek recourse in their only readily available options—either neurochemical drugs that temporarily relieve anxiety, or an indulgence in the druglike effects of mass-media images that bombard and suffocate what she calls "psychic life." Psychic life requires a fulsome sensory response—making sense and making meaning of images, as animals seem to do instinctively. But as Kristeva puts it, we are dispossessed of *anima*, suffer maladies of the soul, and have "run out of imagination."[20]

The druglike effects of mass media after the Deepwater Horizon blowout have yet to be accessed, much like the biological effects of oil and chemical dispersants on marine creatures and the vast array of life forms inhabiting the Gulf ecosystem. What we do know is that the fossil fuel industry invests enormous resources in corporate advertising campaigns to ensure that its own image will remain cleansed of any signs of petrochemical toxicity. The news about even recent petro-disasters gets quickly replaced by newslike testimonials on oil paving the way for economic recovery and prosperity. Consider only the recent bombardment of ads from the American Petroleum Institute in the latest presidential campaign, featuring a carefully staged performance by actors cast as workers who claimed to be "energy voters" and who urged us all to get back on track in a well-oiled economy. Data collected a good two months before the 2010 midterm elections showed that over $150 million had been spent on such staged performances that promoted coal, oil, and gas and that criticized clean-energy alternatives.[21] But all that is nothing compared to the BP media blitz of images dominating commercials before and after the election and especially during televised newscasts, confusing the line between journalism and advertising—as if the two were not already sufficiently blurred. BP tripled its advertising spending after the Gulf oil blowout, investing over $93 million in television and newspaper advertisement in 2010 alone.[22] In a single hour of primetime news reporting on CNN during the high point of its advertising campaign, viewers were typically exposed to BP ads during every single commercial break, sometimes two ads running consecutively. Again and again, national and world news were carefully spliced with the voice and image of an announcer who was cast to deliver the factual news of BP's recovery efforts, and who guided viewers through image after image of pristine Gulf coast beaches, fun fishing adventures, great-tasting seafood, and jobs galore that promised economic salvation to some of the poorest regions in the country. Since the 2012 presidential election, the ads have diminished in frequency but still appear regularly, and still appear predictably during newscasts. To further ensure its control of messages, BP also quickly purchased terms like "oil spill," "volunteer," and "claims" from Google and other search engines in order to direct Internet users to their own website and their own version of the story[23]—a remarkably handy way to distribute information through the simple purchase of language itself.

Such corporate manipulation of the cultural imaginary—recurring pictures of crystal waters washing across white sandy beaches, long-legged herons flying into the sunset, happy tourists consuming mounds of fried seafood, and carefully scripted messages purchased for television and computer screens—at once controls aesthetic response and drugs viewers into cultural anesthesia. This especially pernicious malady of the soul deadens our sensory receptors so that they become incapable of responding to anything more than the images served up and delivered by oil magnates. The effort here is to ensure that the human sensorium becomes as brainwashed as the Gulf itself is washed through with Corexit, that magical dispersant that makes all the oil disappear, as if it never existed at all. Dissolved in this blitz of media images that escapes any weight of the real, oil gets restored to its position of dominance in a restored idyllic Gulf.

In fact, this is all a replay of the 1948 film *The Louisiana Story*, paid for by Standard Oil to produce beautiful images of wetlands undisturbed by oil exploration, and providing a palatable *Louisiana Story* for an industry already beginning to distract attention from its increasing toxic infusions. At about the same time that this film was released, the U.S. Public Health Service was reporting an oily taste in water supplies around the area, potentially generating not simply from oil itself but from the growing petrochemical industry that was setting up its empire on top of former sugar plantations.[24] As petrochemicals replaced sugar as the main addictive commodity of the region, *The Louisiana Story* would need to make way for Cancer Alley and all its chemical releases and bad smells, the only sensations remaining amid an otherwise vague industrial landscape dotted with a few refurbished plantations where tourists are taken on their own media blitz of the past. A corporate-sponsored film is one matter, but even serious works of literature can miss what was really unfolding here in Louisiana. In her sharp reading of Walker Percy's novel *The Moviegoer*, Barbara Eckstein presents us with another version of the malady of the soul. She uses Percy's own diagnostic term, "malaise," to describe a depressive state of dulled sensory response. And she points a finger at physician-writer Percy himself, a native of Louisiana, who situated his Louisiana story in the new suburbs of New Orleans and along the Gulf Coast. In this arguably overrated novel, which won the National Book Award in 1962, Percy gave us a set of upper-class characters consumed in their own depression, the lead character unable to do much more than watch an endless progression of movies. Percy, it turns out, missed the real illness that his characters suffer and that set one of them in pursuit of recurring images. As Eckstein explains, the sickness was not existential malaise, but the actual petrochemical toxins that were in fact infecting the mid-twentieth-century culture of Louisiana in which Percy and his moviegoer resided.[25] The malady here infects even a distinguished author, not to mention physician, who seemed unable to process the images unfolding in this toxic and all-too-real Louisiana story.

Diagnosing and doing something about this malady requires devoted and attentive imaginative work. And given the extraordinary extent of the wreckage caused by human exploitation of fossil fuels, the task necessarily involves us, too, in reimagining the very expanses of time and geography that have marked the terrain of our Angel of History. Ecological writers now speak of "hyperobjects" that exist for durations of time beyond our everyday comprehension and defy our ability to imagine them at all.

Plastics alone may prove to be among the most long-lasting hyperobjects of vinyl chloride extraction and all its hydrocarbon progeny. The chemical dispersant Corexit might even enjoy the status of an anti-image, since its very work is to make oil invisible while obscuring its own toxic content, which remains the subject of intense scientific investigation and controversy. Existing outside the human sensorium, Corexit can still be consumed by small forms of ocean life and can still infuse bodies on up the food chain—where petroleum mixes well, we are told, with the "fat-rich tissues of the liver, brain, kidneys, and ovaries."[26] How do we come to terms with the chemical longevity of PVC and its controversial "end-of-life" scenarios? In her essay about poetic images that seem "beyond imagining," Lynn Keller writes that "we need images that help us think about what is almost unimaginable"[27]—and, let us add, that help us think beyond petroleum in promised pristine waters that are in fact glutted with petroleum's chemically dispersed molecules.

   In an age of eco-disasters, we do need images to navigate our way through the unimaginable. But we also need ways of processing and responding to these images that make them meaningful. It helps, for instance, to re-vision oil as stored sunlight, formed over millennia in sedimentary layers of the earth's crust and not simply as some chemical distillate that is a product of human alchemy. In turn, we might imagine solar panels that also seize the sun's energy as working fast to absorb what oil preserves over millions of years—and ponder ways in which these different yet connected resources might most judiciously be harnessed. We might begin to *see* both the sun and oil in all their elemental formations and technological extensions, and not as substances to be exploited for the latest human addiction. For decades now, feminist theorists in particular have been calling for new understandings of vision itself. Instead of conceptualizing vision as a process in which we look *at* images and objects from some distance, through an assumed controlling gaze, they suggest that vision unfolds in a multidimensional sensorium. We might see and sense objects and others *through* a shared, mutually inhabited sphere in which vision circulates—like sunlight. In her book *Witnessing*, philosopher Kelly Oliver gets right to the heart of the visual dynamics of "bearing witness" to traumatic events—among them, let us now add, bearing witness to creatures horribly damaged by petro-disasters, including us. Her chapter "Toward a New Vision" takes root in the sensed, tactile environments that we all inhabit, infused with "energy, vibrations, particles, and waves" that connect us. Building on the notion of an "*ecological* optics," she urges a vision that sees through and within our shared environments, that gives us "access to each other's sensations," and that emerges as "circulation through the tissues of bodies, the tissues of elements, and the tissues of language."[28]

Something about the swampy wetlands of the Gulf encourages precisely this kind of vision and sensation, what Monique Allewaert describes as a "Swamp Sublime," in which the entanglements that proliferate in these former plantation zones and wetlands disable taxonomies that would distinguish the "human from the animal from the vegetable from the atmospheric," and instead reveal "an assemblage of interpenetrating forces" that she calls an "ecology."[29] Within such a Gulf ecology, and urged on by the very shocking images that emerge from a great petro-disaster on the Gulf, we might begin to imagine the Gulf itself as a site where sensation can be redistributed and dispersed, and where a fulsome ecological sensorium can begin to form on the deep horizons of a fertile, interpenetrating, sensory consciousness. Something is flapping its wings here—in the midst of and maybe even because of all the damage. It may take an animal bearing the weight of the real, if not an angel, to animate a shared sensory imaginary. The fatty tissue that easily absorbs oil might also become the very tissues of bodies and elements through which vision circulates.

Bird will probably die unless there is human intervention. But we must add: *humans will probably become deadened unless there is bird intervention.* The mutual intervention of humans and birds in the life of the Gulf begs the work of angels who inhabit their shared bodies—animal and spirit—bearing the weight of the real and lifting wings for imaginative flight. Imagining pelicans as angels on the Gulf urges on this shared sensory labor and also incites the overwhelming task of reclaiming from oil empires this ninth-largest body of water in the world and its sixteen thousand miles of U.S. shoreline alone. Such an effort requires that those responsible for disasters be held accountable for their actions. In the latest settlement of criminal charges associated with British Petroleum's negligence in the Gulf, BP pleaded guilty and agreed to pay $4.5 billion in fines. More than half this amount will be devoted to the restoration of wetlands in Louisiana and along the Gulf Coast.[30] According to the U.S. Department of Justice, the decision to use such funds for wetland restoration reflected the strategy of Louisiana's own Coastal Master Plan adopted in 2012. For a state that is typically in the pocket of oil companies, this intervention usurps BP's fake media images of a cleansed Gulf, and manages both to incite and fund the long-imagined work of restoration. Among the major interventions charted in the Coastal Plan are river diversion projects aimed at the reclamation of barrier islands, the precise site where the brown pelican roosts.

Such efforts at reclamation also require an almost magical belief that birds do keep fresh the winds for the opening of the morning's eye, every single day—even when the birds themselves are coming back from near extinction. Some humans who dare to save birds from extinction and infuse them into our shared imaginary effect

astonishing change in the space of a mere decade. Others impulsively take the oiled
pelican in hand and get on with the delicate work of cleaning it up. Humans have
taken birds in hand for remarkable imaginative feats. In her Nobel Prize Lecture, Toni
Morrison tells the story of an old blind woman who is taunted by a group of young
people who question her wisdom. They arrive at her door one day, claiming that they
have a bird in hand. Knowing that she is blind, they ask her if she knows whether the
bird is dead or alive. After a long silence, the old woman tells them: "I don't know
whether the bird you are holding is dead or alive, but what I do know is that it is in
your hands. It is in your hands."[31]

The bird in this account may be far from the Gulf, but the world Morrison creates
in her novel *Beloved* is not far removed from the sugar plantations purchased and polluted by big oil for its petrochemical distillations. The apparitional bird-in-hand bears
the weight of both history and imagination in these troubled yet sublime landscapes.
Closer at home in the wetlands of Louisiana, two young girls emerge in a different
story, this one told by New Orleans writer Barb Johnson.[32] We come upon these girls
as they sit atop an oil tank one night, gazing out over an endless vista of oil refineries
leading into a vast and dark Gulf of Mexico. They sneak away from their high school
dance, jump into an old Valiant, and drive to Emerald City in their rough-and-tumble
youthful attempt to escape the dregs of Louisiana oil towns. They are hot to get out of
all this mess. As they sit high on the oil tank, they might appear to be two small angels
presiding over a pile of wreckage. Imagined birds are all around them. One girl leans
toward the other, extending her hand, which "migrates," we are told, like "a small bird
flying into the wind." Enveloping the girls, night rolls out, filled with "dark fields, lit
only by the pale fruit of egrets sleeping in the trees along the bayou." Gazing out over
these bird-infused swamps and the Gulf of Mexico, they wonder what's out there—
"More of this world or maybe another."

The small migrating bird of a hand. The small bird held in the hand and presented
to an old blind woman. Sleeping egrets appearing as fruit in trees and nourishing the
night in these beautiful and poisoned bayous. A mother pelican perched on a flag,
feeding her young. There are images enough here, even in an Emerald City filled with
oil tanks and refineries. But how is it that we might seize the image, take it up and take
it in, make meaning of it, work it and work through it and all the wreckage? As I turned
off from I-10 that day and headed to Geismar in search of petrochemical plants in
Ascension Parish, I thought hard about whether angels could ever ascend here in this
tragic landscape. Its rich sedimentary delta sustained indigenous people for centuries
before the days of sugar and oil, but now that seems another world entirely, beyond
anyone's reach. But in the story, one of these girls grows up and opens a laundromat in
a drug-ridden neighborhood in New Orleans. Here, she and her partner, a woman who

is also a poet, hold together a community of damaged, vulnerable people, as if taking them under their wings. Somehow, long ago, sitting on top an oil tank in Emerald City, she envisioned the image of another world out there—and she seized it. So here, amid the drugs and soulful maladies of a troubled neighborhood, at least two women are doing their part to clean up one soiled spot on this earth.

Meanwhile, somewhere in Ascension Parish, or maybe in a field way down near Murphy Oil Refinery, a great bird has been released from the wreckage of its oiled body. *Released*, as they say in the petrochemical business—but this time not as toxic refuse set loose in the air. This time the bird breathes air. This time the bird lives. *Don't you want to touch it?* And don't you want to clean it all up—not only the birds, who have in fact incited volunteers from across the globe to take flight to the Gulf and help with the cleansing, but as part of an impulsive, spirited response to cleansing and reclaiming imagination itself, for the kind of venture two young girls embraced when they climbed up an oil tank and gazed out across a plundered Gulf, searching for some image of another world.

NOTES

1. Walter Benjamin, "On the Concept of History," *Selected Writings: Volume 4, 1938–1940*, ed. Howard Eiland and Michael W. Jennings (Cambridge: Belknap Press of Harvard University Press, 2003), 389–400. Benjamin's essays on history, where his Angel of History has assumed its memorable position, derive from his immersion both in Marxist materialist traditions and in spiritual/mythical systems of belief. His Angel of History therefore emerges as a figure at once incarnate and mythic, material and imaginative—occupying a body similar to that of the pelican as a winged creature living amid the material wreckage of the Gulf yet also capable of taking flight in the human imagination.

2. Sheryl St. Germain, *Navigating Disaster: Sixteen Essays of Love and a Poem of Despair* (Hammond: Louisiana Literature Press, 2012), 145. All passages from "Midnight Oil" are reproduced with the author's permission.

3. Michael Watts, "A Tale of Two Gulfs: Life, Death, and Dispossession along Two Oil Frontiers," *American Quarterly* 64, no. 3 (2012): 456.

4. On the Gulf of Mexico Exemption, see National Commission on the BP Deepwater Horizon Oil Spill and Offshore Drilling, *Deep Water: The Gulf Oil Disaster and the Future of Offshore Drilling, Report to the President*, January 2011, 62. On the history of oil in the Gulf of Mexico rehearsed here, see Watts's succinct essay, and William R. Freudenburg and Robert Gramling, *Blowout in the Gulf: The BP Oil Disaster and the Future of America* (Cambridge: MIT Press, 2011).

5. Mike Tidwell, *Bayou Farewell: The Rich Life and Tragic Death of Louisiana's Cajun Coast* (New York: Vintage, 2010), 29.

6. Barbara Eckstein, *Sustaining New Orleans: Literature, Local Memory, and the Fate of a City* (New York: Routledge, 2005), 102.

7. Barbara Allen, *Uneasy Alchemy: Citizens and Experts in Louisiana's Chemical Corridor Disputes* (Cambridge: MIT Press, 2003), 72–73.

8. Ibid., 132–50.

9. The term "sugar kings" comes from Hart Crane's nostalgic story "The Land of the Sugar Kings," published in 1958; see Eckstein, *Sustaining New Orleans*, 106. For the actual historical record, see Richard Follett, *The Sugar Masters: Planters and Slaves in Louisiana's Cane World* (Baton Rouge: Louisiana State University Press, 2007).

10. Beverly Wright, "Living and Dying in Louisiana's 'Cancer Alley,'" in *The Quest for Environmental Justice*, ed. Robert D. Bullard (San Francisco: Sierra Club Books, 2005), 96–97.

11. See Allen, *Uneasy Alchemy*, 29, 33, 13.

12. J. Timmons Roberts and Melissa M. Toffolon-Weiss, *Chronicles from the Environmental Justice Frontline* (Cambridge: Cambridge University Press, 2001), 6–7.

13. Ibid., 6.

14. "Special Report: The Spill," *National Geographic*, October 2010, 32–33.

15. Jacques Rancière, *The Politics of Aesthetics: The Distribution of the Sensible*, trans. Gabriel Rockhill (New York: Continuum, 2006).

16. Nicole Cooley, "Four Studies of the Afterlife," *Breach* (Baton Rouge: Louisiana State University Press, 2010), 15.

17. National Commission on the BP Deepwater Horizon Oil Spill and Offshore Drilling, *Deep Water*, 181.

18. See Tidwell, *Bayou Farewell*, 58–62.

19. Dennis Lim, "Birds Do It, Bees Do It (Fill Screens)," *New York Times*, April 1, 2012, 10, 14. Lim's article surveys several studies of animals depicted on screen.

20. Julia Kristeva, "The Soul and the Image," *New Maladies of the Soul*, trans. Ross Guberman (New York: Columbia University Press, 1995), 7–10.

21. Eric Lipton and Clifford Krauss, "Fossil Fuel Industry Ads Dominate TV Campaign," *New York Times*, September 13, 2012.

22. "BP Tripled Its Ad Budget After Oil Spill," *Wall Street Journal*, September 1, 2012.

23. "BP Buys Google Ads for Search Term "Oil Spill," Reuters, June 9, 2010.

24. Eckstein, *Sustaining New Orleans*, 97, 118.

25. See Eckstein, *Sustaining New Orleans*, 98–100, who notes the curious phenomenon of how "Dr. Percy as public intellectual remained separate from public health" and concludes: "Product of his place and time and disposition, he served the mid-century's and the city's sense of irony, but not a recognition of its place within Hydrocarbon Society."

26. National Commission on the BP Deepwater Horizon Oil Spill and Offshore Drilling, *Deep Water*, 180.

27. Lynn Keller, "Beyond Imagining, Imagining Beyond," *PMLA* 127, no. 3 (May 2012): 582. For the concept of "hyperobject," Keller references Timothy Morton, *The Ecological Thought* (Cambridge: Harvard University Press, 2010).

28. Kelly Oliver, *Witnessing: Beyond Recognition* (Minneapolis: University of Minnesota Press, 2001) 193, 174, 223. Feminist criticism has long devoted attention to the dynamics of vision, from early feminist studies of the "gaze" in film, which was invariably projected on women, on through the works of such theorists as Luce Irigaray, Donna Haraway, Teresa Brennan, and many others. For the notion of an "ecological optics," Oliver references J. J. Gibson, "Ecological Optics," *Vision Research* 1 (1961): 253–62, and *The Senses Considered as Perceptual System* (Boston: Houghton Mifflin, 1966).

29. Monique Allewaert, "Swamp Sublime: Ecologies of Resistance in the American Plantation Zone," *PMLA* 123, no. 2 (2008): 340–57.

30. See the article on the website for the Coalition to Restore Coastal Louisiana: http://crcl .org/blog-menu-item/post/bp-settles-deepwater-horizon-disaster-criminal-charges-for-4-5-bil lion-funding-headed-to-louisiana-for-restoration.html.

31. Toni Morrison, "The Nobel Lecture in Literature, 1993" (New York: Alfred A. Knopf, 2007).

32. Barb Johnson, *More of This World or Maybe Another: Stories* (New York: Harper Perennial, 2009), 13–18.

Contributors

GEORGIANA BANITA is assistant professor of North American literature and media at the University of Bamberg and honorary research fellow at the United States Studies Center, University of Sydney. She is the author of *Plotting Justice: Narrative Ethics and Literary Culture after 9/11*.

ROSS BARRETT is assistant professor of art history at the University of South Carolina. He is the author of *Rendering Violence: Riots, Strikes, and Upheaval in Nineteenth-Century American Art*.

FREDERICK BUELL is professor of English at Queens College. He is the author of *W. H. Auden as a Social Poet, Full Summer, National Culture and the New Global System*, and *From Apocalypse to Way of Life: Environmental Crisis in the American Century*.

GERRY CANAVAN is assistant professor of English at Marquette University. He is the coeditor of *Green Planets: Ecology and Science Fiction*.

MELANIE DOHERTY is assistant professor of English at Wesleyan College.

SARAH FROHARDT-LANE is assistant professor of history at Ripon College.

MATTHEW T. HUBER is assistant professor of geography at Syracuse University. He is the author of *Lifeblood: Oil, Freedom, and the Forces of Capital*.

DOLLY JØRGENSEN is an environmental history researcher at Umeå University in Sweden. She is the coeditor of *New Natures: Joining Environmental History with Science and Technology Studies*.

STEPHANIE LEMENAGER is Barbara and Carlisle Moore Distinguished Professor in English and American Literature at the University of Oregon, where she is an affiliate in environmental studies. Her latest book, *Living Oil: Petroleum Culture in the American Century,* was published in 2014. She is the author of *Manifest and Other Destinies: Territorial Fictions of the Nineteenth-Century United States* and the coeditor of *Environmental Criticism for the Twenty-First Century.*

HANNA MUSIOL is lecturer in English at Northeastern University and in American studies at the University of Massachusetts, Boston. She is associate editor of *Cultural Studies: An Anthology.*

CHAD H. PARKER is assistant professor of history at the University of Louisiana at Lafayette.

RUTH SALVAGGIO is professor of English and American studies at the University of North Carolina, Chapel Hill. She is the author of *Hearing Sappho in New Orleans: The Call of Poetry from Congo Square to the Ninth Ward, The Sounds of Feminist Theory*, and *Enlightened Absence: Neoclassical Configurations of the Feminine.*

HEIDI SCOTT is assistant professor of English at Florida International University. She is author of *Chaos and the Microcosm: Literary Ecology in the Nineteenth Century.*

ALLAN STOEKL is professor of French and comparative literature at Penn State University. Most recently, he is the author of *Bataille's Peak: Energy, Religion, and Postsustainability.*

IMRE SZEMAN is Canada Research Chair in Cultural Studies at the University of Alberta. Most recently, he is the coauthor of *After Globalization.*

MICHAEL WATTS is Class of 63 Professor of Geography at the University of California, Berkeley. He is the author of *Silent Violence: Food, Famine, and Peasantry in Northern Nigeria* and editor of *Curse of the Black Gold: 50 Years of Oil in the Niger Delta.*

JENNIFER WENZEL is associate professor in the Department of English and Comparative Literature, and the Department of Middle Eastern, South Asian, and African studies at Columbia University. She is the author *of Bulletproof: Afterlives of Anticolonial Prophecy in South Africa and Beyond.*

SHEENA WILSON is assistant professor and director of the Bilingual Writing Centre at the Campus Saint-Jean of the University of Alberta.

DANIEL WORDEN is associate professor of English at the University of New Mexico. He is the author of *Masculine Style: The American West and Literary Modernism.*

ROCHELLE RAINERI ZUCK is assistant professor of English at the University of Minnesota, Duluth.

CATHERINE ZUROMSKIS is associate professor of art history at the University of New Mexico. She is the author of *Snapshot Photography: The Lives of Images.*

Index